Have you been to our website?

For code downloads, print and e-book bundles, extensive samples from all books, special deals, and our blog, please visit us at:

www.rheinwerk-computing.com

Rheinwerk Computing

The Rheinwerk Computing series offers new and established professionals comprehensive guidance to enrich their skillsets and enhance their career prospects. Our publications are written by the leading experts in their fields. Each book is detailed and hands-on to help readers develop essential, practical skills that they can apply to their daily work.

Explore more of the Rheinwerk Computing library!

Bernd Öggl, Michael Kofler
Git: Project Management for Developers and DevOps Teams
2023, 407 pages, paperback and e-book
www.rheinwerk-computing.com/5555

Bernd Öggl, Michael Kofler
Docker: Practical Guide for Developers and DevOps Teams
2023, 491 pages, paperback and e-book
www.rheinwerk-computing.com/5650

Kevin Welter
Kubernetes: Practical Guide for Developers and DevOps Teams
2024, 401 pages, paperback and e-book
www.rheinwerk-computing.com/5964

Metin Karatas
Developing AI Applications: An Introduction
2024, 402 pages, paperback and e-book
www.rheinwerk-computing.com/5899

Michael Kofler
Scripting: Automation with Bash, PowerShell, and Python
2024, 470 pages, paperback and e-book
www.rheinwerk-computing.com/5851

www.rheinwerk-computing.com

Sujeevan Vijayakumaran

DevOps

Frameworks, Techniques, and Tools

Editor Meagan White
Acquisitions Editor Hareem Shafi
German Edition Editor Christoph Meister
Copyeditor Lauren Miorcec
Cover Design Graham Geary
Photo Credit iStockphoto: 1254474435/© bloodua; Shutterstock: 2151327127/© sakkmesterke
Layout Design Vera Brauner
Production Eric Wyche
Typesetting III-satz, Germany
Printed and bound in Canada, on paper from sustainable sources

ISBN 978-1-4932-2670-2
1st edition 2025
1st German edition published 2024 by Rheinwerk Verlag

© 2025 by:
Rheinwerk Publishing, Inc.
2 Heritage Drive, Suite 305
Quincy, MA 02171
USA
info@rheinwerk-publishing.com

Represented in the E.U. by:
Rheinwerk Verlag GmbH
Rheinwerkallee 4
53227 Bonn
Germany
service@rheinwerk-verlag.de

Library of Congress Cataloging-in-Publication Control Number: 2024046256

All rights reserved. Neither this publication nor any part of it may be copied or reproduced in any form or by any means or translated into another language, without the prior consent of Rheinwerk Publishing.

Rheinwerk Publishing makes no warranties or representations with respect to the content hereof and specifically disclaims any implied warranties of merchantability or fitness for any particular purpose. Rheinwerk Publishing assumes no responsibility for any errors that may appear in this publication.

"Rheinwerk Publishing", "Rheinwerk Computing", and the Rheinwerk Publishing and Rheinwerk Computing logos are registered trademarks of Rheinwerk Verlag GmbH, Bonn, Germany.

All products mentioned in this book are registered or unregistered trademarks of their respective companies.

Contents at a Glance

1	Introduction	15
2	What Is DevOps?	23
3	The Example Company	47
4	Project Management and Planning	61
5	Collaboration when Coding	73
6	Continuous Integration and the Build Process	113
7	Quality Assurance	153
8	Continuous Delivery and Deployment	173
9	Operating the Service	221
10	From Monitoring to Observability	267
11	Security and Compliance	307
12	Successfully Implementing the DevOps Transformation	347
13	DevOps Platforms	377
14	Beyond Culture and Tools	389

Contents

1 Introduction — 15
- 1.1 Culture — 16
- 1.2 Technology — 17
- 1.3 My Path to DevOps and to This Book — 17
- 1.4 Target Audience — 19
- 1.5 The Structure of the Book — 20
- 1.6 Feedback — 20
- 1.7 Thank You! — 21

2 What Is DevOps? — 23
- 2.1 DevOps: The Big Picture — 23
  - 2.1.1 CALMS — 25
  - 2.1.2 The Three Ways — 28
  - 2.1.3 Conclusion on the Three Ways and the CALMS Model — 35
- 2.2 Misunderstandings about DevOps — 36
  - 2.2.1 Too Strong a Focus on Automation — 36
  - 2.2.2 With DevOps, but without Tests! — 37
  - 2.2.3 Incorrect Understanding of Team Structuring — 38
  - 2.2.4 Not Tearing Down All the Walls — 39
  - 2.2.5 Tools over Processes over People — 40
  - 2.2.6 One to One Copies of Working Methods from Other Companies — 41
- 2.3 The DevOps Software Development Lifecycle — 42
- 2.4 Summary — 45

3 The Example Company — 47
- 3.1 nicely-dressed.com — 48
- 3.2 The Development Model — 48
- 3.3 The Business Team: Requirements Analysis — 49

3.4	**The Architecture Team: Design of the Application**	50
3.5	**The Development Teams**	50
	3.5.1 The Development Process	51
	3.5.2 Integrations with Obstacles	51
3.6	**The Quality Assurance Team**	52
3.7	**The Operations Team: The Ops in DevOps**	53
	3.7.1 Manual Build of the Project	55
	3.7.2 Deployment with Obstacles	55
	3.7.3 Alarm from the Monitoring System	56
3.8	**The Infrastructure Team**	56
3.9	**The Security Team**	57
3.10	**Summary**	58

4 Project Management and Planning — 61

4.1	**The First Step: The Agile Mindset**	61
4.2	**Project Management for Everyone?**	64
	4.2.1 Jira Can Do (Almost) Everything	65
	4.2.2 People over Processes	66
	4.2.3 Good Project Management outside of Jira	67
	4.2.4 More than Just a Project Management Tool	69
	4.2.5 Project Management at nicely-dressed.com	71
4.3	**Summary**	72

5 Collaboration when Coding — 73

5.1	**Typical Problems with Managing the Source Code**	73
	5.1.1 Organization of the Code	74
	5.1.2 Isolation for Supposed Security Reasons	75
	5.1.3 Long Development Times Hindering Quick Security Fixes	76
	5.1.4 Development Workflow without a Proper Structure	77
	5.1.5 Big Bang Integrations	77
	5.1.6 Code Reviews Could Help	78
	5.1.7 Technical Debt	79
	5.1.8 High Learning Curve due to Lack of Documentation	80
5.2	**Improve the Organization of the Code**	81

5.3	**There Is No Way around Git**		83
	5.3.1	Git Solutions at a Glance	83
	5.3.2	Development Workflows with Git	84
	5.3.3	Source Code Management at nicely-dressed.com	90
5.4	**Code Reviews and Pair Programming**		91
	5.4.1	Code Reviews	91
	5.4.2	Simplify Code Reviews	97
	5.4.3	Pair Programming	99
5.5	**Inner Sourcing: Sharing Code within the Company**		102
	5.5.1	Open Source	103
	5.5.2	The Path to Inner Sourcing	104
	5.5.3	Advantages of Inner Sourcing	106
	5.5.4	Monorepositories	108
5.6	**Summary**		111

6   Continuous Integration and the Build Process   113

6.1	**Typical Problems in the Build Process**		113
	6.1.1	Onboarding with Stumbling Blocks	114
	6.1.2	Build Difficulties Due to Infrequent Integrations	116
	6.1.3	Only a Few Tests	117
	6.1.4	A Build Server behind Closed Doors	118
6.2	**Modern Build Management**		119
6.3	**Continuous Integration**		122
6.4	**The Continuous Integration Server and the Pipelines**		126
	6.4.1	The Basic Structure of a Pipeline	126
	6.4.2	Scaling and Reproducibility	128
	6.4.3	Declarative Pipelines versus Scripted Pipelines	130
6.5	**Efficient Pipeline Authoring**		132
	6.5.1	Avoid Central Pipelines	132
	6.5.2	Provide Pipeline Building Blocks	133
	6.5.3	Create Visibility	134
6.6	**Overview of Continuous Integration Servers**		134
	6.6.1	Jenkins	136
	6.6.2	GitLab CI/CD	142
	6.6.3	GitHub Actions	146
	6.6.4	Other Continuous Integration Servers and Tools	150
	6.6.5	Continuous Integration at nicely-dressed.com	151
6.7	**Summary**		152

7 Quality Assurance — 153

7.1 Typical Problems with Testing — 153
7.1.1 Teams in Their Silos — 154
7.1.2 Different Understandings of Requirements — 155
7.1.3 The Number of Bugs as a Metric — 156
7.1.4 Is It Fixed Yet? — 157

7.2 Testing as Part of the DevOps Process — 158
7.2.1 Tests in the Build Pipeline — 160
7.2.2 Different Tests for Different Tasks — 162
7.2.3 Automate Tests — 164
7.2.4 Test-Driven Development — 169

7.3 Summary — 171

8 Continuous Delivery and Deployment — 173

8.1 Typical Release Management Problems — 173
8.1.1 Separate Handling of Changes and Documentation — 174
8.1.2 Lengthy Release Process — 175
8.1.3 Automations That Are Not Worthwhile — 176
8.1.4 Hostilities between the Teams — 176
8.1.5 Deployment on Production Systems with Obstacles — 177
8.1.6 Conclusion — 178

8.2 Implementing Continuous Delivery and Deployment — 179
8.2.1 Bringing Development and Operations Together — 179
8.2.2 QA, Staging, and Production Environments — 185
8.2.3 Deployment on Fridays — 190

8.3 Build Management for Deployments — 191
8.3.1 The Question of Version Numbers — 192
8.3.2 Packaging — 193
8.3.3 Containerization — 194
8.3.4 Container Registry and Package Registry — 196

8.4 Rollbacks, Canaries, and Feature Flags — 198
8.4.1 Rollbacks — 199
8.4.2 Step-by-Step Activation Using Blue-Green and Canary Deployments — 200
8.4.3 Feature Flags — 203

8.5	**Deployment Targets**		205
	8.5.1 Orchestrating Deployments with Kubernetes		205
	8.5.2 Orchestrating Deployments at nicely-dressed.com		219
8.6	**Summary**		219

9   Operating the Service

221

9.1	**Typical Problems with Operating Services**		221
	9.1.1 Lengthy Infrastructure Planning		222
	9.1.2 Hardware Exchange with Obstacles		222
	9.1.3 Unfavorable Server Utilization		223
	9.1.4 Common Outages during the Night		224
9.2	**Breaking Up the Highly Coupled Infrastructure Architecture**		226
	9.2.1 Cattle, not Pets		226
	9.2.2 Abstracting the Infrastructure		228
	9.2.3 Containers for Faster Deployments		231
9.3	**Cloud Computing**		233
	9.3.1 What Is the Cloud?		233
	9.3.2 Cloud Models		234
	9.3.3 Service Models		235
	9.3.4 Cloud Native		240
	9.3.5 The Cloud at nicely-dressed.com		242
9.4	**Stronger Collaboration between Development and Operations**		243
	9.4.1 Everyone Should Be Ready		243
	9.4.2 Blameless Post-Mortems		245
	9.4.3 Communication Solutions and ChatOps		247
9.5	**Configuration Management: Everything as Code**		249
	9.5.1 Infrastructure as Code with Terraform		250
	9.5.2 Ansible versus Puppet		253
9.6	**Chaos Engineering**		258
	9.6.1 Making Systems Fail		258
	9.6.2 Chaos Engineering without Chaos, but with a Plan		260
9.7	**Reliability Engineering**		262
	9.7.1 Site Reliability Engineering		263
	9.7.2 Database Reliability Engineering		264
9.8	**Summary**		265

10　From Monitoring to Observability　　267

10.1 No Visibility at nicely-dressed.com　　268
　10.1.1　Service Outages Happen Every Day　　268
　10.1.2　Performance, Performance!　　272
　10.1.3　Logs　　274

10.2 With Insight Comes Foresight　　275
　10.2.1　Observability Engineering　　276
　10.2.2　Insights into Processes with Tracing　　278
　10.2.3　A/B Tests　　279
　10.2.4　Business Monitoring　　281

10.3 Tools for Monitoring, Observability, and Tracing　　282
　10.3.1　Monitor Systems with Icinga and Nagios　　283
　10.3.2　Monitoring with Metrics and Time Series Databases　　284
　10.3.3　Data Visualization with Grafana　　292
　10.3.4　Error Tracking　　294
　10.3.5　Distributed Tracing　　295
　10.3.6　Logging　　296
　10.3.7　Service Meshes　　297
　10.3.8　Observability Platforms　　298
　10.3.9　Monitoring and Observability at nicely-dressed.com　　300

10.4 Availability　　300
　10.4.1　Service Level Agreements　　301
　10.4.2　Service Level Objectives　　302
　10.4.3　Service Level Indicators　　303
　10.4.4　Error Budgets　　303

10.5 Summary　　304

11　Security and Compliance　　307

11.1 Security Disrupts the Agile Waterfall　　308

11.2 DevOps with a Separate Security Team　　310
　11.2.1　To Deploy or Not to Deploy?　　311
　11.2.2　The Search for Undocumented Dependencies　　312
　11.2.3　Frustration and Blocking　　313

11.3 DevSecOps: Building Security into DevOps　　314
　11.3.1　The DevSecOps Team Structure　　315
　11.3.2　Shift Left: Find Errors Earlier　　316
　11.3.3　Inner Sourcing Ensures Formal Security　　317

	11.3.4	Security as an Integral Part of the Development Process	318
	11.3.5	Dealing with Mistakes	319

11.4 Tools for Higher Security ... 321
	11.4.1	Dashboards and Reporting	321
	11.4.2	Pull and Merge Requests	322
	11.4.3	Security Scanners in Detail	322

11.5 Supply Chain Security ... 329
	11.5.1	Attacks on the Supply Chain	330
	11.5.2	Software Bill of Materials	332
	11.5.3	Security of the Build and Deployment Server	332
	11.5.4	Secure User Accounts	333
	11.5.5	No Code Is Good Code	334
	11.5.6	Security at nicely-dressed.com	335

11.6 Compliance ... 335
	11.6.1	Define Compliance Guidelines	336
	11.6.2	Manual Compliance	338
	11.6.3	Fully Automated Compliance	339
	11.6.4	Compliance at nicely-dressed.com	345

11.7 Summary ... 345

12 Successfully Implementing the DevOps Transformation   347

12.1 Introducing a DevOps Culture ... 347
	12.1.1	Bottom-Up or Top-Down?	348
	12.1.2	First Steps in the DevOps Transformation	350

12.2 Making DevOps Success Measurable with DORA Metrics ... 362
	12.2.1	DORA Metric 1: Deployment Frequency	363
	12.2.2	DORA Metric 2: Lead Time	364
	12.2.3	DORA Metric 3: Change Failure Rate	365
	12.2.4	DORA Metric 4: Time to Restore Service	366
	12.2.5	DORA Metric 5: Operational Performance and Reliability	366
	12.2.6	Findings from the State of DevOps Report	367

12.3 Value Stream Mapping ... 368
	12.3.1	The Value Stream of a Pizzeria	369
	12.3.2	The Value Stream at nicely-dressed.com	370
	12.3.3	Implementation of Value Stream Mapping	371

12.4 Summary ... 375

13 DevOps Platforms — 377

13.1 Toolchain Complexity — 378
- 13.1.1 Step 0: Toolchains Grow Historically — 378
- 13.1.2 Step 1: Multiple Toolchains to Increase Maintainability — 379
- 13.1.3 Step 2: Standardized Tools, but Still with a Lot of Duct Tape — 381
- 13.1.4 Step 3: DevOps Platforms — 382

13.2 DevOps Platforms at a Glance — 384
- 13.2.1 GitLab — 384
- 13.2.2 GitHub — 385
- 13.2.3 Azure DevOps — 385
- 13.2.4 Atlassian — 386
- 13.2.5 Other Platforms — 386

13.3 Summary — 387

14 Beyond Culture and Tools — 389

14.1 The Role of AI in DevOps — 389
- 14.1.1 Making Work Easier with AI-Supported Code Generation? — 390
- 14.1.2 More Code Leads to Higher Review Requirements! — 392
- 14.1.3 AI-Supporting Features — 393
- 14.1.4 Data Protection and Privacy — 394
- 14.1.5 It's the Overall Concept That Counts! — 395

14.2 DataOps, MLOps, and AIOps — 397
- 14.2.1 DataOps — 397
- 14.2.2 MLOps — 398
- 14.2.3 AIOps — 398

14.3 DevOps as a Job — 399
- 14.3.1 The Question of DevOps Engineers — 399
- 14.3.2 Soft Skills — 401
- 14.3.3 The Technical DevOps Learning Path — 402

14.4 Summary — 410

The Author — 411

Index — 413

Chapter 1
Introduction

"Nothing is as constant as change."

This quote is now almost 2,000 years old but is more relevant than ever: Product development cycles need to be faster today to meet constantly changing requirements, over which we often have no direct influence. A recent event that illustrates this reality is the COVID-19 pandemic, which highlighted how crucial flexibility is.

The pandemic not only required that many people quickly switch from office jobs to remote work, but it also affected how work was organized in the first place. Suddenly, familiar methods and processes of in-office work had to be adapted to remote working environments from one day to the next. This challenge affected not only pure IT companies but also other organizations, authorities, schools, and associations.

Companies that were able to quickly adapt both their internal workflows and business models to changing conditions gained a significant competitive advantage. The pandemic clearly shows how rapidly requirements can change and how important it is to adapt with agility.

In this constantly changing environment, DevOps culture plays a crucial role. DevOps promotes collaboration between the development (Dev) and operations (Ops) teams to accelerate processes, improve quality, and increase flexibility at the same time. This concept enables companies to adapt effectively to changing requirements and thus operate successfully in a dynamic world.

For example, one notable legal change that probably affected all companies in Germany was the decision on June 3, 2020, to reduce the VAT rate from 19% to 16% for six months as of July 1, 2020. Between this decision and the changeover, there was only just under four weeks to implement, test, and roll out the necessary changes.

Recent events that disrupt business requirements have shown how adaptable companies and organizations can be when the need arises. However, instead of making quick adjustments only in exceptional situations, companies could make their processes more efficient if they made such adjustments more regularly as part of their normal operations. This would benefit not only the companies themselves but also all the users of their services.

And the benefits of regularly adjusting processes doesn't apply only to hip IT start-ups working on apps and websites. In fact, many companies—and even public authorities—are already IT organizations at their core, even if they do not necessarily offer pure software products. For example, a bank's core business is largely based on the software needed to manage bank accounts and services. So, in a sense, a bank is more of a software development company with an additional banking license. It is therefore not surprising that some banks, such as ING, describe themselves in exactly this way.

If you already know a little about IT, the idea of continuously improving processes will sound familiar to you: After all, that's the basic idea behind agile software development. But isn't this book supposed to be about DevOps, not agile software development?

This is precisely where the first good question arises: What is agility, what is DevOps, how are the two connected, and how do they diverge?

But let's take a step back because perhaps the most important question is actually: Why do it at all? Far too often, people follow trends of all kinds and take actions based on what everyone else is doing. However, it makes much more sense to ask why a certain approach is trending and then introduce changes based on this analysis.

This book answers these questions and many others. It is divided into two overarching sections: culture and technology. We start by looking at culture: we cover the concepts, advantages, and disadvantages of DevOps without going into too much detail about individual technical tools. However, we cannot entirely ignore the IT tool landscape. As this landscape is constantly changing, I do not cover too many specific details about each tool; instead, you will find an overview of the state of the art, with a focus on the basic DevOps technology to consider and the concepts that underlie it.

We don't want this book to be dry and theoretical; therefore, we look at the whole topic through the lens of a fictional company, which is presented in Chapter 3. This company has not yet relied on DevOps principles. Chapter by chapter, the company introduces more and more DevOps principles to show what a DevOps transformation can look like. After all, very few companies will be able to start a DevOps transformation on a greenfield.

1.1 Culture

One thing is very important: Understanding DevOps culture is more important than understanding the actual tools. Because DevOps is a culture, it is a way of working together as a team. Many people have a misunderstanding of DevOps, which is why it is important that you understand DevOps culture properly before we look at technical tools in this book.

The term "DevOps" is a portmanteau of "development" and "operations." It is about breaking down the walls between the development team and the operations team, allowing them to work together toward a common goal: the continuous development and provision of the software-based solution.

In this book, I describe what DevOps culture is, what it is not, and where it overlaps with other ways of working. Most importantly, DevOps must be lived throughout the organization, from the top to the bottom of the hierarchy.

Likely the most difficult challenge for you, and especially for your organization, is to flip the switch in your head; implementing DevOps culture requires a change of perspective. It also requires everyone in the organization to adapt to different tasks and perspectives: Developers will sometimes have to be on call, admins should know what a pull request is, and everyone will ultimately be responsible for the security and reliability of the application.

1.2 Technology

Even though DevOps is essentially a culture, and its description alone could fill the pages of this book, I also cover some tools of the DevOps landscape so that you can better understand the current state of the art.

However, I do not discuss every single tool down to the last detail; rather, I discuss the concepts of the tools so that you can understand why they in particular are used in the current DevOps tool landscape. After reading this section, you should also understand why some tools are used less frequently or where and in what combination they can be useful, especially if DevOps is implemented properly.

It is important to understand that the technology is merely a tool for implementing the culture. Good tools support the culture, but even the best tools will not help you if the processes and culture are not lived in the organization.

1.3 My Path to DevOps and to This Book

Many friends, acquaintances, and colleagues have asked me why I was writing a DevOps book, as the tenor is that the topic is already old hat and "almost done." Meanwhile, both smaller and larger companies claim that they are already doing DevOps. The reality, however, is a little different.

My motivation for this project has its roots at the beginning of my IT career. In 2011, I started studying while also working at a medium-sized company and wanted to eventually become a software developer. However, I never really got around to programming back then, as there was much more to do to improve the company's development productivity.

There was almost no automation at all in this company; numerous tasks were done manually. This meant that the various teams lost a lot of time, did not work well together, and did not use the tools for building and testing efficiently. And here I'm writing only about the development workflow: What happened to the software when it was rolled out barely reached the development team because the finished package was essentially thrown "over the wall" to another team from a service provider, which then took over operations. What happened after that? I have no idea because there was never any feedback.

In addition, the development team had no environments in which to test the software close to production. Whether a change to the software worked was always a big question that nobody could really answer. As a stopgap solution, employees set up many small test environments by hand, which had to be cleaned up and reinstalled manually time and time again, wasting a lot of time. The test systems were never really clean.

Many of these shortcomings are probably still typical of small- or medium-sized companies today, especially older companies. The statement "But we've always done it this way!" is heard often enough, and many companies lack the will to constantly improve their processes and learn new approaches, so there are practically no further training measures in such companies. Unfortunately, after a few decades of standstill, the developers' work has been made unnecessarily difficult, severely impacting productivity.

Over time, I introduced automation at this company, simplified the test setup for the development team, created automated test environments, and made everyday development more efficient and enjoyable.

However, I quickly realized that automation is one thing, but its use and acceptance from managers and directors is quite another. Even if most of the automation could have been used with little effort, it was essential that people got to grips with it. This only works if there is time for it and process improvements are also desired.

Even though I was not familiar with DevOps as a term and culture at the time—both were still in their infancy—it quickly became clear to me that this culture must be firmly anchored in the company. There must be the will to constantly improve, to question complex processes, and to cut out any that are deemed useless. The entire organization must actively break down barriers between teams and areas of responsibility so that work can be carried out efficiently.

While I worked at this company, I attended two different universities in succession: first a university of applied sciences and then a university where I completed my master's degree. I completed one or two group projects in which I had to work with a team of students to develop and deploy a software project. I noticed that although my various team members had decent programming skills, hardly anyone had ever worked on a "real" software project.

This meant that although basic development knowledge was available in theory, hardly anyone I worked with on these projects had any idea or practical experience of what modern software development should look like today, from basic tools such as Git and continuous integration tooling to correct reviewing, testing, and deployment procedures.

Although many of these topics were mentioned at certain points in my courses, the students lacked practical experience to bring the individual points together.

A university project like this would actually be a good starting point for trying out DevOps techniques, as it would give you an empty playground where you can try out new processes and tools. Companies tend to be confronted with projects that are older than most students themselves, making it much more difficult to make the leap to new DevOps techniques, as it is difficult to leave the beaten track. If you are currently in training or at university, I hope this book will help you understand the basics of modern software development according to DevOps principles at an early stage.

After my studies, I had the opportunity to work in other companies as a consultant, where I often recognized a similar pattern: entrenched, rigid structures and processes in which different teams throw tasks and problems back and forth between each other instead of working together collaboratively.

This always had to do with a corporate culture that did not enable good cooperation; in some cases, the corporate culture even actively prevented it. Such a culture almost always harmed the end users: Useful functions and urgent bug fixes were often delayed in reaching them. Users were understandably dissatisfied with the products. In private conversations with friends and acquaintances in the IT scene, I often heard similar stories.

I have been working as a solutions architect at GitLab since the spring of 2020 until October 2024. I mainly advised companies from Germany, Austria, and Switzerland to help them implement DevOps with GitLab. This job has made me realize that although DevOps is being used almost everywhere, including at large companies with thousands of employees, the focus is still almost exclusively on the tool landscape and, therefore, on the technology. The culture is adapted only with difficulty, if at all.

All these experiences have made it clear to me that there is still a long way to go in the practical implementation of DevOps. In numerous discussions with customers, people often paint the wrong picture of DevOps or oversimplify it. With this book, I want to simplify the entry and transition into the DevOps world and thus contribute to improvements in work culture, employee satisfaction, and general productivity.

1.4 Target Audience

The book is basically aimed at anyone who has a role in the software development lifecycle, including not only development teams and operations teams, but also the business teams and managers who set the direction. DevOps is relevant for everyone in an organization, and all teams must align themselves with DevOps principles; only then can the implementation be successful.

The book is therefore aimed at two overlapping groups. First, it aims to provide a general, clear description of what DevOps is and how a DevOps transformation can succeed. This part of the book is aimed at anyone who wants to take a closer look at DevOps. These include not only people on cross-functional teams with a focus on development and operations, but also those in management with decision-making powers. After all, DevOps cannot be realistically implemented without support from the top. And this requires a profound understanding of the processes and dynamics in the entire software development lifecycle, from planning to programming and delivery.

Second, the section discussing tools and technology is mainly aimed at those who work on a DevOps team on a daily basis. However, decision-makers can also gain some insight from this section; with a better understanding of the technology, they will be able to better understand the value of the tools. Time and again, I have seen companies and their employees calling for a tool like Kubernetes without having considered whether it would be the right tool for their teams. So before you make a gut decision on which tools and techniques to use, you should ensure that you understand their underlying concepts.

1.5 The Structure of the Book

The next chapter begins with a definition of DevOps. To contextualize this definition, Chapter 3 introduces Nicely Dressed Ltd., an example company used to show you the pitfalls of the old world—of traditional waterfall development procedures—and how you can do things better in the new DevOps world.

The individual stages of the DevOps lifecycle are covered in Chapter 3 through Chapter 11, using Nicely Dressed Ltd. repeatedly as an example throughout. Each chapter contains both a cultural and technical part and includes illustrative examples to describe the DevOps principles so that you also get to know the tools and techniques that are relevant in each stage.

The end of each section includes a summary for reflection to remind you of the most important aspects of the chapter, share a few insights from practice with you, and ideally provide you with ideas that you can use to reflect on your own organization and technology. Hopefully, you will find a few starting points that you can adapt to your organization.

1.6 Feedback

This book is not just about technology. On the contrary, it's about work culture. And that is my biggest challenge. While there is usually a clear solution to technical problems, this is not the case with problems related to culture.

When it comes to changing your work culture, there are a lot of truths, and you can get a lot of things right and a lot more wrong. There are many pitfalls and numerous mistakes that can cause frustration. And unfortunately, there is almost never a universal solution that fits all companies, all teams, and all people. Ultimately, many factors in this book depend on various other factors: the industry, the size of the company, and the existing corporate culture.

And that's exactly why I'm interested in your feedback. If you have any points that you think are missing here or should be dealt with in more depth, send me an email at *mail@svij.org*. I am very interested in what has and has not worked well for you and what you have learned from the process.

1.7 Thank You!

A book like this cannot be realized without active support at various levels—both professional and nonprofessional. Accordingly, I must thank a number of people, and I hope that I have not forgotten anyone.

First and foremost, I would like to thank Dirk Deimeke, my mentor and advisor, who always gave me excellent feedback to help me make the book the way it is. And his help went hand in hand with the support of my editor Christoph Meister, who provided a lot of input to make the book better and better.

I would also like to thank my family, most importantly my extended family, my parents, Rajeevan, Shrani, Maaduri, Max, and Mythili, as well as my nieces and nephews Mithra, Mira, Riaan, and Levi, who unfortunately heard far too little from me while I was busy writing.

I would also like to thank my former colleagues at GitLab, who not only have contributed significantly to my DevOps career but have also given me a great deal of feedback on individual minor and major points of this book: Ralf Gronkowski, Vlad Budica, Kristof Goossens, Alexander Dess, Timo Schuit, Manuel Kraft, Martin Brümmer, Jörn Schneeweisz, Jörg Heilig, Sarina Kraft, Andrea Obermeier, Michael Friedrich, Sander Brienen, Ted Gieschen, Julia Gätjens, Christoph Parschau, Christoph Caspar, Simon Mansfield, Idir Ouhab, Christopher Allenfort, Dominique Top, Stefania Chaplin, James Moverley, Stephen Walters, and Janina Roppelt.

I have learned a lot over the past 15 years in the German Linux and open-source community, mainly from my colleagues at ubuntuusers.de: Torsten Franz, Stefan Betz, Christian Klotz, Jörg Kastning, Sarah Blume, Philipp Schmidt, Christoph Volkert, and Dominik Wagenführ.

Last but not least, I would also like to thank two former colleagues from my first job at otris software AG, Dr. Veit Jahns and Thomas Schmidt.

And of course, a big thank-you to the listeners of my German podcast, TILpod, who have been able to listen to and empathize with the book every month.

Chapter 2
What Is DevOps?

For a long time now, DevOps has no longer been just a fad. Many people do *something* with DevOps and have their own opinions on and understanding of the topic. If you ask them what DevOps is, you often first hear an explanation of what it means and what it is good for, but closer questioning reveals a mixture of superficial knowledge of DevOps and of a loose list of explicit technologies and buzzwords.

Since you are reading this book, you're likely interested in DevOps, have probably heard something about it, and now want to get a clearer picture of whether you can use it in your company or your team. Good idea!

Or you may be a little more skeptical. I'm not a big fan of buzzwords and superlatives, and I recognize that the term *DevOps* could be used in that way. Of course, not everything is awesome and beautiful as soon as you implement DevOps principles. Of course, not everything that lacks DevOps principles is bad. Of course, not everything with "DevOps" written on it is DevOps.

I often can't blame people who may not have a clear idea of DevOps for being skeptical. Nowadays, many companies write "DevOps" on their products or job titles because it makes them sound much more modern. What used to be *agile* and *Scrum* is now DevOps. The main goal of using such labels is to appear to be moving with the times.

Whether the implementation of DevOps in such companies is legitimate or just an illusion can sometimes be seen from the outside, but sometimes not. My favorite example is the role of the DevOps engineer, which every company wanting to become more modern seems to be looking for. This trend is beneficial for those who apply for these jobs, as they usually pay more than similar jobs without DevOps in their titles. Please don't get me wrong: There are many legitimate job listings out there for DevOps engineers! Unfortunately, there are also many listings that use the term whose responsibilities don't really relate to DevOps. This topic is discussed in more detail in Chapter 14, Section 14.3.

So let's dive straight into the topic and take a look at what DevOps actually is.

2.1 DevOps: The Big Picture

DevOps is essentially a work culture. It is not a technology, a tool, or a job title, and no, it is not a team name either, though it is derived from the two terms *development* and *operations*, which are traditionally used to describe the development and operations teams. To be able to work successfully according to DevOps principles, the principles of the DevOps culture must be practiced as comprehensively as possible in a company.

This means that not only should the individual teams be working according to DevOps principles, but the entire company should as well; otherwise, their chances of successfully implementing the DevOps culture are slim. It makes no sense for individual teams to be referred to as DevOps teams, while other parts of the DevOps principles aren't in use in the company. After all, implementing DevOps is about the big picture and not about individual teams.

Also, DevOps is not about the tools that are used. Instead, the tools support the processes, which in turn support the people. It would therefore be the wrong approach to change processes just because you are convinced that you should finally introduce Kubernetes in your company. If you are wondering why I am mentioning this again, it is essential and—although it is well known—is ignored time and again.

While agile software development is a good basis, DevOps is about looking at the entire *software development lifecycle*. This lifecycle begins with project planning, continues with development and the rollout to end users, and then starts the planning process over again using feedback from operations for future development work. In fact, you could also use the term *software delivery lifecycle*—after all, development without delivery is pretty pointless.

Implementing DevOps in a company is basically about significantly expanding the company's agile software development process. Agile software development aims to ensure that not everything is planned precisely in advance, but rather that work is carried out in shorter increments to allow the teams to be more flexible with changing requirements. A more in-depth introduction to agile software development is provided in Chapter 4, Section 4.1.

Although DevOps is supposed to be a living culture, you can of course approach it theoretically and try to summarize it with definitions. However, you must be aware that these definitions only ever represent ideals and abstractions that you must adapt to your specific environment.

Nevertheless, it is useful to know these principles, as they form a good framework that you can use as a guide during implementation. Two DevOps models have become established in practice—namely, the *CALMS model* and the *Three Ways*.

I will refer to both models again and again throughout the book, returning to their main points repeatedly so that it is clear how each model relates to DevOps principles. For now, let's get into the main points of the models. Both CALMS and the Three Ways approach DevOps in different ways.

2.1.1 CALMS

The CALMS model contains the guiding principles of DevOps. But first, let's take a step back: What does CALMS stand for? CALMS is an acronym of five values:

- Culture
- Automation
- Lean
- Measurement
- Sharing

The lean value is often considered optional, so the model is sometimes referred to as CAMS with CALMS as an extension.

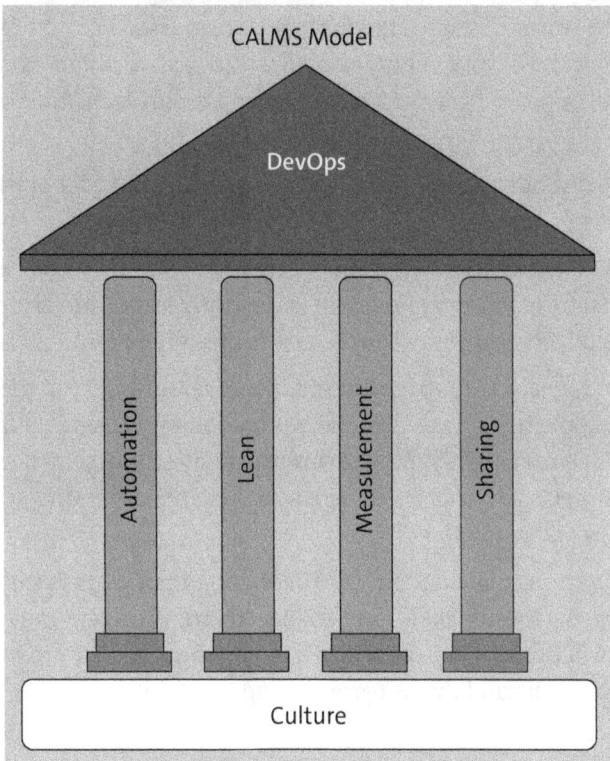

Figure 2.1 The CALMS Model

The CALMS model will help you with your DevOps transformation, as it emphasizes that all DevOps principles must be introduced equally and therefore grow together. The principles should stand on the foundation of culture; otherwise, it won't work.

Here, I will describe each CALMS value, but relatively superficially. In the following chapters, I will repeatedly refer to the individual components of the CALMS model to explain more directly how they relate to DevOps and where they can be found in the CALMS model.

Culture

Culture refers to work culture: How does collaboration take place in a team and within an organization, regardless of the organization's different roles and areas of responsibility?

Put simply, an organization that works according to DevOps principles essentially has the development and operations teams working together and not, as is so often the case, competing against each other. Instead of both teams working in silos and being isolated from each other by large, high walls, they work toward a common goal.

But this is where the first problem arises: The developers have to develop new features with as few errors as possible. The focus of the operations team is on the secure and stable use of the software written by the development team. In traditional work contexts in which these teams are siloed, these two roles can come in conflict with each other.

The culture of the company is crucial, as all the people across teams must be brought together not only organizationally, but also in terms of goals. It is important for both teams to fulfill both sets of goals.

A traditional development team that is siloed from the operations team may not be able to smoothly roll out new features when deployment is too complicated and error prone. Instead, teamwork and collaboration between the two teams is required.

A key aspect of DevOps culture is to strive for frequent rollouts of releases to keep the time from error correction to rollout as short as possible. Automation is necessary for this task. And frequent rollouts can occur only if the teams really have the confidence to roll out the software. Other aspects also play a role here, such as extensive testing to be able to publish changes safely and regularly.

And that is precisely why it is important that not only do these two groups work in this culture, but so do supervisors, decision-makers, and management. Everyone must agree to accept mistakes that will occur along the way. Changing the work culture in this way cannot be achieved overnight, and that is perfectly fine.

Automation

The automation value simply means that as many tasks as possible should be automated. Basically, the aim is to increase the speed of development work by automating as much as possible so that no or very few manual steps have to be carried out. This includes automation of tests at all levels, building the artifacts and deploying them, and monitoring after the deployment has been carried out.

It is important that the software is not rolled out blindly, but that it is well tested and rolled out automatically with a certain degree of self-confidence and self-assurance.

The problem, however, is that many companies equate DevOps only with automation. So you often hear, "We have a continuous integration/continuous development (CI/CD) pipeline, so that means we do DevOps!" This is a big mistake because those who care only about automation and less about the other aspects of DevOps are essentially automating the errors throughout the complete software delivery lifecycle.

Although automation is a good step in the right direction—after all, a CI/CD pipeline is an elementary component of DevOps at the technical level—it is important to emphasize that DevOps does not just consist of automation and a CI/CD pipeline. And simply having such a pipeline does not help if you do not use it regularly (i.e., several times a day).

Lean

Although the lean value originated in the manufacturing industry, it is now also used in many other areas, including software development. We know the most famous lean model from Toyota's industrial production. The lean value is considered optional in the CALMS model. However, for the sake of completeness, and because it includes valid and useful points, I cover it here in the book.

Lean basically concerns the elimination of waste. Everything that is garbage or waste can and should be removed. In concrete terms, this means that everything that does not add value should be removed.

Another important component of lean is continuous improvement, increasing benefits for customers and focusing on long-term goals. Other components of the lean value are the people who can and should bring about continuous improvement by encouraging them to do so.

In practice, adhering to the lean value means that you have so much confidence in the team and the software that you can also run well-tested experiments on production systems. This should get you to your goal much faster if you know what works and what doesn't without having to spend a lot of time discussing it.

Even if lean did not originally come from IT, it is very much in line with IT and, as a result, with DevOps principles.

Measurement

In general, the measurement value is about metrics, essentially using quantifiable data to determine what works and what doesn't work in the development process. Metrics can be large or small scale.

A classic example of a metric that can be tracked is the *lead time*, or how long it takes until a code change (such as a bug fix) is both implemented and rolled out on the production system and thus goes live.

Part of the lead time is the *processing time*, or the time from the developer's commit until the change is rolled out on the production system. If processing time is measured, you can make a better statement about how good the processes around the CI/CD pipeline are in order to be able to roll out changes quickly to end users, not only in an emergency but also under normal circumstances.

A basic distinction is made between process metrics, operational metrics, and business metrics so that, ideally, many relevant aspects are regularly kept in view and evaluated.

Over the course of the book, you will learn about many other metrics that are important in the DevOps world, such as the DORA metrics, discussed in Chapter 12.

Sharing

Sharing primarily means that everyone is able to learn from each other. This does not mean that new technologies should be disseminated within the organization and that everyone should suddenly be an expert in every technology. Rather, it means that problems should be documented and solutions should be shared within and outside the organization so that no one makes the same mistakes again.

This includes both minor issues and major problems that would have led to lengthy downtime. It is important that the existence of errors is fundamentally accepted and that we learn from them. And that acceptance should apply across team boundaries. This philosophy has an effect on culture: Only by dealing with mistakes in a healthy way can minor and major problems be shared within your own team or the wider world without fear.

Conclusion on the CALMS model

The CALMS model is a good foundation for understanding and categorizing the core features of DevOps. It is a good guide for introducing cultural changes in your organization. Note, however, that like many of the points covered in this book, the model is not a one-size-fits-all solution.

Also note that the CALMS model cannot be used to measure the success of DevOps; the DORA metrics exist for this purpose and are discussed in more detail in Chapter 12.

2.1.2 The Three Ways

In contrast to CALMS, the Three Ways model is based on the idea that all DevOps principles can be derived from three basic ideas. There is some overlap with the CALMS values, albeit from a slightly different perspective and therefore with a different focus.

Basically, the Three Ways model is focused on improving the flow from the development team to the operations team—and therefore from the business to the customer. This includes the feedback loop back from the customer to the business, but also regular feedback loops throughout the entire process.

Throughout the rest of the book, as with the CALMS model, I will refer to the Three Ways regularly so that it is clear how their individual aspects, methods, and ways of working relate to DevOps.

The Three Ways were largely developed by Gene Kim, who presented them in a blog post in August 2012 (*https://itrevolution.com/articles/the-three-ways-principles-underpinning-devops/*).

This post served as the basis for his book *The DevOps Handbook*.

The First Way: Systems Thinking

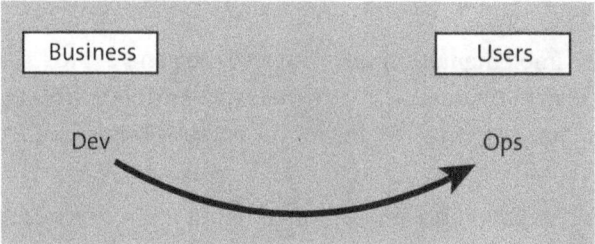

Figure 2.2 The First Way

The First Way is about understanding the entire development process as a system and as a flow. Systems are usually complex and consist of many different components that, depending on the organization, have to be developed and provided by many different teams. It is important to understand the system as a whole and the flow between the individual components of the system. One aspect of the First Way is that changes should flow as smoothly as possible from the development team to the operations team. In the traditional way of working, this is usually not so smooth: Far too often there are hurdles that lead to delays and have to be overcome manually.

Ultimately, the work of the development team can be equated with the business; after all, it is this team that provides the value that the system's customers want to use. And this is possible only if the application is provided accordingly. This in turn goes hand in hand with the swift and smooth path from the development team to the operations team. In concrete terms, this means that changes should be delivered as quickly as possible so that the added value can be used quickly.

Also, the size of the changes must be adjusted to ensure a smooth process. If you want to bring huge changes that are made over the course of six months of work to the customer, you may have several problems on your hands; for example, many changes may not have been properly tested or implemented. And, as happens far too often with this kind of work process, further improvements may already have been implemented based on these changes, which would complicate the deployment. That's why it helps to keep the changes as small as possible to simplify the flow from the development team to the operations team and thus from business to customer.

In addition, the work that is done must be made visible to everyone. Far too often with traditional workflows, work is not visible to everyone, both within the team and outside it. This means that progress is often hidden. In such cases, the team may try to conceal slow progress and may be nervous that possible errors will be discovered by others.

This is particularly problematic between teams that are dependent on each other. The quality assurance (QA) team should be able to see what progress is being made so that quality deficiencies can be rectified at an early stage. In the same way, the operations team should be able to see at an early stage what is relevant for operations so that questions about the infrastructure do not have to be revisited once the changes are completed.

This concerns not only the actual implementation work but also the general progress regarding the planning. The status of individual tasks is crucial here: Are they currently being worked on? Are team members currently waiting for the completion of another dependency?

A planning board can be useful for making sure everyone on the team and across teams can keep track of the work. An *issue board*, often implemented according to the Kanban method, is a classic example. A simple Kanban board contains three columns with work packages that are assigned to either "To do," "In Progress," or "Done" (see Figure 2.3). This ensures good and simple visibility. In practice, several different levels are usually used to make cases such as "In Review," "Waiting," or "Blocked" quickly visible.

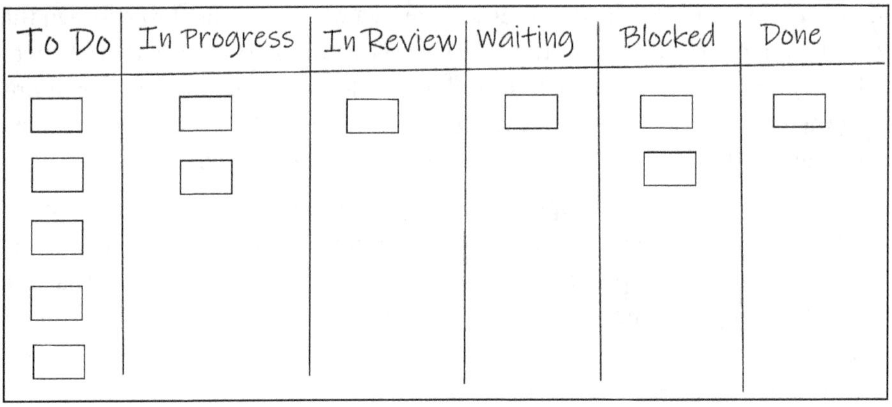

Figure 2.3 Example Kanban Board

The basic rule is to always keep the work visible, as this is the only way to identify problems at an early stage! This is important both within teams and between different teams.

Above all, keeping the work visible helps you to see whether too much is being developed at the same time. With long development times in the traditional approach, there are many tasks that have to be implemented over a very long period of time. It is simply not possible to do everything at the same time. And since work and time resources are limited, your team should not work on too many projects at the same time.

Therefore, how much work is happening in parallel must be visible. Having too many work packages in the "To Do" status, perhaps even several tasks that need to be completed by one person, is a good indicator that the developers are overloaded, simply have too many tasks at the same time, or do not have the proper information on task prioritization.

An issue board can also help by visualizing how many tasks are currently in a certain status or with a certain person. This analysis goes hand in hand with the visualization of work statuses. This allows you to pay attention to the second aspect: limiting the work that is currently being carried out to ensure flow.

The whole process will almost never run completely smoothly, which is why it is always important to remove unnecessary processes and streamline the work. And this applies to the entire value chain, including eliminating unnecessary handovers and approvals.

However, much more important than streamlining processes is to build the systems in such a way that the entire system is understood as well as possible, even if not everything is thoroughly understandable down to the last detail. This is precisely why the First Way is also called *systems thinking*—the entire system should be considered. The main focus is on both reducing complexity and making the system comprehensible. The two go hand in hand.

Only by understanding the path of the product throughout the process as a whole system can your development teams identify at an early stage problems that would otherwise have remained hidden. Otherwise, the experience gained by the operations team during the operation of the service would not reach the developers, who in turn would not be able to benefit from these learnings.

Of course, the company is just one example: Systems thinking means that all components and their various dependencies must be taken into account, understood, and monitored. This is possible only if the big picture is fundamentally understood as a system and that you don't get tangled up in isolated solutions from different areas of expertise.

A common practice is to have the business team draw up the requirements, which are passed on to the architecture team, which then hands over the task packages to the development teams, which then, once the work has been completed, passes the work on to the QA team and then to the operations team, which then puts the software into operation.

There are transitions between each of these teams that need to be coordinated and where approvals take place. These transitions take an incredible amount of time, but a lot of knowledge is needed early on. This is a major problem that I'll return to and address repeatedly throughout the book.

These transitions, including any decisions and approvals that must occur during these transitions, must be reduced to only what is necessary. For example, approvals that come from higher up in the organizational structure are often required even though these decision-makers cannot make a truly well-founded statement about whether the given tasks have been completed as desired. This information usually comes from the teams themselves anyway, so why not hand these approval tasks over to the teams themselves?

Handovers can be reduced and simplified by using automation to make recurring tasks, which would otherwise only lead to further waiting times, simpler and more efficient. The most obvious example is automation of deployments, which not only saves work, but also prevents potential errors.

Smaller tasks, such as providing access to data for systems when new employees start, could also be automated. Operations teams often have to manually create accounts and activate access, tasks that are good options for automation!

The topic of automation goes hand in hand with the next point: During development and deployment, there are and will always be bottlenecks that should be uncovered and eliminated to keep the flow between from the development team to the operations team, and thus from business to customer, smooth. Any approach that could help should be introduced to facilitate the flow. The goal is to recognize and remove recurring problems and hurdles.

The Second Way: Feedback

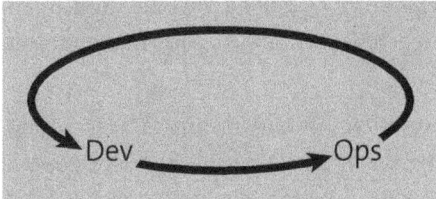

Figure 2.4 The Second Way

The First Way is primarily about understanding the entire system to find ways to optimize the flow from the development team to the operations team and thus from business to customer. The Second Way is exactly the opposite: The needs of the operations team and the customers should flow to the development team and thus should inform business decisions.

The Second Way is primarily about keeping the incoming feedback from customers and operations constant and accepting it quickly. This approach is directly related to the First Way, because rapid and constant feedback is possible only if users are provided with frequent changes, which can be achieved reliably only if the steps taken to get there are small.

One prerequisite for this approach is being able to work with complex systems as safely as possible. This is not an easy task in principle, as, unsurprisingly, it is difficult to gain an overview of complex systems. One solution is to set up an interdisciplinary team so that each person has a broad view of the system but with individual specializations.

With this approach, problems that inevitably occur in the production system are used as a basis for initiating improvements so that development runs more smoothly in the next iteration. Knowledge gained from this process must be shared within the team so that everyone can understand the complex system and deal with errors.

As modules from different teams work together in a complex system, problems must be made visible so that they can be worked on together: Nothing should be hidden or ignored.

Automated testing is a central aspect of being able to work safely with a system. It is important not only to test the central processes but also to consider issues that could occur outside the expected functioning of the system, known as *edge cases*. Regular feedback also helps in identifying such cases, but not if the system is deployed only in very rare cases. Problems that are relevant to the software design, for example, would be discovered only at a later stage, which would make a correction more time-consuming and therefore much more expensive.

The feedback loop of rolled-out code is ultimately the slowest form of feedback. It is therefore important to use automated tests that identify errors as quickly as possible; that way, errors that become apparent during these tests will not lead to problems for customers and will not need to be collected and tracked via bug reports or issue trackers.

Automated tests are available at various levels (I will go into more detail in Chapter 7):

- **Unit tests**
  These tests are executed locally during and after programming to provide quick and easy feedback.
- **Integration tests**
  These tests check interactions between several components.
- **System tests**
  These tests check the entire system; therefore, they run slower than other tests.
- **Acceptance tests**
  These tests check whether the result meets the requirements.

As always, it is important to automate as much as possible and keep tests as simple as possible.

Early feedback on changes made is essential so that problems are not pushed further and that the roots of the problems are addressed. It is about avoiding and reducing *technical debt*.

Rapid feedback is gained not only from using the various testing types and carrying out frequent deployments but also from reviews from colleagues from the team or from other teams. The goal of conducting reviews is to find opportunities to improve the code and to gather knowledge from the environment in which the code will run from the team.

The Third Way: Continuous Learning

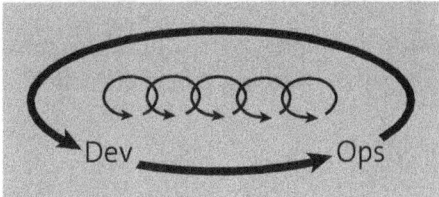

Figure 2.5 The Third Way

The Third Way revolves around continuous learning. Here, your main focus is on being open to constantly learning about new techniques, processes, and tools. Of course, learning is not an end in itself; rather, it is an approach to solving problems. You should always consider which problems currently exist and which errors occur again and again. You should not simply accept or ignore these stumbling blocks, but rather see them as an opportunity for improvement. This approach must be deeply rooted in the culture of the organization.

This approach requires a good team culture around managing errors. I have observed teams that have an unhealthy error culture. You can understand a team's error culture by asking, How does it deal with mistakes when they happen? Are mistakes allowed to happen at all? Does the team conduct experiments on the live system to gain insight into whether it is behaving as intended?

Companies with a poor error culture can be recognized by their many silos. Teams in such companies only ever think about their own way of working and their own problems. Inconsistencies in collaboration with other teams are not corrected at the source; instead, these teams build workarounds.

Inconsistencies occur when teams do not know how other teams work or when they are completely isolated from and have no communication with each other. Also, important decisions and background information may not be shared and may be treated like private secrets.

This type of error culture leads to major problems when errors do occur, such as one that causes major downtime. For example, an organization with such a culture may quickly seek a guilty party to blame errors on instead of analyzing the root cause of the actual problem and dealing with it. This finger-pointing should be avoided. Instead, every effort should be made to understand the problem together and work out solutions to prevent it from happening again in the future.

Mistakes should not be covered up but actively shared so that everyone can learn something from them. No one should be afraid to make mistakes, but that doesn't mean everyone should be careless either. Finding the right balance between accepting that mistakes will happen and expecting your team to be careful not to make them is the great art here.

How errors are reported is also particularly important to a good error culture. This is even a step before the actual error handling. Are errors taken seriously, or are they ignored? Are teams criticized when errors are discovered?

One aspect that I have already touched on in the last few sections is the organizational structure of the company and the way in which the various teams handle information. You often see that many organizations completely seal off their teams from each other so that neither code nor project management can be viewed. Information is treated as top secret, even within the organization, so nobody can see any of this information.

And yet there is often no reason to do so! This approach usually reinforces the silo mentality, simply due to the organizational structure of information. Generally speaking, transparency creates far more advantages than disadvantages. An important point that I will return to again and again over the course of this book is the importance of transparency in all aspects of the software development lifecycle, because without it, close collaboration is largely impossible.

However, learning is not just an organizational aspect. For example, developers can learn from a living object; by regularly rolling out their own changes, even if they are very small, they can test small parts of potentially larger changes and then incorporate the findings into further development.

However, these types of learning activities must also be covered and supported accordingly by management personnel; a typical development team is unlikely to carry out experiments on live systems (even if the developments are well tested and, therefore, safe) if these experiments are neither approved nor wanted by the team's managers.

Finally, a culture of continuous learning also means that information—such as new findings and information on how individual teams have rectified both minor and major errors—should be shared throughout the entire organization. This information-sharing can be done through blog posts or presentations, for example. Such activities can help to prevent future errors or inefficiencies, as the various teams can learn from how other teams have dealt with certain problems.

2.1.3 Conclusion on the Three Ways and the CALMS Model

And that brings us to the end of our discussion on the Three Ways and CALMS. By now, you should recognize that these two DevOps models are very similar but approach aspects of DevOps with a different focus. The principles of both models will run through the entire book. I have deliberately avoided using too many examples to keep the introduction as short as possible. Throughout the rest of the book, you will regularly encounter, and should hopefully recognize, the aspects and principles of DevOps presented here.

> **Note: Reflection**
>
> In this subsection, you have learned about the CALMS model and the Three Ways. One of the main problems with many organizations' uses of the CALMS model is that they rely purely on automation and invest little to nothing in the other aspects of the model.
>
> Reflect on your team and your organization. Do you recognize which aspects of your organization can be handled better and where there is a need for optimization? The next sections of this chapter will go into greater depth on how DevOps is often misunderstood.

2.2 Misunderstandings about DevOps

Now that you know the basic ideas behind DevOps, I'm sure you'll agree that it all sounds perfect: shorter paths to deployment, simpler processes, more communication, better handling of errors. Of course everyone wants these things for their organization; nobody would disagree here.

But this list of virtues and good ideas is not enough; the theory must be put into practice, meaning that these virtues and ideas must be lived and become part of the DNA of your team and the organization. In fact, this is much easier said than done, because, naturally, things often go wrong when implementing DevOps.

Typically, things go wrong when a company does not understand DevOps as a whole or incorporates only partial aspects of DevOps: No, a joint Slack channel for the development and operations teams is not enough to implement DevOps.

In this section, I discuss the common mistakes and misunderstandings of DevOps that you should be aware of so that you can recognize how our fictitious company, introduced in the next chapter, correctly puts DevOps into practice.

2.2.1 Too Strong a Focus on Automation

Every now and then I see companies and teams that are introducing DevOps into their workflows but focus too much on automation, either ignoring the other aspects of DevOps or focusing far too little on them. In the end, these companies and teams have often only achieved the A in CALMS, have not established the right culture, and have ended up with a lot of waste. They do not track metrics, and sharing is still a foreign concept. Figure 2.6 shows that they have not erected a solid building in this way.

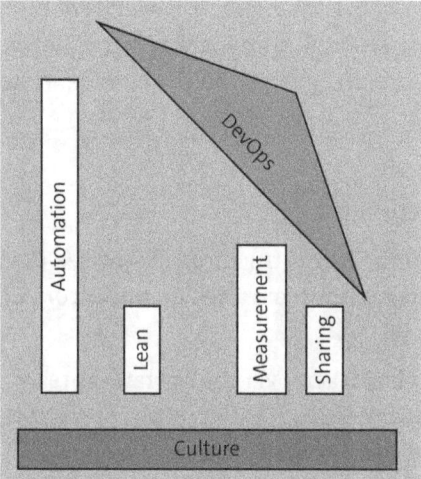

Figure 2.6 Imbalance in DevOps Due to Strong Focus on Automation

In these scenarios, the companies hire DevOps engineers, who write and maintain a CI/CD pipeline, and that's it. Although a CI/CD pipeline is an elementary component of the technical part of DevOps, it does little to change the other aspects; the teams continue to work in exactly the same way as they did before.

I often see CI/CD pipelines that can be run fully automatically in theory but only after a senior manager who may not have been involved in the technical implementation or who cannot assess the quality of the project has approved the deployment. This means that an important aspect is missing—namely, the assumption of responsibility by the team.

The result is the worst of both worlds: cumbersome manual processes and additional technical overhead due to the tools of a CI/CD pipeline, which now has to be maintained in addition to the actual product.

2.2.2 With DevOps, but without Tests!

One prejudice that is often heard is that DevOps produces much worse code, as the changes are rolled out far too quickly. And the assumption is that if something is produced quickly, then it must be of low quality. Proponents of this view assume that DevOps means poorer QA, as new code is simply thrown directly onto production environments, according to the motto "users can test whether everything works."

That is, such a program is seen as typical banana software: The program develops with the customer.

That's not true, of course. If a DevOps workflow is done correctly, the opposite is the case—namely, that testing is much more structured and targeted in teams that work according to DevOps principles, as a great deal is automated.

At the end of a development line following DevOps, many more tests will have been carried out than in traditional development cycles. So the point here is that all levers are set in motion in order to have a secure parachute, giving you the confidence to jump (or, roll out the changes).

2.2.3 Incorrect Understanding of Team Structuring

A classic misunderstanding about DevOps concerns team structuring. Some see DevOps as simply a combined development and operations team. This is almost correct, but with various restrictions.

Others see DevOps as a development team that has taken over the operations team's responsibilities; in this view, admins are no longer needed! That's not DevOps, either. Admins are still necessary in a DevOps workflow. However, the tasks and responsibilities of all the people involved change in a DevOps environment; they no longer work in the traditional way (see Figure 2.7).

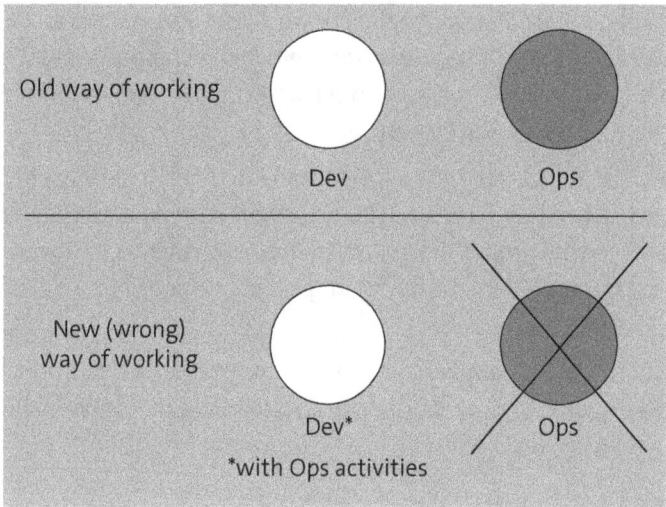

Figure 2.7 Incorrect View of DevOps as Elimination of Operations

In some cases, companies put in place separate DevOps teams that sit between the development and operations teams, and on a permanent basis. During a company's transition to DevOps, such a team could be good idea, but certainly not in the long term because that is not the purpose of DevOps!

There are numerous other misinterpretations regarding team structuring that I will address throughout the book.

In essence, a proper approach to team structuring is basically about bringing the development team and the operations team together and improving the strengths and eliminating the weaknesses of the respective roles. But the question is, how can these teams be combined? Which other teams or roles should be included? For example, the QA team, the security team, the business team (i.e., the people who ultimately make the business decisions), and even the finance team could be involved.

2.2.4 Not Tearing Down All the Walls

Working according to DevOps principles does not mean that *all* old processes and habits have to be thrown out the window. This is precisely the next problem that often becomes apparent. In the DevOps context, people often talk about breaking down the wall, tearing down the fences, or removing the silos.

It is therefore a question of breaking down the barriers between the different teams. A typical mistake is that, as shown in Figure 2.8, the walls between the development team and the operations team are torn down, but not between the other teams outside of engineering.

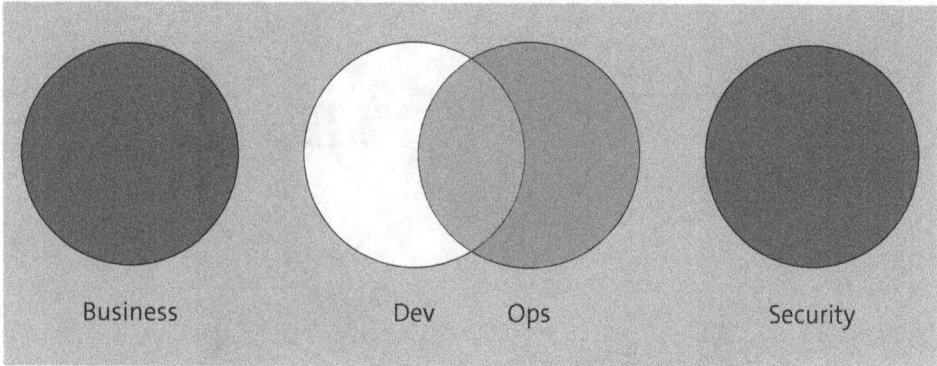

Figure 2.8 Siloed Business and Security Teams

Cultural change can be achieved only when everyone changes together, and this applies to all roles and areas of responsibility in the organization. Managers should also implement business decisions based on feedback from the development and operations teams, as they are best positioned to assess what works and what causes problems.

This idea applies to finances and budgeting as well: The lower the barrier between the decision-making level, which decides on budgets, and the technology level, the better that informed decisions can be made. For example, consider a team that is working on migrating a data center to a public cloud. Details of such a project need to be considered by both the technicians and managers; the project would result in a different billing model, so the pros and cons making the migration need to be discussed. No department can make this decision alone.

Therefore, introducing DevOps involves bringing together not only the technical roles but also all the other areas of responsibility. A DevOps culture that is practiced only in the machine room is doomed to fail.

2.2.5 Tools over Processes over People

The next problem is that many companies introducing DevOps start with purchasing tools, then adapt their processes to DevOps principles, and then finally start selecting people. You need everything—tools, technology, processes, and people—but you need to approach the implementation of these things in exactly the opposite way.

The principle is always people over processes over tools. This means that people come before processes and processes before tools, as mentioned at the beginning of this chapter. This guiding principle comes from Scrum and has an important meaning in DevOps (see Figure 2.9).

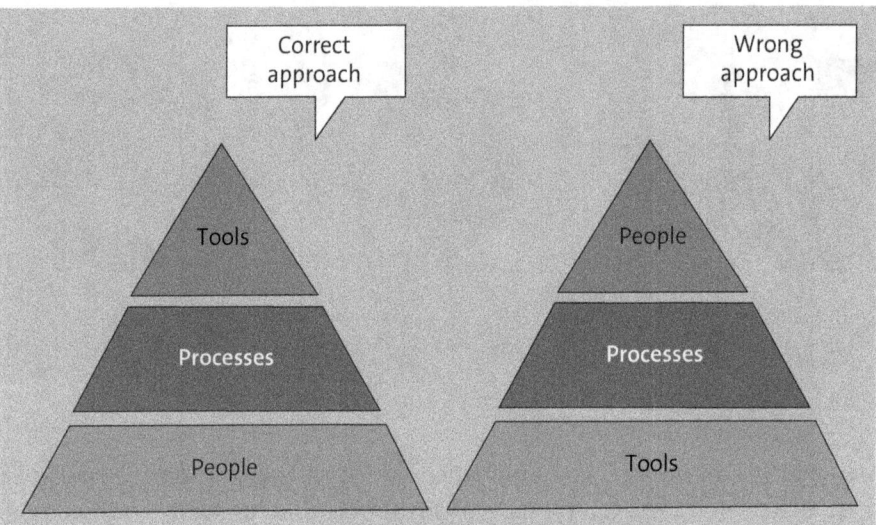

Figure 2.9 People as the Foundation of DevOps

The tools you select are important and will help you and your team implement DevOps principles, but they are useless if your processes are poor. And even if your processes are world-leading, they won't help if the mentality of the people on the team or in the organization is still wrong.

So the first step is to select the people in your organization's teams. Only then should you work on adapting your processes to DevOps, and only then should you select the tools. While it's great if a company has a CI/CD pipeline to integrate and roll out changes, it doesn't help much if every intermediate step has to be approved by a number of managers. Obtaining several approvals for a deployment not only takes quite a long time, but it also shows that you don't trust the team and, therefore, the people running it, even though they do the main work and are best positioned to assess the progress and quality of the project. Autonomous and independent teams are a core characteristic of DevOps!

But this trust also has to be built up slowly. And that is neither very easy nor can it be done overnight. Trust can be developed only with support from the very top. If there is a lack of understanding and trust, a well-functioning organization that works according to DevOps principles can almost be ruined.

How to select people for a DevOps implementation is touched on repeatedly in the following chapters, but Chapter 12 provides a deeper look.

2.2.6 One to One Copies of Working Methods from Other Companies

In the spirit of DevOps—as mutual learning is one of its core principles—many companies and organizations share their experiences of adapting DevOps at conferences and in blog posts. This information is useful for companies looking to make their own transitions to DevOps.

However, it is not useful to copy the working methods, structures, or tools of other companies without adjusting them to your own needs and environment. Every organization is different. Each has its own challenges that need to be overcome. And many have their own compliance guidelines, some of which are prescribed by law.

You will not find a simple checklist in this book that teaches you how to introduce DevOps in a few short steps. It would be nice if it were that simple, but it's not just a matter of a few general points that you can tick off of a checklist, giving you a certified DevOps environment.

Rather, I intend for this book to give you an idea of what a DevOps culture involves. Nevertheless, at the beginning of Chapter 12, I provide a description of what the first steps of a DevOps transformation can look like. But, as already mentioned, don't take this as a checklist, but rather as inspiration for your own transformation!

It is important to learn many aspects from other companies without copying their approach one to one. I often hear in conversations that people take certain actions simply because, after all, Google, Amazon, Netflix, or some American start-up do it that way.

However, very few companies are in any way comparable with such big companies. Many factors are involved, such as company size, company age (and, therefore, the age of the codebase), the company culture and willingness to change, and, above all, the industry. A company like Google must and can work differently than a German car manufacturer, a Swiss bank, or a US public sector organization, and this is not only due to legal regulations. Several industries with their own challenges collide here.

Don't get me wrong: We can and should learn a lot from the big tech companies! However, you shouldn't try to copy their approaches, as your company's needs and culture are likely very different; instead, you should adapt these companies' approaches wisely to your own context.

Many of the points in this section of the chapter may not yet be immediately clear to you, as I have not provided concrete examples to help you visualize them. However, I revisit these points again and again throughout the book, which should make them much clearer.

> **Note: Reflection**
> You will no doubt have recognized one or two points in this section from either your own company or from other companies.
>
> If you want to actively drive the DevOps transformation at your company forward, consider writing down one or two of the misconceptions you hear about. When you're communicating the overall DevOps vision, which is discussed in more detail in Chapter 11, you will then have a few misconceptions you've heard firsthand that you can reference and refute.
>
> An initial opinion that needs to be dispelled is to work according to the motto, "Never touch a running system!" Changes are always necessary, and they should be made without fear, but not blindly. An open culture of error is essential for this and must be constantly fostered.
>
> This is where metrics can help you to make fact-based decisions. Of course, metrics have to be determined first. Reflect on what information you would like to have in order to gain better insight into and a better feel for the application.

2.3 The DevOps Software Development Lifecycle

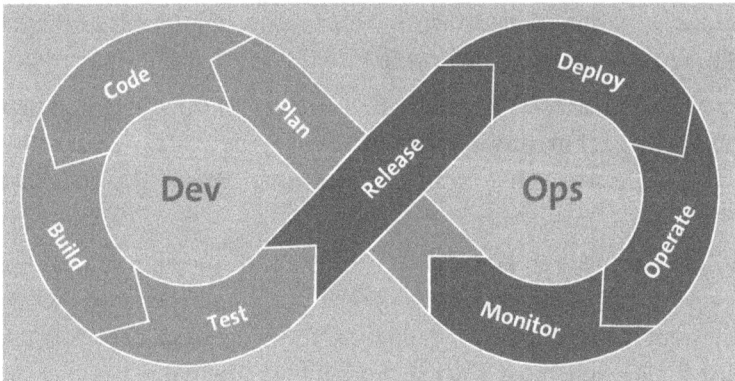

Figure 2.10 The Infinite DevOps Software Development Lifecycle

The software development lifecycle is depicted in Figure 2.10 as an infinity sign (i.e. a horizontal figure eight). This depiction communicates that the software development lifecycle is a continuous process that is never complete as long as the software is in use. It consists of various components that go hand in hand. Many aspects overlap and cannot be clearly defined, as you will see throughout this book.

It is important to emphasize that this lifecycle should be kept as short as possible in order to implement DevOps principles with all their advantages.

The first part of this book focuses on individual aspects of the DevOps lifecycle. Using the example company that you will get to know in the next chapter, I will show you how DevOps principles can be introduced for each sub-step and what benefits result from doing so.

Note that the lifecycle represents a flow of information and activities and that it starts again and again with the shortest possible iterations. This contrasts with a waterfall-like development model in which each stage is worked through continuously and in which development is actually considered completed. The stages are as follows:

- **Plan**

  Projects usually start with planning. The planning process in a waterfall model is often very detailed, comprehensive, and inflexible; everything is planned in advance, and an attempt is made to avoid having to add any changes later.

  Such an approach doesn't align at all with the DevOps world. Agile software development concepts such as Scrum and Kanban are now very widespread and an integral part of everyday work according to DevOps principles. Planning steps according to the agile way of working are a basic principle in the DevOps world. Chapter 4 provides some insight into agile software development planning in the context of DevOps.

- **Code**

  Source code is written when the team is programming the project. You might be thinking, "Okay, but I'm already doing that; what makes DevOps coding any different?"

  And this is exactly where there is a lot to consider that can go wrong. Not only can many mistakes be made when planning projects, but the foundations can and must also be laid for programming and the processes surrounding it so that the process fits in well with the other components of the DevOps lifecycle. It's about collaboration within the development team, but also with other departments. This is examined in more detail in Chapter 5.

- **Build**

  Source code is useful only if it can be built into artifacts that can be easily published. This step is therefore about the importance of build tools and processes. The goal is reproducible automatic builds that can be built from source without manual intervention. This also includes managing dependencies with the build tool of choice and optimizing build time to keep development productivity high. More detail is provided in Chapter 6.

- **Testing**

    While traditional companies often still have completely separate QA teams, the cross-functional teams of a company organized around DevOps principles are responsible for unit and integration tests.

    The aim is to move the tests as far forward as possible in the development process and integrate them into the pipeline in such a way that there is early feedback on whether they fail. Automation and collaboration between the formerly standalone development and QA teams play a major role here. You guessed it: This topic is the focus of Chapter 7.

- **Release**

    The result of the pure development process is the finished artifact. There are many different creative solutions for publishing artifacts. The purpose is to build a standardized package that is also reproducible. This is the basis for the next steps to enable smooth deployment in the deployment process.

- **Deploy**

    Once a software package has been built, the previously written and built artifacts are published. Release management, which is often simply referred to as *deployment*, is about how the application is rolled out to end users. If you do it right, this process works so well that you can experiment without major risks and don't have to shy away from releases.

    As the release and deploy steps of the software development lifecycle are closely linked, both topics are covered together in Chapter 8.

- **Operate**

    After the project has been written, built, and deployed, it still needs to be operated. This stage is about what to do if services fail partially or completely and how to deal with the errors. Chapter 9 deals with the operation of the application.

- **Monitoring**

    Monitoring is closely related to operations; it helps inform the team of problems as quickly as possible. Because all the various roles work on the same team, a great deal of synergy can be created so that release and monitoring can take place virtually simultaneously.

    This also means that you can run various tests and experiments on the production environment with real users in order to learn from these findings, which can then influence the next development cycle. In Chapter 10, I describe the core elements of monitoring.

- **Security**

    The last aspect of this list, but not the last aspect in the DevOps lifecycle, is security. As can be seen in Figure 2.10, the topic of security must run through the entire lifecycle, as security standards and principles should be integrated everywhere and cannot (as is so often the case) be added at the end. After all, why include an operations team but leave out the security team?

It is important that security and data protection issues are considered from the outset in all parts of the development and deployment process. Security is not an aspect that can be added at a specific point; it must be an integral part of the entire system. Security is discussed in Chapter 11.

Finally, Chapter 12 covers how you can implement the principles of DevOps culture in your own environment. DevOps platforms, which I briefly introduce in Chapter 13, can help with this.

> **Note: Reflection**
>
> The individual parts of the software development lifecycle are important. It is important to remember that they are inextricably linked. You may be thinking, "Yes, well, I'm already doing the code, build, and test steps of the software development lifecycle, so I'm already halfway there!" That's good, but it's only half the truth.
>
> A focus on coding, building, and testing is common. In practice, however, it turns out that operations is often neglected or even outsourced to other teams and that security is often overlooked.

2.4 Summary

In this chapter you got a short overview of what DevOps is and why it's always important to understand and follow the CALMS model and The Three Ways. Additionally, you should have a better understanding of what DevOps is not and may have even identified some issues in your current organization.

Chapter 3
The Example Company

It is now time to introduce the example company we'll use throughout the rest of the book to demonstrate how DevOps can be implemented in an organization. First, in this chapter let's take a look at how the company currently works without DevOps. In the following chapters, I show how the company applies individual aspects of each stage of the DevOps lifecycle and the advantages they bring.

This practical example will serve to show how various DevOps principles improve a company's productivity, collaboration, and maintenance.

The easiest way to introduce DevOps principles is to start on a greenfield, meaning that no major changeover or migration is necessary and all DevOps principles can be implemented immediately.

However, very few companies will be able to start on a greenfield; most companies will have to work with old code, established processes, and possibly even antiquated tools as they implement DevOps principles. It's often not just the code that's been hanging around for a while but also the mindset of the people who work there; they may say, "We've always done it this way, and it has always worked!"

And that is precisely the biggest challenge! Oftentimes, these companies have good reasons for working the way they always have. However, whether it will be possible to continue working in the same way in the future is another matter. Organizations can be sluggish, due to not only the people who work there but also the code, which often comes with high technical debt.

Nicely Dressed Ltd. should serve only as an example, but you may discover some similarities between its problems and the problems faced by your own organization. Even so, the examples and solutions discussed in this book cannot and should not be copied one to one.

As already stated, the DevOps model is not a model solution that you can impose on your company in just a few steps. Instead, the goal of presenting this model is to give you some ideas for how to introduce DevOps into your company, outlining some common problems and discussing some solutions to those problems using DevOps. Your task is then to adapt these approaches to the reality of your environment. It is always important that you look at the big picture or—as the Three Ways would put it—always think of the whole system.

3 The Example Company

3.1 nicely-dressed.com

Nicely Dressed Ltd was founded at the end of the 1990s and is therefore over 20 years old. It earns its money by selling clothes both online at *nicely-dressed.com* and in brick-and-mortar stores. The online shop is a complete in-house development, and there is still some source code from the early days. Some parts of it are over 20 years old.

It is hardly surprising that the code has grown historically and, therefore, has technical debt that would certainly be avoided if the online store were completely rewritten according to the current state of the art. In the time since the online store was first developed, software development practices have changed again and again, as have the technologies used.

In practice, however, the code works, and the online store generates good, but not huge, profits. So far, management has been satisfied by the sales numbers.

But over time, sales, and therefore profits, began to shrink, so the company decided to place a stronger focus on the online store. In the current economic situation, running a physical store is no easy task, but online stores are also facing stiff competition.

Although the online store is basically working and changes are constantly being implemented and rolled out, there are also some major problems. Employees are voicing concerns about problems and limitations in the development process that are leading to widespread dissatisfaction.

3.2 The Development Model

The first problem that many employees in the company have recognized is that although the company introduced an agile framework to the development process a few years ago, it is still very much based on the waterfall model (see Figure 3.1).

Officially, development of *nicely-dressed.com* does not work according to the waterfall model, but essentially only the implementation phase was transferred to the agile framework. The rest continued to be strictly separated, so the advantages of agile software development could not unfold.

The main problem for the development team is that everything is planned in detail at a very early stage and then designed in the analysis and design phases of development. This process usually takes several weeks or even months. Only then can the work start for the development team, which then has to implement changes over several months. There is no coordination or even a feedback loop between the implementation and the design processes.

With this approach, many ideas decided on in the analysis and design phases turn out to be unrealistic and cannot be implemented in time. Other requirements have to be coordinated, planned, and then developed several months in advance; if some of those requirements change in the meantime with regard to the specifications or the framework conditions, the work turns out to be useless.

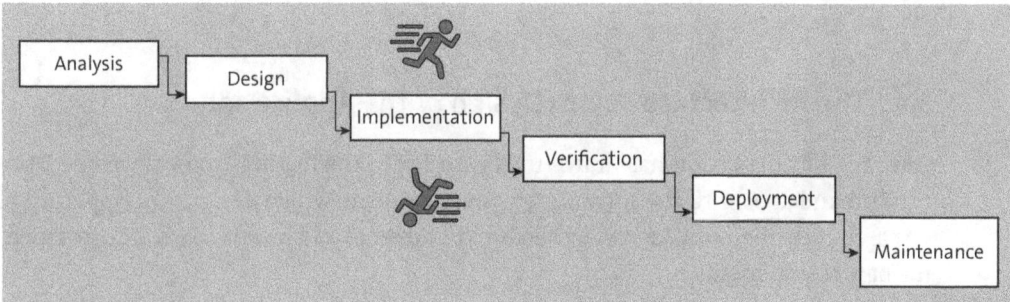

Figure 3.1 Company's Agile-ized Waterfall Model and Its Stages

At the end of a sprint in a typical agile environment, the software should be ready to be rolled out. However, the *nicely-dressed.com* development team tends to take a few sprints to complete its implementation work, and the work then passes to the QA team, which mainly carries out manual tests. As soon as the implementation work is accepted, the software is deployed. Only then is the work completed and the system goes into maintenance. There are approximately six months between the requirements phase and the live operation of the implemented changes. Each phase of the software development lifecycle is the responsibility of a separate team, and the teams communicate very little with each other. The members of the respective teams do not know exactly what the subsequent teams are doing, nor do they know what impact their work has on the activities of the other teams.

3.3 The Business Team: Requirements Analysis

At the beginning of the development phase, the business team at *nicely-dressed.com* decides what needs to be developed in the following months. Essentially, the members of this team carry out the requirements analysis and then draw up a plan as to which new features are needed and where optimization is necessary.

The team is also responsible for deciding what features could be removed. The team works closely with the marketing team, as some functions will be launched together with various campaigns.

The core objective of the business team is to define the requirements that will help drive more customers to the store and thus boost sales.

However, the team carries out the requirements analysis essentially without any knowledge of the technical relationships and the status of the codebase, and it rarely works directly with the architecture and development teams. As a result, they cannot foresee the effort and interactions that the planned changes will require.

Also, there are hardly any metrics that measure the effects of changes in a detailed and flexible manner. Only the sales figures are generated once at the end of the quarter, so it becomes apparent only at a late stage how successful the previous months were.

3.4 The Architecture Team: Design of the Application

Once the business team has drawn up a wish list of the features to be developed, the plan is handed over to the software architecture team, which designs the application. The architects then consider what needs to be adapted and how it will be done to meet the new requirements.

Although close cooperation with the developers is essential in this stage, as their experience is crucial to design an application that can be successfully implemented, this cooperation is not happening; the developers are still busy fine-tuning work for the last job, which now urgently needs to be deployed to the live environment. They therefore do not have time to provide input on the upcoming changes.

3.5 The Development Teams

Once the architecture team has (supposedly) completed the architecture for the new release, it passes the task on to the various software development teams. Over the following months, these teams develop the desired features and correct any errors. Several teams are working on different parts of the software. For example, there is a team for most of the store, one for the billing backend to ensure that billing works properly, and teams for the applications for the various mobile operating systems.

There is no direct cooperation between these different teams. The teams work in their own chambers and talk to people from the other teams only when necessary. The teams therefore only ever receive the supposedly finished work from the respective predecessor team.

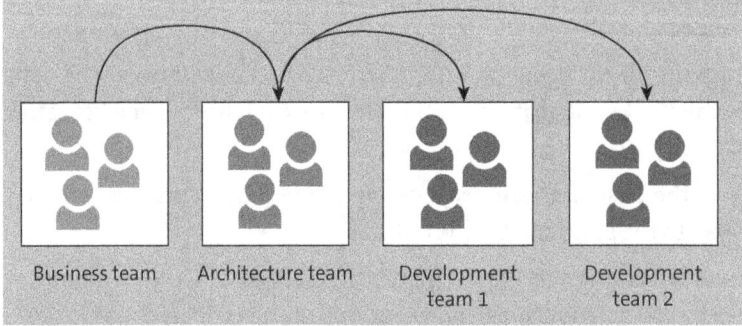

Figure 3.2 Workflow from the Business Team to the Architecture Team to the Development Teams

3.5.1 The Development Process

Here, we'll look at an example workflow that a developer for *nicely-dressed.com* may experience. Before the implementation work begins, the developer grabs the following ticket: "The first offers for the Christmas season should be available in time for the first week of Advent, and there should be further offers displayed on the start page for each subsequent week of Advent."

As all the development work has to be completed in the first six months and it is still January when the work on the ticket starts, the developer still has plenty of time. The ticket contains a lot more information, such as when which offers should be displayed and when they should be removed.

The developer starts working on the part of the ticket regarding the offers for the first week of Advent. The source code developed by her team is managed in a Subversion repository. The developer works locally with a working copy, just like her other colleagues, as is the norm with Subversion.

The developer implements the feature for the offers for the first week of Advent over a period of several weeks. During this time, there is no further check out of the repository; the changes written by the other members of the development team completing their own tickets are therefore not downloaded again.

In the meantime, the developers have questions as to exactly how some of the features were intended, as the documentation of the requirements analysis and the architecture phase is unclear. These requests for clarification go to the specialist staff. Unfortunately, after some time, the developers do not receive any concrete answers because the decision-makers can no longer remember exactly what they had in mind. Their planning phase is long over, so the context in which these features were discussed is missing. This means that further work needs to be done before the implementation phase can continue.

While the Christmas offer feature continues to be implemented over several weeks, other colleagues are working diligently on other parts of the online store. So while our developer is implementing the Christmas offers feature, someone else on the team is working on a major refactoring of the code to remove some legacy issues. This work will also extend over a longer period of time. However, they will be implemented, integrated, and checked into the repository before the Christmas offers feature is finalized.

3.5.2 Integrations with Obstacles

Everyone on the team now has a problem when they want to integrate their code. Because everyone uses a branch in the repository, they don't always pick up the changes from the other developers during the implementation work. In source code management, a branch is a development line in which the source code is contained accordingly.

The developers manually test the features they worked on during development and can check their part of the application relatively easily. At the end of their own development work, however, they have to download the changes from other developers that have accumulated in the meantime. Unfortunately, this almost always leads to conflicts in the code.

The developer who implemented the Christmas offers feature now has to deal with numerous conflicts, as the refactoring affects the code she touched. She now has to adapt her work to the changes, which will take a few more days.

She then has to repeat the whole process over again because even more changes were made in the meantime, which fortunately were not quite as big as the previous ones were. But this was only possible because she agreed internally within the team that she had to integrate her work. In the meantime, the other developers have paused their own work so that she won't have to continue redoing her work again and again.

When the integration is finally successful, the team immediately encounters another problem: Individual developers report that they can no longer build the web store. The work of the rest of the team stands still, because if it is no longer certain that the project can be built, it can no longer be developed in a meaningful way.

Fortunately, it did not turn out to be a major problem: A new dependency was introduced that had not been documented. Once all team members have downloaded this dependency, the problem is solved. This solution just needs to be communicated to everyone, which, again, takes some time.

Deployment to the production environment is scheduled for a few months later. The deployment to production is generally carried out at the end of each quarter. Before we look at this topic, we need to take a look at what happens when the development team and its sub-teams have completed their work on the new implementations.

3.6 The Quality Assurance Team

Before deployment is carried out, the QA team must check the development team's work. This step also repeatedly leads to conflicts, which are not only technical but also of human nature.

The goal and motivation of the development team is basically to develop new features and make necessary bug fixes. The developers are not particularly interested in what the QA team does. The QA team's goal is to find as many bugs as possible, which the development teams then have to correct.

The goals of the teams are therefore contradictory: The more errors the QA team finds, the more work the development team have to face. This causes the development team to have a bad attitude toward the QA team. The development team doesn't know exactly what the QA team is doing to find the bugs and doesn't really care; they are supposed to correct problems and develop features, not find bugs.

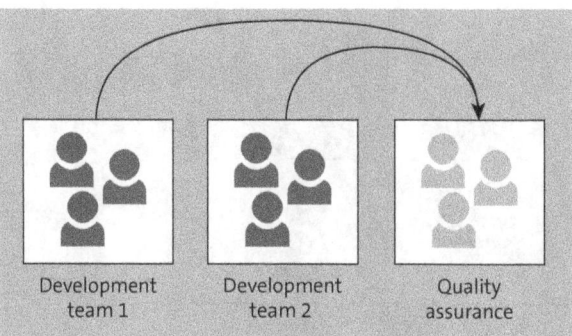

Figure 3.3 Workflow from Development to QA

The QA team carries out various types of tests, some of which are manual and some of which are automated. The automated scripts are often distributed on the computers of individual team members rather than being stored in the source code repository next to the actual web store project. The individual QA team members find errors in the web store and report them to the development team and are then initially off the hook.

Several problems arise from this process: First of all, the teams are completely separate from each other, both organizationally and professionally. They do not directly collaborate, as they only ever throw errors over the fence from one team to the other, as you can see in Figure 3.1. Ultimately, this means that fundamental problems that could be addressed and resolved more easily with closer collaboration are not even mentioned.

Another problem is that the QA phase, in which the QA team mainly works, takes place completely after the implementation phase. Therefore, the individual members of the development team receive feedback on features a few months after implementation.

As a result, a lot of problems are passed back and forth between the teams that could be solved much earlier if the software development cycle were faster. Furthermore, the source code for the tests is stored in a separate repository, which means that the QA team corrects the tests only every three months (i.e., every time they work on the project again and need to adapt them to the changes in the application). Also, the tests run on a specific computer that only certain people have access to. All the scripts and the tooling around them were installed manually once on this computer, but they have been sparsely documented.

3.7 The Operations Team: The Ops in DevOps

The operations team puts the application into operation and supports it. This team makes up the Ops in DevOps. The team's focus is on ensuring that the application is always running and available to customers. Because without a running online store, there is no revenue for the company.

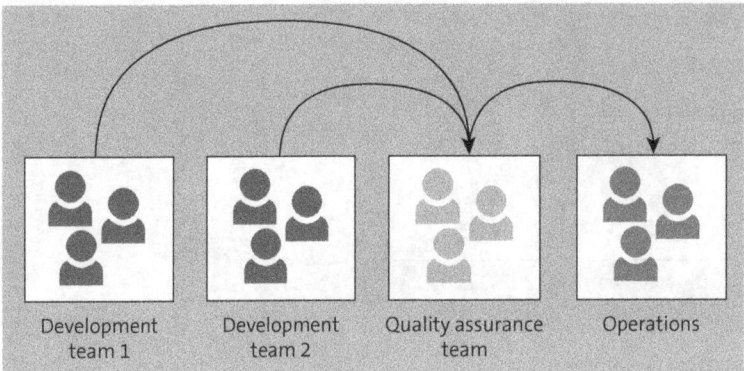

Figure 3.4 Workflow from Development to QA to Operations

> **Note: Admins and the Company**
>
> So far, I have been using the term *operations team*. You may be wondering why I haven't mentioned the familiar system administration team. That is a legitimate question! Many people use the terms operations team and administration team interchangeably. However, this is not entirely correct, even if many aspects of the teams' roles overlap.
>
> When I write about the operations team here, it is the team that runs and operates the application.
>
> Administrators plan and manage the infrastructure on which an application runs, including servers, networks, and data centers. Administrator tasks often also include installing the operating system, managing the virtualization solution, and importing updates from various systems. The role and its tasks are therefore also strategic. Depending on the company, the areas of responsibility are fluid and overlap. Sometimes security is part of the role, and sometimes there are independent network administrators.
>
> I am mainly focusing on the role of the operations team in this book because this team operates the application that is developed by the development team. I call those who manage the hardware the infrastructure team. There is also the platform team, which is the link between the infrastructure team and the operations team and works on the platform on which the application runs.
>
> The smaller the company, the more the responsibilities of these teams overlap; therefore, the role of the system administrator could be broader. So don't be surprised if you ever encounter system administrators who run applications but also manage their underlying systems. However, even though some companies may have broader and overlapping team responsibilities, I use these more precise terms to distinguish between these teams' typical roles.

In addition to the pure operation of the online store, the operations team is also responsible for importing new deployments. The operations team therefore comes into play once the development and QA phases have been successfully completed every six months. The general flow can be seen in Figure 3.4.

The operations team supports the activities of the various development teams. The more changes are made, the more work the operations team has. The development team does two things directly for the operations team:

- The developers build the application as a package to make it available to the operations team.
- The developers update the documentation each time the application is built so that the operations team can adapt certain application-dependent conditions.

However, as with the development and QA teams, there are also high walls between the development team and the operations team.

3.7.1 Manual Build of the Project

After QA has approved a release, the development team gets back to work and builds the package. The development team uses a server for this, on which the project already exists and all locally required dependencies are preinstalled. However, this server is used only for the release (i.e., usually only every six months).

Before the release is built, a lot has to be done manually: The development team must check again which new dependencies have been added and then install them. The team then builds all components of the online store individually. Finally, the team copies the respective packages that are built from the individual components together once by hand for the final build of the project.

A common problem here is that sometimes the team needs to omit individual components during the build, as changes are not incorporated into every component with every release. However, the operations team needs all components once again for each deployment.

3.7.2 Deployment with Obstacles

The operations team receives the finished package after a certain waiting period. The deployment is to be imported at the deployment date at the beginning of the second quarter. It also contains the feature for the Christmas offers. However, when the deployment is installed during the defined maintenance window, several errors are found.

The deployment initially fails because something could not be found. After a little checking, it turns out that a new dependency that was introduced had not been documented for the operations team. As the deployment is installed at night to minimize the loss of sales in the event of possible outages, the store is offline for the time being.

Since no one from the development team is available, the operations team is in the dark for a while until it finds and corrects the problem—and fortunately for *nicely-dressed.com*, this problem can be corrected.

3.7.3 Alarm from the Monitoring System

A few hours after deployment, the monitoring system reports that the application is offline. After a little research, it turns out that the new version of the web store is filling up the file system and causing the web store to crash. As the operations team has hardly any knowledge of the application, it first has to find where the problem is located and how to deal with it, as it is still late at night and no developer is online.

Fortunately, it turns out that there is a simple solution: The team simply needs to write a script to continuously delete the log file that is filling up the file system. The next day, a description of the problem is sent to the developers, and they look into how to fix the problem. However, as development has continued to progress since the last deployment was finished, a new deployment is needed, which would unfortunately also release new features that are not yet ready for the public.

Additionally, the process with all its handovers and stumbling blocks is so cumbersome that nobody wants to deploy again. They know that new problems with new bugs are waiting, which everyone wants to avoid. Instead, the employees decide that the temporary quick fix should remain in place until the next release.

So instead of quickly correcting a small problem and rolling out the change, the teams postpone it. Another dependency is added (namely, the operations team's script that deletes the log file) instead of correcting the error and making the system simpler. It's only a matter of time before someone misses the log file for a system analysis and nobody knows exactly why this information is no longer collected.

3.8 The Infrastructure Team

The role of the infrastructure team is straightforward, but just as important as the other teams discussed so far: It provides the infrastructure for the operations team. In concrete terms, this involves the operation of the data center, including the acquisition, provision, and repair of server hardware. The infrastructure team therefore provides the solid foundation on which the basic operating system is installed so that the operations team can use it, as shown in Figure 3.5.

In the case of *nicely-dressed.com*, there are also walls between the operations team and the infrastructure team, as some problems and challenges often need to be discussed. For example, if hardware needs to be replaced, deployment of a release must not take place—information the development team needs to know.

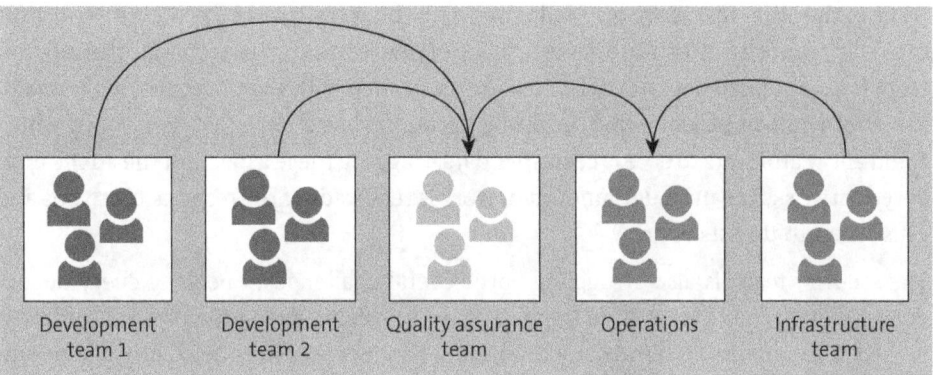

Figure 3.5 Workflow from the Infrastructure Team to Operations

3.9 The Security Team

Finally, there is the security team. The role of this team is to ensure that the application is secure and contains as few known security vulnerabilities as possible. This means that the team has to take care of both the security of the web store and the infrastructure on which it runs. As you can see in Figure 3.6, they have to maintain a relationship with every team.

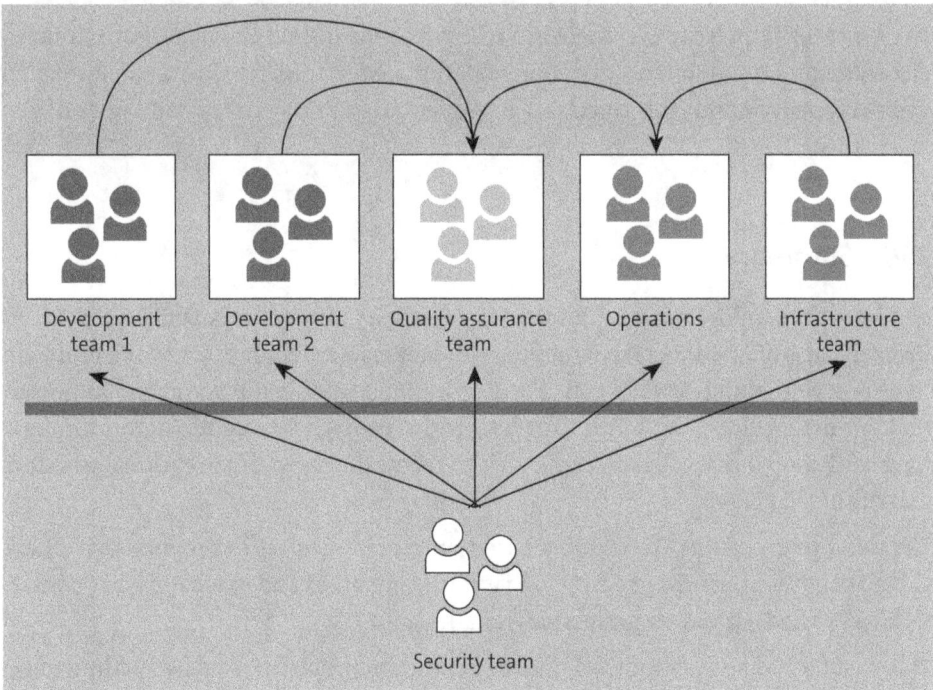

Figure 3.6 Relationship between the Security Team and All Other Teams

None of the other teams are on good terms with the security team, as security concerns often block processes and the development of new features. For example, the security team flags outdated dependencies that are associated with major security vulnerabilities and serious vulnerabilities in the application that enable SQL injections, which could allow attackers to access customer data. Flagging these aspects of the application as security risks mean a huge amount of work is required for the responsible teams and cause their to-do lists to grow.

The security team is also struggling with several challenges: Checking every dependency of every application is an impossibility, especially when little or no automation is used. The situation is hardly any different when it comes to code quality and potential security problems. Even a risk as simple as SQL injections cannot be completely ruled out if you have to check every corner of the online store by hand. So instead of thoroughly and regularly auditing the application, the security team often takes only a cursory look at the application. Although it uses some tools to perform various analyses, they are used exclusively by the security team itself, which then passes the results on to the operations team and the development team. The security team writes long lists of detected problems without knowing the exact context in which those problems exist. The team distributes tasks to the other teams, nags the others to implement solutions to detected problems, and can veto almost all decisions with the irrefutable argument that the security of the application is affected, which can slow down the work of the other teams.

From a security perspective, there are many reasons to pull the handbrake in many places, such as if serious security vulnerabilities are found shortly before or even after deployment. However, this could cause frequent disruptions to the daily work of several teams.

3.10 Summary

You now have an overview of the working methods of the teams working on Nicely Dressed Ltd.'s online shop, the structure of the teams (see Figure 3.7), and the problems they face; you should now have an idea of how things work in this company. Some may find this description to be a little exaggerated. In reality, however, it is not uncommon for a company to lack collaboration between teams due to its organizational structure and culture in general.

The description of this company is based on a great deal of experience that I have gained over the course of my career in various companies and based on many conversations I've had with other professionals in the industry.

This chapter has already touched on some common problems. So now it's time to get started: Let's take a closer look at the individual problems this company faces to find out where and how DevOps can help.

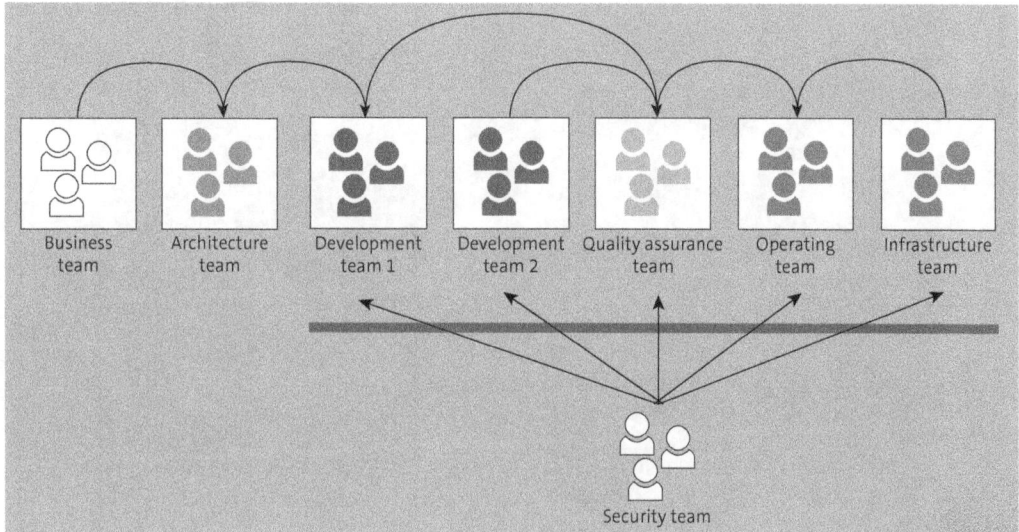

Figure 3.7 The Entire Team Structure

In this chapter, you may have recognized one or two problems from your organization. If you work in technology, you can adapt the culture only to a certain extent. However, the higher up in the hierarchy you are, the more likely you are to have the opportunity to actively shape the culture.

Chapter 4
Project Management and Planning

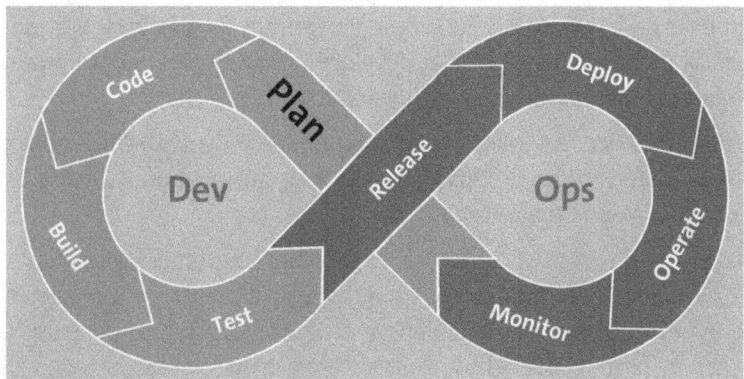

Figure 4.1 Plan Phase: Requirements, Goals, and Tasks Are Collected, Managed, and Defined

As a first step, it is worth taking a closer look at what can be optimized and improved during project planning alone.

4.1 The First Step: The Agile Mindset

Many companies are already working with an agile framework these days. Adopting such a framework is the foundation that needs to be laid in order to adopt DevOps principles, and it is an important step away from the previously common waterfall model, which is no longer common today for a good reason.

The waterfall model involves a lengthy planning period before the given project is implemented and tested by various teams over a period of six months. Requirements that change over time may be skillfully ignored and simply may not be taken into account, as the waterfall model is not very flexible.

The problem is that, in many cases, companies attempt to transition to an agile framework but fail to completely implement it throughout its processes and culture. Often, only the pure development work is agile-ized. What is missing is a view of the entire value stream, which ultimately delivers the value of the product to the end user. Unfortunately, this often results in an *agile waterfall*: Development work is carried out in sprints, according to the agile framework, but the delivery is not made available to end users at all.

Such a process begins with the requirements analysis and specification, followed by the design phase and then the implementation phase. Only then is the product tested and then delivered. Such an approach means that any errors uncovered in the QA phase, for example, would have to be corrected in the design or implementation phase, which would have already finished a long time ago. At each stage, the software is thrown over the wall to the next team, and the teams communicate inadequately with each other.

The required period of six months between requirements planning and the finalization of the requirements tends to be too long. Not only are requirements typically constantly changing, but the deployment on the target platform could also undergo serious changes, as it is also typically being further developed, such as through hardware replacements or updates to the operating system version.

As already stated, the period between planning and deployment is usually around six months. If you look at a work package, this means that up to six months can pass between the creation of a task and the roll-out of the change.

As previously explained, this period of time is called lead time, and there is also *cycle time* (see Figure 4.2), which corresponds to the time that elapses between the start of work on a task and its completion, including deployment. Both are typical metrics in the DevOps world that are used to assess how well DevOps is being implemented.

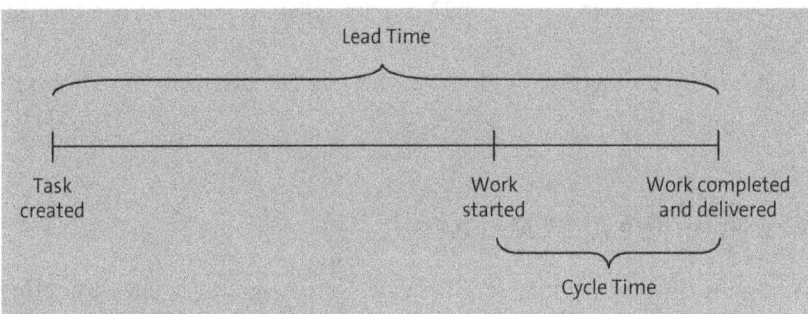

Figure 4.2 Lead Time and Cycle Time

One of the main pillars of DevOps is keeping lead times as short as possible. With a short lead time, changes can be rolled out as quickly as possible, which can be very helpful in many areas, such as in being able to react quickly to changing market conditions, to close security vulnerabilities, or to quickly implement bug fixes. This pillar is shown in the first way of the Three Ways.

However, before you can implement a short lead time, you need to look at project planning. Many companies have replaced the waterfall model with agile methods for good reasons. The main goal of such methods is to achieve faster and more efficient development with a higher quality of software.

Agile software development began in the early 2000s with the publication of the *Agile Manifesto*, which defines the 12 basic principles of agile software development and can be found at *https://agilemanifesto.org/*.

The principles of agile software development and DevOps are similar and overlap. Among other similarities, both assert that making changes to a project is fundamentally good and emphasize the importance of publishing releases frequently, strengthening collaboration within and between teams, and reflecting on and adapting software and people.

> **Note: The 12 Basic Principles of Agile Software Development**
>
> The 12 basic principles of agile software development according to the *Agile Manifesto* are as follows:
>
> 1. Our highest priority is to satisfy the customer through early and continuous delivery of valuable software.
> 2. Welcome changing requirements, even late in development. Agile processes harness change for the customer's competitive advantage.
> 3. Deliver working software frequently, from a couple of weeks to a couple of months, with a preference to the shorter timescale.
> 4. Business people and developers must work together daily throughout the project.
> 5. Build projects around motivated individuals. Give them the environment and support they need, and trust them to get the job done.
> 6. The most efficient and effective method of conveying information to and within a development team is face-to-face conversation.
> 7. Working software is the primary measure of progress.
> 8. Agile processes promote sustainable development. The sponsors, developers, and users should be able to maintain a constant pace indefinitely.
> 9. Continuous attention to technical excellence and good design enhances agility.
> 10. Simplicity—the art of maximizing the amount of work not done—is essential.
> 11. The best architectures, requirements, and designs emerge from self-organizing teams.
> 12. At regular intervals, the team reflects on how to become more effective, then tunes and adjusts its behavior accordingly.

Agile software development is widespread and well known in many companies these days, which is why I won't go into it in too much depth here. Although the concept is already over 20 years old, the adaptation of agile software development has gained momentum over time; it can be said that today, it is generally well established, unlike DevOps.

In agile software development, work is carried out in sprints that are kept as short as possible. Instead of following a six-month development cycle, as is the case for the waterfall method, work is usually carried out in two weeks, during which planned developments are to be implemented.

But what is the difference between agile software development and DevOps? Basically, you can say that DevOps is a supplement to agile software development. However, as already briefly touched on, the latter mainly focuses on development and almost ignores everything that comes afterwards. Although the manifesto emphasizes the importance of frequent releases and deployments, it contains no further mention of exactly how this frequency can be achieved.

But that's exactly what's so exciting! What can we learn from the deployed release? How do the end users use it? By utilizing various metrics, much more information can be gained, which can be useful for the next iteration.

Let's go back to project management: The agile waterfall approach is a first step that can pave the way for DevOps. A lead time of six months is definitely too long. Instead, development should take place in faster cycles to allow teams to react more quickly to changing requirements.

At the end of a sprint, the software should be ready to be delivered not only in theory but also in practice. The key to being able to achieve such lead times is closer collaboration between and involvement of all relevant stakeholders. They should sit together at one table instead of throwing their work results over the wall between each other, as is typically the case in a waterfall environment.

> **Note: Scrum**
>
> Agile project management is often equated with Scrum. This is not correct. Scrum is *one* option for agile project management. There are others, such as Kanban and the Scaled Agile Framework.

However, an agile framework is only the first part of DevOps, and it is also only a small part that must be further adapted in the subsequent steps of the process.

> **Note: Reflection**
>
> Agility is the first step into the DevOps world. If you have not previously developed software using an agile framework, then adapting DevOps principles will likely not be possible. Fortunately, the likelihood of failing at this stage is relatively low, as many companies now work using agile methods.

4.2 Project Management for Everyone?

The first phase in the software development lifecycle is the planning phase. So let's take a look at how project management works from a DevOps perspective.

Numerous commercial and noncommercial tools for project management are available on the market. The largely undisputed top tool in agile project management is Jira, developed by Atlassian. Almost everyone reading this book has likely used Jira at some point.

Jira is the industry standard for project management tasks. It is embedded deeply in many processes in many organizations—with all its advantages and disadvantages. Depending on who you ask, this is either a curse or a blessing. Project managers, product owners, agile coaches, and Scrum masters rave about Jira and often don't know much about other similar products. In fact, because of how widely used Jira is among Scrum masters, Scrum is often equated with Jira.

On the other hand, if you ask technicians what they think of Jira, you tend to hear a lot of skepticism. For some, Jira is a necessary evil that is simply part of project management. Others simply hate it and are happy if they don't have to touch it. Some of the more extreme opinions of the latter can be read on *https://ifuckinghatejira.com/*. The main criticism is that Jira is a management tool and, in the hands of poor project management, can cause more effort and annoyance than benefit. It tends to bureaucratize and complicate development work.

But let's stick to the facts: Jira can be used in many ways. As a comprehensive project management tool, it can be used as an issue tracker or as a workflow tool. You can use it to collect tasks in the form of tickets and assign them to members of your team. My focus here is primarily on issue tracking for agile project management.

4.2.1 Jira Can Do (Almost) Everything

A great strength of Jira is that you can customize the software down to the smallest detail and that there are numerous plug-ins with which you can expand the range of functions. This is the case for many similar tools, and regardless of the tool you're using, it can be both a blessing and a curse.

The curse is that every installation can be heavily modified with home remedies and plug-ins. For an engineer who simply wants to look at the issues and track their progress, every project can look different because everything can be customized. This may seem like an advantage at first, but I now largely see it as a disadvantage. If there are too many ways to arrive at a solution, you have to find your way around again and again.

It's also not uncommon for a company to have not just one but several Jira instances, all of which are maintained by different teams. Each instance and each project within each instance may work very differently due to the use of plug-ins and hardwired workflows. For people who are not involved in these projects, how things work may be completely opaque. Anyone who needs to jump back and forth between projects would have to repeatedly reacquaint themselves with different workflows.

Also, initiatives such as inner sourcing don't work well if there are too many Jira instances: Too many accesses with the right authorizations have to be created and managed, and the effort required to find your way around somewhere else is too high. Regardless of the specific project management tool, this is an anti-pattern, especially in the context of DevOps culture, as the creation of tickets should be as *simple* as possible. The aim should always be for the tool to support the processes and, therefore, the people—true to the motto, people over processes over tools.

Therefore, if you use software for project management, whether it is Jira or another tool, make sure that customizations do not get out of hand. Otherwise, you will end up with more management work than before. The simpler and more uniform the solution is designed, the better. (Exceptions prove the rule, of course, but they must be justified).

4.2.2 People over Processes

Jira is a powerful tool. But with great power comes great responsibility, so use it wisely.

Jira makes it very easy to introduce complex processes. These processes can quickly become bureaucratic, which slows down the development flow. Rules and conditions for project progress can be quickly formulated in Jira, and by overusing such rules, you may find that your development teams are only busy ticking off tasks and rather than focusing on writing code and actually working. From a management perspective, the rules look necessary and sensible, but they make the processes more cumbersome and error prone. The great art of project management is to cut all unnecessary overhead so that only the really important rules can be implemented.

But that's not all: Project management reflects the work steps of your team members. Formalizing these steps too much could create the very rigid structures that flexible and agile project management is supposed to abolish.

A common workflow, for example, is that a ticket always changes from one predefined status to the next—for example, from "Planning" to "Backlog" via "Ready for Development," "In Development," "In Test," "In QA," and so on, until it eventually ends up in the "Released to Production" status before it is closed.

Do you notice something?

With this process, we have reinforced the very silo mentality that we actually want to abolish: When something is in planning, the developers often have nothing to say. When the work step moves on to development, the QA team waits. Ultimately, the responsibility lies with the operations team, which has to take care of deployment.

Not only is this workflow unfavorable from a pure DevOps point of view, but it often also leads to complicated workflows, as it is not easy to see which tasks follow which status and what to do if something has to be run back in the process.

So if you cast processes into rigid specifications using project management software, you run the risk that these structures are reflected throughout the system and ultimately in the product you are developing. And that is generally not the best way to go.

> **Note: Conway's Law**
> This insight is not new, and we will look at it in more detail in Chapter 5, Section 5.1. In a nutshell, it involves a thought by the American computer scientist Melvin Conway: The systems that an organization builds are always a figure of its communication structure. If the group working on a piece of software is divided into three teams, the resulting program will also consist of three parts. The individual parts of the software correspond to the parts of the group.

Therefore, make sure that your project management tool—be it Jira or a similar tool—does not force your team to follow a certain complex workflow. Of course, it is sometimes necessary to standardize processes and monitor compliance, but it is important to find the right balance between a flexible, perhaps even creative use of the tools and processes that are as standardized as possible. My advice is to look at whether all the complex processes in project management are really needed in exactly the same way.

The simpler processes are, the more likely it is that information and important insights will come to light that might otherwise be hidden. After all, the people on your teams want to work with the tools, not against them.

If project management isn't working at all and you feel like you're just struggling with the tools, you may find yourself wanting to change tools: "If we used a different tool instead of Jira, we wouldn't have these problems." In my experience, however, this is true only in very few cases. The errors and shortcomings of Jira exist in other tools as well, and poor processes can be implemented with any tool.

In addition, migrating to another tool is quite time-consuming. Understandably, very few organizations are prepared to undertake such a migration. This is not surprising, as project management tools are often deeply integrated into the company. So instead of overturning well-established processes with the tool, the better approach is to rethink how you can organize your project management with Jira, or whatever project management tool you're using.

4.2.3 Good Project Management outside of Jira

In project management, both with and without Jira, it is possible to avoid uncontrolled growth with a few simple steps and structures so that tools can be used as efficiently as possible:

1. **Creating and ensuring visibility**
   One of the most important DevOps principles is creating and ensuring the visibility

of information. The various teams should be able to not only view the tickets of other teams but also comment on them and create them. On a technical level, this prevents silos from forming between different teams.

Your teams should also be able to access further information from third-party tools. For example, even nontechnical personnel should also be able to see pull requests and merge requests to make the progress of work visible.

2. **Using templates**

   Templates are helpful to ensure that the most important information is recorded in a ticket. A template can provide a structure so that those who create tickets do not forget the necessary information. For bug reports, for example, template sections for information such as what the observed error is, how it can be reproduced, and what the expected behavior is could be included to guide ticket creation, reducing the proportion of tickets with useless descriptions such as "does not work."

3. **Staying close to the standard**

   Don't reinvent the wheel. The closer you stay to the standard use of the project management tool you use, the better. This will not only reduce the complexity of ticket management, but it will also help to reduce potential problems during upgrades. In addition, it will facilitate the process of starting new or different projects.

4. **Using labels for organization**

   Jira has a feature allowing you to create your own ticket types. For example, you could create tasks, bugs, or stories. Make sure that you create as few types as possible and that you use no or very few custom fields. The basic idea is to ensure that employees can use the system as efficiently as possible and with as little complexity as possible.

   Labels can help with managing multiple ticket types and maintaining efficiency. They can be changed and inserted in a ticket quite quickly. However, using labels requires careful consideration as to how they can be defined and used. Templates can be useful here, as they would allow you to select predefined labels. Labels also allow the tickets to be categorized, and they can be quickly adjusted after they have been created.

5. **Focusing on simple workflows**

   Jira is unpopular for a reason: A frequent point of criticism is the ability to create custom complex workflows that require many approval steps and other transitions from one status to another. If workflows are unintuitive and confusing, they become tedious to use. Make sure that you simplify unnecessarily complex workflows to increase user-friendliness.

6. **Keeping the size of tickets as small as possible**

   Make sure that the scope of tickets is kept as small as possible. Keeping ticket sizes small will allow you to implement DevOps principles more quickly, regardless of the tool you're using; many smaller iterations can be implemented more quickly and require less clarification, as it would already be clear at the start of the work what needs to be done. Break down tasks and feedback into the smallest possible parts that can be processed step by step.

7. **Using metrics**
   Following the measurement value of the CALMS model, it is important to pay attention to which metrics are provided by the project management tool you're using and regularly check how they are trending. Project management tools offer a lot of data that can be used to measure productivity. If you notice that the values are deteriorating, investigate why this may be the case. The trend is generally more important here than the specific figure, because every company starts from a different point.

But don't confuse metrics with targets. If the number of bug reports decreases and there are fewer tickets with the status "critical," this is, of course, a good sign. But never set a target to have fewer tickets; such a target will likely only encourage your team to no longer document problems or to combine many issues into one ticket.

4.2.4 More than Just a Project Management Tool

If you want to dive deep into the secrets of project management, you will find plenty of tips and advice on the Internet and in other books. Though you'll find a lot of information on project management, only you can say what really works for you and your situation. You'll also find a lot of advice specifically on Jira, half of which contradicts itself. What some people swear by is considered a mistake by others. Gladly, that's not the topic of this book! From a DevOps perspective, regardless of the specific tool you're using, the most important thing is that the team's work and the progress being made on the project are visible, ideally to everyone.

From a DevOps perspective, it is completely irrelevant which approach—for example, Scrum or Kanban—your team is using to carry out its work. In order for all relevant parties to have full visibility into the work being done, it is much more important that all the tools in use can be integrated with each other and with the entire software development lifecycle and that everyone can participate.

Regardless of which specific tools you are using, integrating all of the tools can be a struggle, as all information can't always be easily shared between tools and it is not always easy to create a ticket in a project management tool from other sources. In addition, many people often lack access to systems because information is compartmentalized. This makes collaboration difficult!

This point becomes clearer with specific examples. A solved problem is the question of whether the team is currently working on a ticket. In other words, it is a question of whether there are already commits for the given ticket (i.e., whether code has already been actively written). In Jira, a ticket will have in its history the commits for that ticket or will contain additional links to the corresponding feature branch and pull or merge request. The current development work can be accessed with just a few clicks.

But that's only half the story. And it answers only the most obvious questions. What about security vulnerabilities? If a security vulnerability is displayed on the dashboard, you want to create a ticket immediately based on it, but you also want to see directly whether a ticket already exists or whether your team has already processed this case. This information affects several people at once, because managers, developers, and the security team all want to know about and monitor the status of that ticket. And this can be done only if everyone has access to the same information. Therefore, your project management approach must be well integrated into the security scanners, even though at first glance they appear to have nothing to do with each other.

Another example is the handling of feature flags. Chapter 8, Section 8.4 deals with the example of the company Knight Capital, which was (among other things) driven to ruin by the inadvertent reuse of an outdated feature flag.

If feature flags are used, the ticket that each individual feature flag is attached to should be immediately visible in order to track both the implementation and the subsequent cleanup. Therefore, visibility must not be limited to the ticket that requested the feature; the ticket must also integrate the tool that monitors the feature flags. This is not necessarily a feature that is common in today's tool landscape, but hopefully you can see where I'm going with this.

We can continue with these examples almost indefinitely. What I'm trying to say is that it's important to look not only at the actual project management tool but also at how it can be integrated into the entire DevOps lifecycle so that everyone has full visibility at all levels and that everything is traceable. I don't know of a perfect solution for this; every tool has its advantages but is always missing something.

Instead of integrating these features into Jira (or the project management tool of your choice) so that everything is visible in one place, my advice is use the project management capabilities of GitLab or GitHub, especially for smaller projects. By using these platforms, you do not integrate the technical aspects into the project management tool, but you use the features of the coding tools to display and manage the progress of the projects.

Both platforms have been offering functions for this purpose for some time, and more capabilities have been developed in recent years, particularly for the paid versions of these platforms. The focus has been primarily on software development projects. While Jira, Asana, and similar tools are about general project management, completely independent of software development, the GitLab and GitHub platforms focus on code management. The ability to flexibly set labels without necessarily having to define workflows can help to keep project management simple. For teams that work according to DevOps principles, a project management tool that works flexibly with features like labels and milestones is much easier to use, as it is specifically geared toward software development processes and allows you to easily filter work by individual tickets.

All in all, these platforms simplify the development process. The disadvantage is that most project managers first have to learn how to use these tools. As is so often the case, there is potential for conflict here; as always, everyone involved must be brought on board. Those responsible must recognize these problems and make appropriate decisions. Even simple project management tools used directly on the source code will not solve all problems, as they tend to be far too simplified.

However, this section ultimately also shows that culture is also important when selecting the right tool, and the right tool with the right settings supports the culture and processes—or not, if it is not implemented correctly.

4.2.5 Project Management at nicely-dressed.com

For *nicely-dressed.com*, too, the aim is to modernize and simplify the technologies used. Up until now, a lot has been managed with Jira. The company has numerous different Jira instances; the development teams have a shared Jira instance, but the QA and the operations teams have separate instances. In addition, the company has other instances that revolve around the administration of the company itself. Each instance had a different set of plug-ins, making it cumbersome to use and maintain.

As a first step, the project management tool for the individual teams should be set up cross-functionally so that all departments have access to it. The aim is to facilitate collaboration through transparency. However, in the first step, the project management approach for the administration of the company itself will not change, as the company intends to retain the existing processes. These processes could be easily mapped in a classic project management tool such as Jira.

Project management features are not the only important criteria for the new tool to be selected. Instead, the number of tools should be reduced, and the tools should be close to the code and the CI/CD pipelines. Accordingly, it makes sense to take a closer look at GitHub and GitLab for project management at *nicely-dressed.com*.

Some might argue that a disadvantage of both tools is that they rely heavily on labels, which are quite flexible but do not offer a simple and straightforward solution as to how the structure in the projects should be set up.

> **Note: Reflection**
>
> Replacing or introducing a project management tool is no easy task. As you might have guessed, I'm not a big fan of complex setups, at least not when it comes to developing and operating software.
>
> In many organizations, replacing the central project management tool is not easy, and for good reasons. From a DevOps perspective, two things are important: visibility for all relevant people and ease of use.

> Visibility is not limited to making the ticket system accessible. It is more about ensuring that the relevant information also becomes part of the development and operational processes so that the relevant information is available quickly and easily. See what you can integrate and optimize so that the tool is helpful in everyday life and does not get in the way.

4.3 Summary

This chapter dealt with how a DevOps approach can help optimize the planning phase of project management. Long before your team transitions to DevOps, your teams have to start working with agile software development. Without agility, there is essentially no DevOps, because it is the logical next step.

Efficient and transparent project management is the cornerstone of a successful development and operational process. It is important that feedback from previous development cycles and information from operations is continuously incorporated in the smallest possible iterations.

Chapter 5
Collaboration when Coding

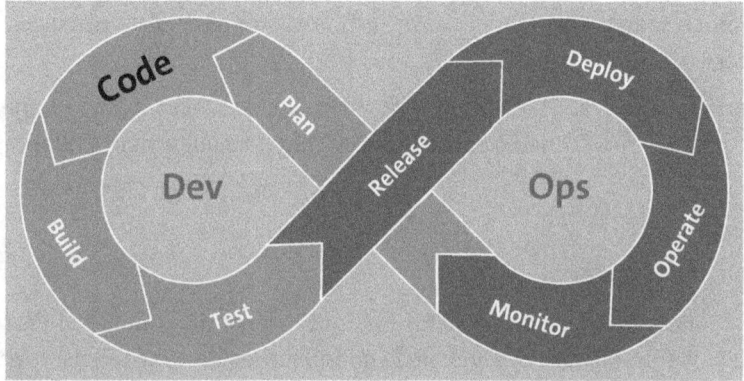

Figure 5.1 Code: In the Development Phase, Code Is Written and Checked into the Repository

Now that we have introduced the general concept of DevOps in Chapter 2 and the planning phase of the DevOps lifecycle in Chapter 4, in this chapter we discuss the next phase of the DevOps lifecycle: development, in which developers write the code.

At first glance, development appears to be a solely technical part of the DevOps lifecycle. This is true to a certain extent, of course, but this phase actually involves much more than that. It also concerns the following:

- How is the development workflow organized?
- How does the team work collaboratively?

These questions have little to do with the technology or the programming languages, frameworks, and tools used. They form the cornerstone for effective and efficient collaboration in the project, as all other phases of the DevOps lifecycle are based on the code.

In this chapter, we look at the various DevOps principles mainly from a programming perspective.

5.1 Typical Problems with Managing the Source Code

Let's first take a look at the source code and the coding workflow at *nicely-dressed.com*. The development team at *nicely-dressed.com* face some challenges, which can essentially be divided into the following three points:

5 Collaboration when Coding

- Organization of the code
- The development workflow, in which developers commit code without any structure
- Technical debt, creating a high learning curve for new developers

In the early stages of the development process, these problems hinder collaboration between the teams. In addition, the organizational weaknesses lead to many avoidable weaknesses in the actual product.

Of course, bugs and programming errors can never be completely avoided, but the shortcomings in source code management cause such errors to multiply particularly strongly and permeate the entire project.

In the following sections, I will take a closer look at the challenges faced by the *nicely-dressed.com* development team, focusing particularly on the consequences of this poor working method.

5.1.1 Organization of the Code

In addition to the architecture of the software, the organization of the source code has a strong impact on how the software is developed and influences how the teams work together. The source code of *nicely-dressed.com* is versioned in a Subversion repository. The exact tool used does not play a major role here; a much bigger question is how the source code is managed and how the repositories are set up.

Figure 5.2 shows the fundamental problem that exists in many companies: The various teams with their different areas of responsibility work in isolation in their own repositories. It is not possible to view the code of other teams or even collaborate on the same source code.

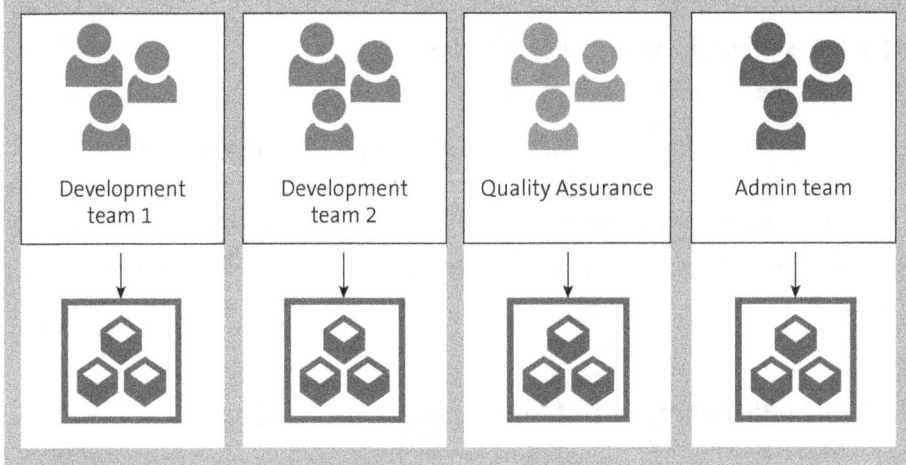

Figure 5.2 High Walls between Teams and Limited Visibility of Work

As you can see, collaboration between the individual teams is not really possible because the individual projects and teams work only in their own silos.

Also, each team can change only files in their own repository; the teams have no insight into the other teams' work. Therefore, errors that involve other teams' code are fixed by workarounds rather than by a fix at the real source.

Conway's Law

The structure of the teams actually has a major impact on the entire product. According to Conway's law (formulated by computer programmer Dr. Melvin Conway in 1968), the architecture of software depends on and is a reflection of the structure of the organization building it. And this is not meant in a positive way—it should be seen as a problem.

If your company has four software development teams, then your product will be developed as four individual projects, even though they actually belong together. How well these four parts fit together in the final product depends on whether these teams work well and closely together. If two teams do not communicate well while working on a given part, that part of the product will have problems of some sort. This is not surprising, as it is always easiest to communicate and work together within a team.

It's very difficult to escape Conway's law. It takes a great deal of management effort to coordinate several teams in such a way that the communication structures have no impact on the work; perhaps it is simply impossible, because if you really tried, the overhead would be incredibly high, as you would be more concerned with coordination than with the actual work.

We can take Conway's law not only as a warning, but also as a lesson: The simpler and less complicated the communication structures are when designing and implementing a project, the better the overall result will be.

The question that arises here is, How can we organize and structure team collaboration in such a way that it benefits the software product? Breaking down the walls between departments to create cross-functional, flexible teams is a step in the right direction.

5.1.2 Isolation for Supposed Security Reasons

Let's take a look at the consequences of the compartmentalization of the teams at *nicely-dressed.com*. Which teams are we talking about?

First, there are the development teams, who develop their own parts of the *nicely-dressed.com* online store behind closed doors in their own little rooms. And then there is the QA team, which has versioned the source code of its automations separately from the online store project. The QA team has no insight into the work of the development teams.

This way of working did not evolve by chance; rather, it is the result of adherence to the *need-to-know principle*, meaning that only those who need to work with the source code are allowed to see it. Managers established this way of working to prevent certain security issues; for example, in the event of a system break-in via a user account of the QA or development team, a malicious actor could access the entire source code of the entire company and find vulnerabilities in the software that could allow them to take further malicious actions, such as accessing customer data.

But the managers weren't concerned just with external attacks. Another reason for the lockdown was to make it more difficult for employees with malicious intentions to access the data. The background to this decision was therefore a deep mistrust of their own employees, especially those outside their own teams.

Such security concerns are real threats and cannot be completely dismissed, but they can often be prevented with a little configuration work in source code management system. In fact, the team leaders had a much more important reason for compartmentalizing their teams, one that was not openly communicated: They didn't want other teams to discover possible security vulnerabilities, inferior code quality, and other problems in the code.

First, they found such code quality issues very embarrassing, and it might have meant trouble with their superiors. Furthermore, they wanted to avoid possible interpersonal points of attack from other teams.

Due to the compartmentalization of the teams and the general sense of mistrust, the atmosphere at *nicely-dressed.com* has become tense, and neither work nor communication is open. There is no real transparency, and when something does go wrong, of course, everyone practices a lot of finger-pointing because nobody wants to know about problems of any kind.

> **Attention: Security Is Important**
> Conversely, the fact that the status of the projects should be visible does not mean that anyone can make changes without any restrictions! We will look at security in Chapter 11.

5.1.3 Long Development Times Hindering Quick Security Fixes

The need-to-know principle and secrecy in the working environment actually makes it more difficult to implement quick security fixes. Work continues as normal on many teams, even though the developers already know that there are many security vulnerabilities. However, the developers do not bother fixing many of these vulnerabilities because the changes are not due to be rolled out for at least six months anyway. After all, there is still plenty of time until then, so it can be dealt with later. And between releases, it is virtually impossible to fix security-critical bugs quickly and flexibly.

Sometimes, of course, a fix for a security vulnerability has to be rolled out, but many important steps, such as QA, are simply skipped. In addition, a complex branching strategy is used to get changes into the production environment. However, only a few long-standing employees have a clear view of this strategy.

Instead, some teams often try to make at least the worst problems as harmless as possible with some unsightly solutions. Or to fix an ugly bug quickly before the next release, they use less-than-ideal solutions, such as hard coding an IP address.

In other cases, the access data to other systems has to be hardwired to ensure everything still works. Of course, nobody is supposed to see such measures, especially not the other teams, so they also make sure to block access to the repository so that nobody will notice.

The fact that the QA team also works with its test scripts completely separately from the actual development team also makes collaboration much more difficult. Although the code of both teams belongs together and is even dependent on each other, the silos between the teams are maintained. The QA team is never able to look into the source code of the individual projects of the online store, which is why many errors are found only much later. The QA process and these problems are discussed in more detail in Chapter 7.

5.1.4 Development Workflow without a Proper Structure

In addition to the problems with source code management, the lack of a proper development workflow is also noticeable at *nicely-dressed.com*: Each person and each team integrates their code as they see fit. I have already explained why this is the case in Chapter 3, in which we discussed the company structure of *nicely-dressed.com*.

The main problem is that the development work for many new functions of the new online store always takes a very long time. The reason for this is that huge task packages are defined in the planning phase, which are only slightly refined in the sprint planning. Therefore, the development team integrates its changes only at the end of the development work, sometimes over several sprints.

5.1.5 Big Bang Integrations

This means that all changes are inevitably *big bang integrations*: Instead of small, manageable chunks, the results of weeks or months of work have to be integrated into the main branch of the project. These changes are very extensive, as they require changes to many lines of code, which makes integration into the main project noticeably more difficult. What this could look like is shown in Figure 5.3.

And as many large changes have to be integrated at the same time, all team members have to download the code from the repository again and adapt their own changes again with each completed integration, meaning there is a lot of waiting.

5 Collaboration when Coding

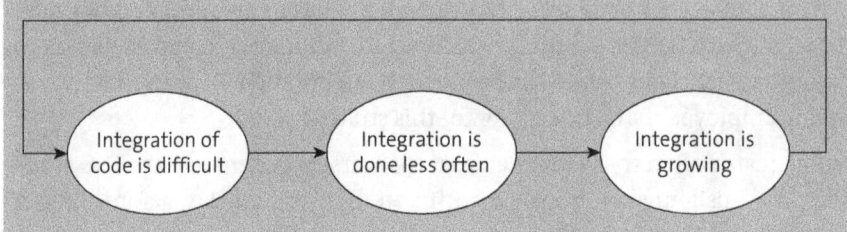

Figure 5.3 The Vicious Cycle of Big Bang Integrations

This ultimately increases the frustration among the developers with each integration, as merging the remaining changes is very laborious and error prone. As a result, a lot of work has to be started from scratch, as major refactoring steps that were not agreed on are carried out anyway.

Not only is frustration a major problem here, but the development work is also delayed quite a lot, as there are always delays due to necessary local corrections. And this ultimately has an impact on the other teams, who depend on timely handovers in order to deliver their parts on time.

The delays that arise in the process always cause trouble for the QA team. As a result, many errors remain undetected; the development teams often take the blame for these errors because the delays that prevent the QA team from properly completing the QA phase ultimately originate from the development teams. The development teams themselves, on the other hand, only recognize in some cases that the large and often imprecise and unspecific task packages from the previous teams are also a problem.

The developers often try to skillfully circumvent the problems so that their source code can be integrated as quickly as possible. To do so, the developers often repeatedly copy and adapt the existing code so that the integration runs more quickly. However, the developers rarely clean up the duplicate code, resulting in further errors.

So far we have looked only at integrations within a single project and therefore a single team. The problems on the individual teams cascade when the individual projects have to be integrated at the end of development in order to get the entire online store up and running.

5.1.6 Code Reviews Could Help

Every now and then, someone at *nicely-dressed.com* introduces the idea of code reviews. The development teams have always developed and integrated their code without having anyone look over the changes. The purpose of code reviews is obvious: By checking each other's work, the developers could ultimately avoid some errors and could accelerate the training of new developers as well.

However, many of the developers greatly disapprove of this innovation. They already need too much time for development work and integration; there is simply not enough time if someone has to sit next to you and keep track of the changes that had been implemented over weeks or months. And they already had the QA team to ensure the quality of the code anyway. That team should make sure that the changes don't cause any damage.

All these problems lead to even more technical debt, as the developers end up developing a lot of low-quality source code. Furthermore, as explained when we introduced the company in Chapter 3, the developers can't always build the project because integration is not easily possible, especially after the long development times of individual features.

5.1.7 Technical Debt

The term *technical debt* describes the price that is paid when teams take shortcuts and make compromises in order to deliver code quickly instead of implementing optimal solutions over the long term. Many of you have surely experienced this: A deadline is approaching, the (working) day was long, and you just want to be done. Instead of thoroughly thinking through a change or a new feature and testing it accordingly, you change the code only to the extent that it somehow works. Integration into other parts of the project, comprehensible documentation, and long-term maintainability are all problems for future you, or for the next team member who has to touch these lines of code again later.

Technical debt has a lot in common with financial debt: It can pile up and later become so unmanageable that the project cannot be continued. On the other hand, debts (or loans) are not bad per se. Sometimes it can be necessary or even sensible to shift expenses into the future to be better positioned to act *now* and react quickly to problems. However, you must always be aware that these debts will have to be paid off later.

To manage technical debt and reduce its negative effects, it is important to document any implemented workarounds that led to the debt. Such documentation should include a description of how high these debts are and what needs to be done to reduce them.

A common cause of technical debt at *nicely-dressed.com* is duplicate code. Instead of properly encapsulating a functionality that is used repeatedly throughout the source code in a separate class or method—which would require access to the code of other teams and, therefore, to different repositories—developers often copy and paste the same code in different places and make some small adjustments.

This duplicate code negatively affects the quality of the project. The developers use duplicate code not only to avoid integration difficulties as much as possible but also to test new implementations and keep the old one for the time being. The main idea behind this decision is that by keeping the old version, the code would already be available and would not have to be restored. However, this ignores the fact that version management programs exist precisely to be able to restore older versions.

The duplicate code not only makes working on the project hell in the long run, it also makes onboarding of new team members very time-consuming and, therefore, expensive. It takes several months before new team members are ready to start, and they only have a rough overview of the codebase. Another problem is that access to the source code is only assigned manually.

5.1.8 High Learning Curve due to Lack of Documentation

There is also hardly any documentation on setting up the development environment. This will be discussed in more detail in the next chapter. In the development phase, it is it is important for developers to familiarize themselves with the code. Due to the high hurdles, it is difficult for newcomers at *nicely-dressed.com* to get started quickly, as even the more experienced team colleagues had little time. This is mainly due to the numerous delays that occur.

The issue is exacerbated by the long development times of the individual features: For newcomers making their first contributions to the code, the learning curve is very steep and the hurdles are very high. The integration problems that constantly arise make the whole situation even more difficult, as it is not easy to get started locally either.

It is then always necessary for a more experienced person from the team to introduce newcomers to the project, but this is also a challenge because the experienced developers already have little time to work on their own features. As a result, the induction process for new hires takes quite a long time, as there are no good and simple structures for learning from each other.

The tooling in use generally reinforces this working method. As previously mentioned, the developers use Subversion, which allows branches to be used to carry out parallel development work and to merge and thus integrate it on completion. However, the *nicely-dressed.com* developers find this process very laborious, so they never use branches.

But even if it weren't difficult, the other problems, which are more cultural and organizational in nature, occur regardless. Ultimately, this is just another example of the fact that it's first about the people, then the processes, and then the tooling—not the other way around.

> **Note: Reflection**
> In this section, we have mainly focused on management of the source code. Large changes in particular, which are to be integrated without a code review, make short and fast releases impossible and make exploitable security vulnerabilities likely.

> Reflect on the issues that also occur for you and those that perhaps do not. The lack of an efficient code review and working with large integrations are typical mistakes that I see in discussions with customers.

5.2 Improve the Organization of the Code

Development is the second phase in the DevOps lifecycle and comes directly after planning. The culture and sharing values of the CALMS model are therefore especially relevant here.

The root cause of the problems at *nicely-dressed.com* is the culture. The problems have nothing to do with competence in programming, the choice of programming language, or use of a tech stack that does not meet the latest standards. Instead, a lot of problems are hidden and a lot are simply ignored. This has led to several problems, which I hope I have described clearly in the previous sections. Perhaps you will recognize at least some of them in projects you have already worked on.

All these problems mean that nobody understands the whole system; after all, the source code is developed behind closed doors. They also prevent faster and more frequent changes from being delivered to customers, according to the First Way.

In order to improve both the culture and the sharing of information within an organization, various aspects need to be optimized. First of all, you need to look at how the storage and management of source code can be improved.

The first step is of a more technical nature: The source code is to be transferred from Subversion to Git. While Subversion is a central version management system, Git relies on a decentralized and distributed approach so that the entire history is stored not only on a central server but also locally on each clone (i.e., on the developers' devices). There are other distributed version management programs such as Mercurial or Bazaar, but these are not very relevant today.

However, replacing the version control system alone does not bring any essential improvement to the situation. The errors from managing the source code in Subversion can be repeated in Git. It is only worth switching if you also take a look at the structure and releases of the source code within the company. You need to think carefully about how collaboration can be fundamentally improved!

At *nicely-dressed.com*, the individual projects are each to be transferred to a Git repository. These projects will no longer be completely closed off from each other as they were before; they will now be made public internally to enable inner sourcing. The basic idea of inner sourcing is that not everything within the company should be developed behind closed doors, allowing people from other teams to contribute to the project in the event of problems. The topic of inner sourcing is discussed in more detail in

Section 5.5. The aim is to ensure that everyone in the organization can participate in a project if needed.

As part of the changeover, the various repositories of the different teams, which belong to one overall project, will also be merged. This means that the source code of the *nicely-dressed.com* online store, together with the tests that come from the QA team, will be stored in a common location (see Figure 5.4).

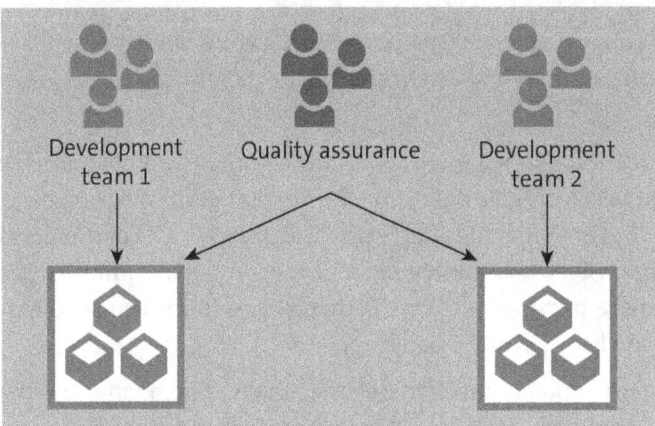

Figure 5.4 Repository Structure of Development and QA Teams

The ultimate goal is to establish a *single source of truth*, one source in which all relevant files for the project are stored. This includes the source code of the project itself, the tests, and aspects from the operation of the application.

The advantages of a single source of truth are clear: The teams have insight into the work of the other teams. Tests can be written directly together with the code that is currently being developed. Previously, the tests were developed completely separately from each other, as no insight was possible. Collaboration between the development teams and the QA team is discussed in more detail in Chapter 7.

> **Note: Reflection**
> Organization and visibility of the code are the be-all and end-all for all subsequent aspects of the DevOps software development lifecycle.
>
> My advice is to improve transparency. The next step ideally happens quite naturally: The teams should recognize the synergies when the projects are brought together to work on the same goal.
>
> There is a common sticking point, though, and I have seen this in practice more often than I would like: There is often a policy that prevents projects from being opened up. The larger the company, the more difficult it is to change this policy. However, the smaller the organization, the easier it should be to convince the right people to do so.

5.3 There Is No Way around Git

You will have noticed that I consider there to be (almost) no alternative to using Git as a version management program. Git is a modern distributed version management system that has been established on the market for years. Virtually everyone who works in IT should not only know Git but also be able to use it. The closer you are to development and operations, the more familiar you need to be with its use.

Many often emphasize the underlying and rather complicated object model when discussing the advantages of Git; its object model makes Git more efficient on a technical level compared to older version management systems. Unfortunately, however, the solution is presented before it is even clear what problem it is supposed to solve. Git's efficient object model is certainly important, but it is much more important to emphasize that Git provides more productivity, visibility, and efficiency than older systems. Because it is a distributed system, users can create new code branches and recombine them on a regular basis. This also makes code review much easier, which is one of the key points of Section 5.4.

5.3.1 Git Solutions at a Glance

The main platforms for source code management are GitHub, GitLab, and, with decreasing importance, Bitbucket. While GitHub is backed by Microsoft, GitLab is an independent company. Bitbucket, like Jira, is backed by Atlassian. While GitHub and Bitbucket are closed source, GitLab is open core with a large open-source base and offers a proprietary license with additional features, which are particularly relevant for companies.

The convenience of the GitHub platform is probably one of the main reasons why Git ultimately prevailed as the most widely used version management systems. Although all Git solutions enable the fast, simple, and efficient creation of branches to drive parallel development, it was only with GitHub's pull requests (or, merge requests; in practice both terms describe the same process) that this workflow became so convenient and easy to use that GitHub primarily became established in open-source communities.

GitLab and Bitbucket have followed suit and also implemented pull requests. Since then, pull requests and code review have been widely used and are available in all Git platforms. A specific problem is being solved here. What the exact implementation looks like is negligible.

> **Note: Other Tools**
> Gitea, a very lightweight source code management tool, is often used for smaller projects. It is a fork of Gogs. There is now another fork of Gitea called Forgejo.

> Of course, there are other tools for central source code management. However, they are much smaller and offer little-to-no integration into all the other stages of the DevOps software development lifecycle. They are also mostly open-source projects that cannot offer professional support, which often rules them out for corporate customers.

Yes, there are other source code management tools apart from Git. Some companies still use Subversion, but mostly for historical reasons. Some still use Mercurial, but it has virtually died out since the ecosystem around Git has developed so well.

Therefore, the focus of this section will be on GitHub, GitLab, and Bitbucket, though only marginally, as many organizations are migrating away from Bitbucket. GitLab and GitHub are used most often, as Bitbucket has seen little further development, and integration into the DevOps lifecycle is better and deeper with the other two platforms.

5.3.2 Development Workflows with Git

Choosing the right DevOps platform should not be taken lightly. In the following sections, you will learn about some aspects that you should consider when making this choice.

At the most basic level, you should implement a smart development workflow and move away from traditional unstructured development, which is often plagued by too many individual solutions, too many delays, and too much technical debt that has piled up due to a lack of processes.

There are many different development workflows. Some are simpler, while others are more complex. The right workflow for a given company or team depends on how many different versions and variants of the software need to be maintained. For example, for a typical web shop such as *nicely-dressed.com*, which consists of only one version in production that is operated within the company, the requirements are quite low. Therefore, a Git workflow that is as simple as possible is ideal here.

However, when multiple versions need to be supported, which is not uncommon, a more complex workflow is likely needed. And if the software is installed directly on the customer's premises, this has an even greater impact on the entire development process, especially regarding the delivery of new versions.

Workflows

An important aspect for effective software development is the right Git workflow, or how changes to the code are managed. The question is always how to deal with different versions: What code is currently being used productively? What code is currently being tested? What code is currently being tweaked? These different versions should never be mixed up—it would be pretty bad if untested code that is still being worked on ended up in production. It's not always easy to maintain an overview of the various versions of the code, and it is even more difficult when several people are working on it at the same time.

There are numerous solutions for this! Some (like Gitflow) are more complex, while others (such as GitHub Flow and GitLab Flow) are much simpler. A lot can be written about this topic, but let's start with a brief introduction so that you can understand which workflow comes with which advantages and disadvantages.

First, the right workflow depends heavily on the given projects. For example, do many development lines have to be maintained in parallel? Are there many old versions that need to be updated and available over a long period of time? Or does everything quickly flow back to one large main line? Simple release management is an argument in favor of keeping source code management as simple as possible.

The same applies to deployment: Not every project can do "real" CI/CD, as not every company develops applications where both the development work and the deployment are completely under the company's control. If instead a company requires a more static approach (i.e., the company builds the package but the customers have to roll it out themselves), many Git features are less relevant. There are many companies that are not doing regular web development but deliver software to customers who have to roll out the software in their own data centers or directly on embedded hardware, such as in cars.

In the following sections, I will discuss the workflows GitHub Flow, GitLab Flow, and Gitflow, which are used in many projects. Gitflow in particular has long served as a best practice workflow when it comes to the introduction and use of Git. However, I'm not a big fan of Gitflow for various reasons and only use it when it can't be avoided, but more on that in the following Gitflow section.

It is important that the workflows are easy for everyone on the team to understand and, therefore, easy to use. Today's IT world is already complex enough, so you shouldn't create unnecessary complexity here either. In concrete terms, this means that you should work with as few long-lasting branches as possible and keep the feature branches as short and small as possible. This is the only way to enable DevOps principles for a fast flow from the business requirements to the customer.

Trunk-Based Development

Let's first take a look at the simplest type of workflow: *trunk-based development*. With this type of workflow, the entire team works directly on the main development branch. The idea is not to work with branches, but to integrate all changes directly into the *trunk*, the main line from which there are no deviations. To ensure that the project can be built, a preintegration build is carried out beforehand so that only buildable additions are integrated.

However, trunk-based development does not scale in larger projects with many participants, so this approach is not an alternative if there are constant merge conflicts and changes cannot be integrated.

The main difference between trunk-based development and the old waterfall model with big bang integrations is that with trunk-based development, the implementation does not run entirely on its own but is usually carried out as part of *pair programming*. This means that several people are working together on an implementation, which results in an instant peer review. The build is also carried out locally, so it can be validated that the project is being built and can be shared with the others.

Although there are actually no branches in trunk-based development, short-lived branches are often used to verify the build on the CI server. This is particularly helpful if pair programming cannot be carried out. More detail on trunk-based development can be found on the website *https://trunkbaseddevelopment.com/*.

GitHub Flow

Trunk-based development is not suitable if several people are working on different features at the same time and need to integrate them into the main branch. Instead, the team would need feature branches, such as those offered by GitHub Flow. With feature branches, there is only one long-lived branch, main (formerly, master). When new features are developed or bugs are fixed, a short-lived feature branch is created based on this branch, which flows directly back into the main branch after a code review (see Figure 5.5). The changes are then rolled out directly to the production systems using the configured pipeline. This means that short-lived branches are always created and merged into the main development branch as quickly as possible. The changes should be kept as small as possible.

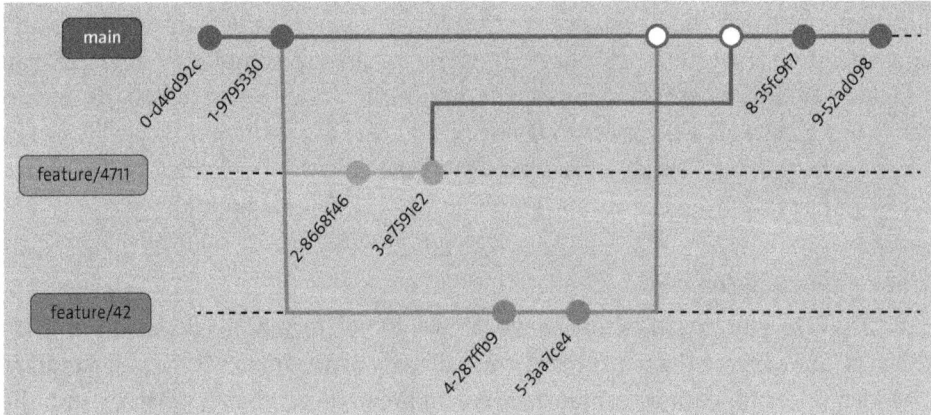

Figure 5.5 GitHub Flow

The exact instructions for the actual implementation of the workflow can be found in the GitHub documentation (*https://docs.github.com/en/get-started/quickstart/github-flow/*).

The simplicity of the workflow is demonstrated by the fact that it can be implemented with just a few commands.

First of all, you must ensure that the main branch is up to date. To do this, first switch to the branch and then use a pull request to make the change available locally:

```
$ git switch main
$ git pull
```

Then, all that remains is to create the new branch and make the changes. Here, we are creating a new feature with the name foobar:

```
$ git switch -c feature/foobar
```

Finally, we only need to push the branch, which would trigger a code review. We discuss code reviews in greater depth shortly.

```
$ git push origin feature/foobar
```

And that's the GitHub Flow workflow! It works well for small web applications that you maintain yourself, but it's *too* simple for anything beyond that. It doesn't even include a staging environment in which you can deploy your test versions.

GitLab Flow

GitLab Flow addresses these limitations and picks up where GitHub Flow leaves off. One limitation of GitHub Flow is that it assumes it will always be possible to deploy to production systems. Filling in this gap, GitLab Flow introduces another long-lived branch called production that always reflects the state of the production environment.

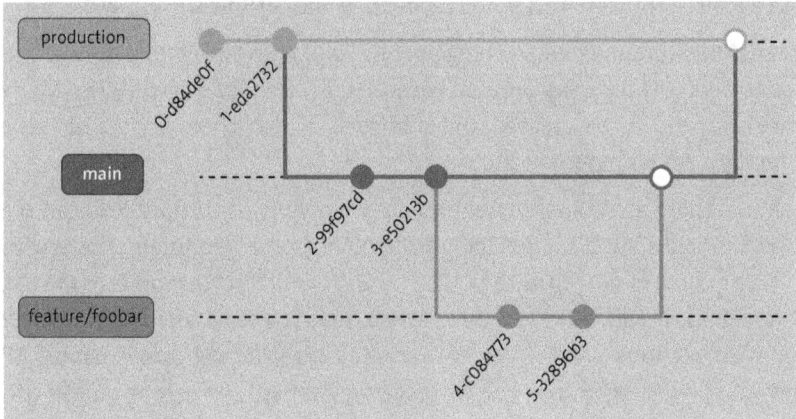

Figure 5.6 GitLab Flow

The usual development work also takes place in this workflow in short-lived feature branches, which are merged into the main branch after code review. The main difference is that deployment to the production environment does not take place automatically after each merge. If a deployment is to take place, the developer needs to explicitly merge it from `main` to `production`. This process provides greater control over the status of the deployment. Figure 5.6 should give you a better understanding of GitLab Flow.

This flow could have some restrictions depending on the requirements of the project, such as a staging environment in which the code is tested before it is rolled out to the production environment. But such a restriction can be mitigated by inserting a `pre-production` branch between `main` and `production`, which would serve as an intermediate step providing greater control over staging and production deployments.

GitLab Flow also provides capabilities for release branches. This type of branch is particularly relevant for software that is delivered to and then rolled out by customers. In this case, `production` and `pre-production` branches are not necessarily required, but you create release branches à la `1.1-stable`, `1.2-stable`, and so on based on the `main` branch. The usual development work continues to run against the main `branch`, on the basis of which the release branches are created. Cherry-picking can be used for bug fixes for older releases.

Complete information about GitLab Flow and the exact commands and instructions can be found in the GitLab documentation (*https://docs.gitlab.com/ee/topics/gitlab_flow.html#introduction-to-gitlab-flow/*).

Gitflow

In 2010, Vincent Driessen published a guide to Git on his blog: *https://nvie.com/posts/a-successful-git-branching-model/*.

That was in the early days of Git, when it was not yet entirely clear how to use Git's useful functions in a consistent way. This guide resulted in the Gitflow workflow.

However, Vincent Driessen updated the blog post in 2020 with a note explaining that he no longer considers Gitflow to be optimal because it is often far too complex. Many long-lived branches need to be created and managed, which then also need to be merged into the right development branches.

As Figure 5.7 shows, there are many, many branches available in Gitflow. You can see some long-lived branches such as `develop`, into which the development flows with short-lived feature branches, as well as the `master` branch, which corresponds to the last published version. In between, however, there are various release branches and hotfix branches, used if corrections need to be incorporated into the various versions. All these branches must be merged into the correct branches so that errors can be corrected not only in older versions but also in the current development line.

I am deliberately not going into all the details here, as they are beyond the scope of this introduction. If you would like to find out more about this workflow, you can find all the information in the blog post.

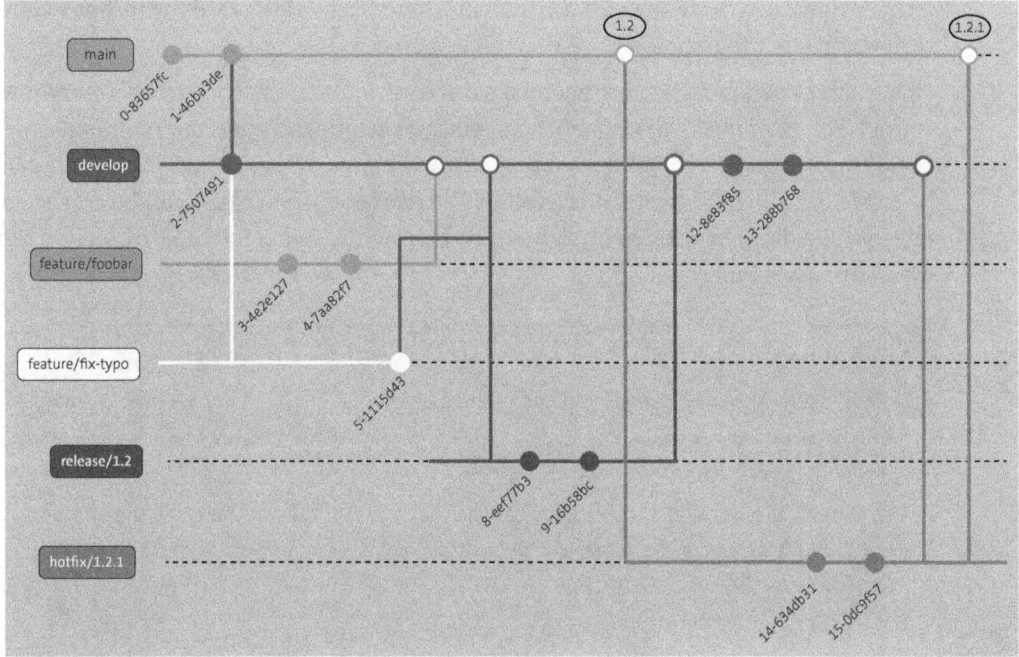

Figure 5.7 Gitflow

Nowadays, many companies rely on a workflow that is close to Gitflow, with some modifications: Typically, they work with several long-lived branches as well as feature and release branches. After a code review, they merge the feature branches into the main development branch, which is usually called develop. When a release is due, they create a release branch based on the develop branch, on which they make the last corrections before the code is merged to the main or master branch and back to the develop branch.

The complexity of Gitflow is necessary for complex setups in which many different version threads need to be maintained and developed simultaneously, which is the case for many projects in industry. The main disadvantage, however, is that this approach makes the work very complicated, as branches often have to be forked and fed back into the correct original branches; developers using this approach have to be extremely careful not to forget or confuse anything. As far as the requirements of the project allow it, simpler workflows in which code reviews can be straightforward and efficient are ideal.

Much more important than understanding the actual implementation of the development workflow, however, is to look at the benefits of using a workflow from a DevOps perspective. Generally, if all the methods of the workflow are implemented properly, the quality of the software will improve. Furthermore, workflows allow development teams to notice faulty and duplicate implementations of any kind in the code much earlier.

In addition, developers on teams that use workflows like this can learn from each other, which also contributes to improved code quality and increased productivity in the long term. We discuss these benefits further in Section 5.4, which deals with both code review and pair programming.

Finally, other people from external teams can contribute to the code when a workflow is used. For example, members of the operations team who have the right expertise could trace errors directly in the source code and perhaps even correct them. This can be particularly helpful from an operational perspective if the cause can be corrected directly, avoiding the need to implement workarounds. This is the topic of Section 5.5, which discusses inner sourcing.

> **Note: Reflection**
>
> A clearly defined workflow is important so that everyone involved knows how development is carried out in order to avoid inefficient uncontrolled growth. Conveniently, teams can define workflows for themselves.
>
> The simpler the workflow, the better for the team. Nothing is more annoying than an unnecessarily complicated workflow. A simple workflow is also an important starting point for code reviews and CI/CD.

5.3.3 Source Code Management at nicely-dressed.com

To modernize the source code management, all those involved in the website development have decided to introduce a stringent code review process, which is to be carried out without high levels of complexity.

This new code review process doesn't really work well with Subversion, so the teams have decided to switch to a new Git-based platform. The workflow needs to be as simple as possible, so GitLab Flow is the obvious choice, as Gitflow is too complicated for the purposes of *nicely-dressed.com* and GitHub Flow is too simple.

> **Note: Reflection**
>
> If you are not yet using a smart development workflow, I recommend working on implementing one as soon as possible. If you are using one that is far too complex, such as Gitflow, then consider whether it wouldn't be better to simplify it. Unnecessarily complex workflows should be avoided.
>
> This also includes other minor factors that I have not gone into. For example, there are many cases in which each member of a team is working with their own fork of the repository and opens pull or merge requests from there. Managing multiple repositories is usually unnecessary, so even in large teams it is better for everyone to work with short-lived branches on the central repository. The rest is taken care of by the repository settings.
>
> Also, don't forget that source code management and the processes around it have a strong influence on the CI/CD server. Ideally, you should manage your source code with as few tools as possible.

5.4 Code Reviews and Pair Programming

The previous sections briefly discussed some of the advantages and disadvantages of code reviews and pair programming. However, these exercises are absolutely central components of a high-quality development workflow, so we discuss them in greater depth in this section.

5.4.1 Code Reviews

Conducting code reviews, of code completed by both the development and operations teams, is fundamentally sensible. If code reviews are missing and everyone touches only their own work, productivity and product quality will suffer. Work may be carried out twice or three times (e.g., in order to fix mistakes that could have been caught by a colleague conducting a code review), changes may be incompatible and require additional work, and errors may creep in because requirements and work steps are not well coordinated.

Code reviews are a simple way to improve collaboration, team knowledge, and code quality. The basic idea behind code review is that every line of code that is to be added to the main development branch of the project is reviewed by at least one other person. This is worthwhile for both high- and low-quality projects, as ideally only improvements to the quality of the code are added. Code review ultimately facilitates communication on the code and promotes a shared development culture.

The initial situation does not have to be as bad as in the XKCD comic that you can see in Figure 5.8. Of course, code reviews ensure that really bad code is filtered out, but they also help improve the skills and knowledge of the team, especially with input from more experienced team members. Everyone can always learn something when participating in code reviews.

Figure 5.8 XKCD Comic about Code Reviews (Source: https://xkcd.com/1513/)

Code review has long been common practice in organizations developing open-source projects. However, I have unfortunately seen that it is not such a common standard at other organizations. The most common reason I hear is, "We have so much to do. We don't have time for code review!", followed by, "But our customers don't benefit from code reviews. They only make our rollouts to customers take longer!"

Apart from a few exceptions, I consider both of these reasons to be lame excuses. As already stated at the beginning of this book, DevOps must be lived throughout the entire organization, and, accordingly, also by the people at the top of the hierarchy. This means that code review must be carried out fully, which takes a little time at the beginning if you are not used to it.

Organizations implementing a code review process should take time should to identify any future problems at an early stage instead of allowing them to lead to high technical debt over time, which would ultimately leave even less time for code review. And good code quality ultimately also leads to a better product for the customer—that is, fewer bugs make it into the finished product and errors can be corrected more quickly, as fewer legacy issues are dragged along the development process.

> **Note: Checklist for Good Code Reviews**
> 1. Keep the changes as small as possible to keep reviews efficient.
> 2. Describe not only *what* you have changed, but also *why*. Bonus points are awarded if you also describe why other solutions were *not* practicable.
> 3. Write meaningful commit messages using the guidelines in point 2.
> 4. Make sure that the changes are always buildable and, if necessary, carry out a rebase.

Code Review within the Development Team

To make the whole code review process a little clearer, let's look at an example.

A developer started at *nicely-dressed.com* only a few weeks ago and is working on her first work package. She is not very familiar with the outdated codebase and realizes that her work may not have solved the problem in the most optimal way. She is not sure whether there is an easier way with the existing code.

Without a code review, she would probably have simply checked the code into the repository. The main thing is that it works. What else can you do?

Well, quite simply, you can ask for help.

To ensure that both her code is of high quality and the implementation is efficient, she asks a senior developer for a review. This developer has been with the company for a while, knows the code very well, and knows of some pitfalls that could lead to problems.

The senior developer gives her new colleague a few tips on what she could do differently to optimize the code. And lo and behold, the senior developer finds errors in the existing code that could lead to problems later on if other parts of the code are not also revised directly as part of a refactoring (see Figure 5.9).

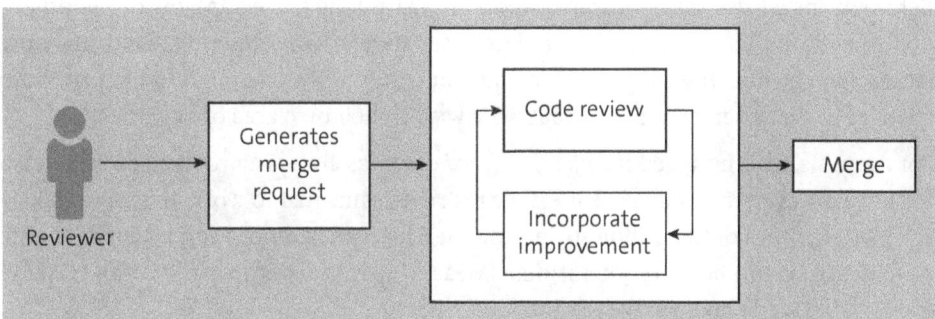

Figure 5.9 Simplified Code Review Process

The advantage of code review is clear: Reviewers can identify and rectify potential problems at an early stage. Importantly, reviews should be carried out soon after the code is written; that way, the developer will know why they solved the problem in that way. A few weeks later, and the developer would be out of the context of the issue and would first have to familiarize themselves with the specific issue again. There is a high risk that something will be forgotten.

But there is another advantage: The developer being reviewed can get to know the codebase much better and faster and learn what they should consider for the next implementation. And the developer carrying out the review can see the code from different perspectives, allowing them to possibly find problems that only people with fresh eyes can recognize. They may even be able to better recognize future challenges for their own implementation work at an early stage.

The argument that code review takes too long is hardly valid here: In traditional waterfall workflows, changes are rolled out only every six months anyway, which regularly leads to problems. This time is saved if possible sources of errors can be eliminated earlier, such as through code review. And the project will benefit in the long term if the code quality is improved and potential sources of error are gradually eliminated.

You will certainly be familiar with these internal reviews within the development team. Code reviews can be even more valuable if colleagues from other teams with different skills are also involved. As the development and operations teams move closer together during a transition to a DevOps environment, code reviews are an easy way to gather feedback between different areas of responsibility.

Code Review between Development and Operations

DevOps does not mean an operations team is unnecessary. Both the development and operations sides are still both equally needed. The truth is somewhere in the middle: Developers are now taking on more responsibility for operating the products, and those who previously worked only on the operations side are getting to know the applications better. And all this is happening on one team.

Before the *nicely-dressed.com* teams started transitioning to DevOps, the development and operations teams worked completely separately; consequently, they not only lacked insight into the other team's code, but they were also unable to contribute changes to the other team's code that were within their own area of expertise.

For example, loading speed for *nicely-dressed.com* was always a problem, as is the case for virtually every complex website. One task the teams had to work on was to reduce the loading times of the online store. However, metrics had not been available, so the problem could not be better quantified. (Measuring metrics is an aspect of the CALMS model, but I won't go into this in more detail here.)

As part of the development work, it was concluded that the project needed a caching server to cache a lot of data so that content could be retrieved and delivered more quickly. The development team was asked to provide the caching server. The team had to provide the infrastructure and needed clear instructions on how the server should be configured.

During the implementation work on the development team's side, however, the developers always needed to "quickly" make a few adjustments. At that point, there was no promising opportunity to work together on the same code.

The operations team had no insight into the code and could not see the configuration parameters that the developers had set. The exchange did not work; they could not help each other. It took far too long for the changes to go through. And even then, the completed work was always in need of improvement.

However, now that the company has started transitioning to a DevOps environment, opening up the code, bringing the teams together, and introducing code review, this process has significantly improved. Instead of ping-ponging the work between the teams, a process with a lot of trial and error, the operations team can work on projects such as the caching server with the development team directly, especially with the new review process. Development work has been significantly accelerated.

This applies not only to changes to the source code of the online store but also to configuration changes to the server infrastructure, where both teams can suggest changes to each other.

Code Review between Development and QA

DevOps is not just about bringing development and operations together; it is about integrating all departments that work on a product, including the QA team. Again, code review between the development and QA teams is valuable because it takes place between different areas of work. For example, when the development team has to reorganize the structure of tests, they need to receive the necessary input from those who are primarily responsible for QA.

And the reverse is also true: From a QA perspective, a review by a developer is always useful to ensure that the changes make sense. For example, based on their knowledge of the codebase, developers can indicate where the tests could be improved or where there are gaps in test cases.

Speeding Up Reviews

A core aspect of DevOps is that regular feedback and equally regular incremental updates of the product to be delivered can make work more agile. To ensure that code reviews are done in a timely manner, there are a few rules to follow. Code reviews should be made as quickly and efficiently as possible. After all, changes should be introduced quickly; otherwise, you won't have won anything.

- **Avoid allowing reviews to be delayed.**

  A common problem with the introduction of reviews is that code review requests are often left lying around. Code reviews may be delayed a few days, sometimes even weeks. This is common because developers are more focused on ensuring that their own work is completed quickly. So if reviews are not being completed as requested, your developers should take the initiative and find someone else to complete review requests that have been laying around if necessary. Know that there are tools that can simplify the whole process.

  Fixed deadlines and regulated processes that organize the review process can help ensure that code reviews are done in a timely manner. You should never have to wait long for a finished change to be reviewed.

- **Foster a culture that does not allow blame!**

  Absolutely central to the success of code reviews is that developers should not be blamed for problems found in their code or for problems their code may have caused. Developers are not automatically to blame for such problems. And similarly, reviewers shouldn't be blamed if bugs in code they reviewed make it into production; after all, someone else could have found the bug.

  Discouraging blame creates a productive atmosphere around the review process! Nobody should be afraid of code reviews. However, it should also be clear that no obviously bad or unfinished code should be submitted. The code review process is generally not an additional instance in which problems are solved for which you have not found an answer yourself. These steps should be completed beforehand, as they otherwise block and complicate the review process.

- **Keep changes manageable and small!**

  In order for reviews to be effective, the changes under review should be as manageable as possible. The smaller the changes, the faster they can and will be reviewed. This is not surprising; the easier it is for a person to carry out a review, the more likely they are to do it. A change involving five lines of code can be reviewed faster than a change involving several thousand lines; such a change would take a lot of time, and it may not be possible to meaningfully review it at all.

Another advantage of smaller review chunks is that they can be integrated and rolled out much more frequently and tend to cause fewer serious problems, as they help avoid large big bang integrations. They prevent not only problems on production systems but also conflicts caused by several people working on the same or overlapping lines of code. The faster changes are integrated, the less pain the implementing developer and the team as a whole suffer.

- **Implement reviewer roulette.**
In terms of collaboration and knowledge transfer, it is important that the same people do not always verify the same parts of the code. Having the same people review the same parts of the code may be useful in some cases of complex issues that are better reviewed by senior developers and those who know those parts of the codebase particularly well. But in order to build up more experts on the team, there should be a good mix of developers who touch various parts of the codebase.

  I often see that the same people always ask their favorite colleagues to review their code because they hardly make any comments and wave the code through with little effort. In these situations, the developers gain nothing, and they could also introduce new disadvantages into the process.

- **Encourage constructive and helpful feedback**
The comments given for reviews should be meaningful and comprehensible, regardless of the developer's level of knowledge. Comments such as "This is bad code" may be correct, but they don't help; they don't make it clear what the problem is, and there is no suggestion on how to fix it.

  The tone of the language used is therefore important. Reviewers should avoid making accusations or using language that portrays the developer who made the changes under review as "stupid." Even comments such as "This is quite simple!" and the word "obviously" could come across as condescending, which is not conducive to harmonious cooperation.

In summary, introducing a code review process is a simple way to encourage collaboration between the individual areas of responsibility so that the teams can move closer together and that knowledge can be shared. In other words, having code reviews keeps with the spirit of the sharing value in the CALMS model.

Code reviews improve processes within the team, ensure higher product quality, and simplify support within the team when the application is rolled out. And last but not least, code reviews ensure a better culture on your teams: Smart reviewing almost automatically creates a shared commitment and ownership of the product, as your teams are not working just in their own corner of the code.

5.4.2 Simplify Code Reviews

One of the most important reasons for code reviews is to provide very early feedback to the developer introducing the change. This feedback comes from both the reviewer on the maintainer team and from automated tests. These automated tests are executed as part of the pipeline, so they should run automatically as part of the code review process without the need for manual intervention to start them for a feature branch. It should be directly visible in the code review whether the pipeline run was successful.

Ideally, these automated tests should go one step further: It should be immediately apparent by simple means what impact the changes made have on the project. This includes, for example, whether the quality of the code, which is defined by standardized quality characteristics, has changed. Depending on the application, it may also be useful to carry out quick load or performance tests to see whether the changes lead to performance changes, whether minor or major.

These tests should always run as quickly as possible, as long waiting times are counterproductive. Ideally, the source code management platform should also support the definition of various quality gates as part of the review in order to block merges if needed. These gates are particularly important for preventing security vulnerabilities from being merged; decisions can always be made as to whether deployment to production environments should be blocked due to new security vulnerabilities.

The visibility of the pipeline in the code review process means that everyone can see directly whether the software has been built and the various tests have been successfully completed before anyone starts a review.

Source code management platforms also have various smaller features related to code review to simplify the process. For example, depending on the platform used, reviewers can note suggestions for additions to individual lines in order to correct errors or add optimizations. The developer under review can then review these changes and accept them directly so that they become part of the merge or pull request (see Figure 5.10).

This feature may be overlooked, but can be very helpful, as it can speed up the review process considerably. A good platform offers metrics that show how quickly reviews are carried out. The smaller the changes under review are, the faster the review process should generally be; a faster review process is also in line with the goal of being able to integrate and roll out changes quickly, even if they are not yet fully functional.

However, do not forget to consider the entire system when evaluating a partial solution. This is also the case here! The best system for source code management is of little use if too much time and effort have to be invested in half-hearted integrations with other tools.

5 Collaboration when Coding

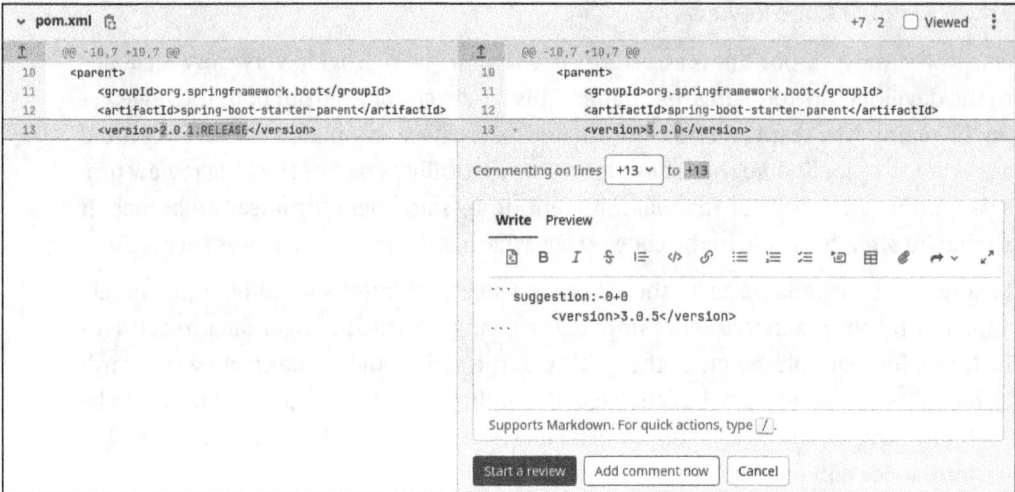

Figure 5.10 GitLab Example of Change Proposal in Merge Request

Approval Rules for the Process

Another useful concept is the definition of rules around who is and is not allowed to accept and approve merge or pull requests. Depending on the size of the project, it makes sense to define at least one approver for changes. Although an approval process would negatively impact the speed at which changes can be completed, it is a simple way of enforcing reviews.

Depending on the industry, such approval rules may even be required by law. The banking and finance industry in particular has such laws, which are checked by auditors. Approval rules can help ensure your company is complying with such laws. We discuss compliance in more detail in Chapter 11, Section 11.6.

The CODEOWNERS file in the source code management program, which defines the owners of the individual files and paths in the respective repository, can be helpful. This file makes it easy to control which changes to which code need to be checked by which people in order to ensure quality. Thanks to the simple integration, no additional effort is required to find the appropriate reviewers; this is done automatically. This is particularly relevant when many projects are managed in a repository.

> **Note: Codeowners File Availability**
> Depending on which Git hosting solution you use, the CODEOWNERS function might be an enterprise feature.

A CODEOWNERS file can look like this:

```
src/ @dev-team
docs/ @docs-team
kubernetes/ @ops-team
```

In this example, changes to the src directory would require the development team to carry out a review. However, if changes were made only to docs or kubernetes, the development team would not have to carry out a review.

While implementing a review process is important, it is also important to ensure it is short. For example, consider a situation in which a review is approved, but the merge takes place only once the pipeline has been successfully completed.

For larger projects with many almost simultaneous merges, however, reviews should be processed sequentially. This means that they should be included in a *merge train*—sometimes also called a *merge queue*—in which the pipeline is run through once before each merge and the merge is canceled if the pipeline fails. If no problems occur, merges are performed continuously but not all at the same time. The main problem with simultaneous merges is that changes that were considered unproblematic when a merge or pull request was opened may no longer be at the time of the actual merge.

To Squash or Not to Squash

Commits give you a history of your project, as every step is visible. If you follow the guidelines I have given (i.e., to keep commits frequent and small), the history of your project should look very small and, therefore, perhaps confusing. One solution to a confusing commit history can be to *squash* commits—that is, to combine and merge multiple commits into a single commit. This usually happens during a merge; most commonly used Git platforms offer the ability to perform a squash followed by a merge.

This process has advantages and disadvantages. I prefer to be able to easily track which changes have been made by a commit, so I prefer to keep commits as small as possible to be able to see in detail which changes were committed and when.

However, this process also has disadvantages. For example, in the event of an error in one commit that was squashed together with other commits, you would have to check and undo several commits, which can be cumbersome. The answer to the question of whether you should squash commits or not is probably that it depends very much on what exactly the change looks like.

5.4.3 Pair Programming

Pair programming is another method that teams can implement either as an alternative or in addition to code reviews to help increase code quality. Although it can actually be used independently of DevOps, pair programming follows many of the basic ideas of DevOps culture.

The idea behind pair programming is simple: Two developers work together on the same problem in the same code. They look at the existing code at the same time and develop the solution to the requirement or bug together.

5 Collaboration when Coding

A common assumption is that two developers deliver twice as much code as just one person. However, this is true only for very simple tasks that are simply coded down. The more complex the problem is, the more likely it is that two people in a team can solve a problem faster than two lone wolves. Because even if you tackle a complex task alone, you have to explain the thought processes to yourself.

> **Note: Rubber Duck Debugging**
>
> A widely used technique in this context is *rubber duck debugging*. Are you faced with a complicated problem or an error that you don't understand? Grab a rubber duck and explain the challenge to it, but slowly and in small steps, as the duck is not a programming professional. By explaining the problem as simply as possible, you will usually arrive at the solution yourself. Pair programming goes one step further and replaces the duck with an intelligent colleague.

In pair programming, one developer explains the current challenge and their own thought processes to the other. One developer writes the code and thinks about the next lines, while the other typically thinks a few steps ahead. This is particularly helpful when other parts of the source code need to be looked at in parallel.

The advantages are obvious: For example, developers can spot potential problems more quickly when talking about them together than if they were conducting an asynchronous code review. Possible errors can be corrected more quickly if you are already in the flow of thought during pair programming.

Figure 5.11 Pair Programming

Pair programming is also good for team culture, as it allows team members to share knowledge (see Figure 5.11). In a session between a junior and a senior developer, both can learn from each other. The junior developer can get to know the codebase and domain knowledge, while the senior developer can get a fresh look at old problems. In a duo, junior developers may find it easier to ask questions, and interpersonal relationships can be improved.

In my experience, pair programming also has another positive side effect: Two people can often concentrate better together than when they work alone. For example, people tend to be able to ignore nonurgent direct messages and emails when working together; when they are alone, they are more likely to reply to them immediately. Therefore, pair programming can help improve and speed up the coding process.

> **Note: Mob Programming**
>
> With pair programming, both people do not necessarily have to be sitting in front of the same computer; they could also use the usual video conferencing solutions.
>
> And with video conferencing, developers can not only work in pairs, but they can also engage in "mob programming" with several people. You can find an e-book on mob programming at *https://remotemobprogramming.org/*.

Pair programming therefore allows developers to write better code more quickly. And after a pair programming session, a separate code review may no longer be absolutely necessary, as two sets of eyes have already looked at the code. However, pair programming and regular code reviews do not have to be mutually exclusive: A development team could use regular code reviews for simpler tasks that can be implemented quickly by one developer and pair programming for heavier tasks that two developers could tackle together.

And of course, in an environment with a strong DevOps culture, pair programming doesn't necessarily have to be with two developers. It may well be that the problem can be solved much better with a representative of the operations, QA, or security team.

> **Note: Reflection**
>
> Please do not work without code reviews. With code reviews, your teams can incorporate the process of addressing bugs and technical debt and of sharing knowledge directly into everyday development.
>
> The introduction of code reviews is possible within your own team alone but will probably require some convincing. Skeptical colleagues may believe that code reviews will result in too much and too slow development. Note that this criticism may actually be true when you first implement a code review process, which is why it is essential to focus on an efficient workflow with as few changes as possible.
>
> Pair programming can also be introduced within your own team with just a few hurdles. It is helpful if you use it to work on complex issues, regardless of whether experienced or less experienced people are involved.

5.5 Inner Sourcing: Sharing Code within the Company

If you have followed the story of *nicely-dressed.com* so far, you will have noticed that many problems arise from the fact that the development, QA, and operations teams do not cooperate well together. But that is not the whole story. Problems also arise due to lack of cooperation between the individual development teams. We could even say that they have tended to work against each other so far.

Although it is a single online store, the codebase of *nicely-dressed.com* consists of different parts that are developed by different teams. For example, one team works on the backend for the billing, another is responsible for the search functionality, and yet another works on the module that shows customers relevant offers.

Previously, the teams all worked completely separately from each other; each team had its own repository with its own access points that only the team could access.

As has already been discussed, such an approach doesn't look too bad at first glance. The teams don't have to worry about other people reading their code and getting bad ideas. However, this organization results in some significant disadvantages. The teams inevitably have to work together because the different parts of the online store call each other via APIs. This would not be a problem in itself if there were complete, up-to-date and detailed documentation.

But really, have you ever seen internal API documentation that is complete, up to date, and, above all, helpful? That's why there are always problems, because occasionally something doesn't work the way it was documented, or a developer isn't be able to find some information in another team's documentation and therefore doesn't know how to put the different parts together.

So far, the collaboration between teams at *nicely-dressed.com* has looked like this: One team would create new tickets in another team's systems. The aim was to discuss errors and problems. However, as the entire development process followed the waterfall model, there were hardly any opportunities to add bug reports, feature requests, or missing documentation on short notice and in an agile manner. Problems remained unresolved.

For a ticket to be processed at all, the relevant team had to compile comprehensive problem descriptions and diagnoses. A team getting a lot of tickets from another often questioned whether the other team was actually calling the API correctly.

This was a lengthy and exhausting process for everyone involved; teams often closed tickets involving problems in external code parts without comment because they preferred to build their own workarounds instead of continuing to struggle with communication.

So instead of fixing the actual problem, the teams would just temporarily fix the symptoms. This in turn led to further technical debt in the source code; temporary fixes are not as stable.

These problems in any organization are serious enough to make working on new features or fixing bugs almost impossible. But even if an organization were to simplify the processes with better technology, such as by introducing a new ticket system, it wouldn't fix the problems stemming the poor working culture. For example, in organizations with a toxic culture, teams may point fingers or blame their own challenges on the mistakes of other teams: "As long as they don't process the ticket over there, I can't continue working."

And since no one wants to be seen as the brakeman or even be responsible for the sluggish progress on other teams, this can lead to a situation in which mistakes are no longer acknowledged or, even worse, sold as features that don't need to be corrected. After all, it's human nature to present your work in a better light and preserve the company's reputation.

These are problems of corporate culture that cannot be solved technically. Instead, a rethink is required, using inner sourcing. Inner sourcing is not much more than the use of practices from the open-source movement, but within a company.

5.5.1 Open Source

Before we take a closer look at inner sourcing, let's discuss the practices and values are that are practiced in open-source projects. These practices are relatively close to DevOps principles, as both place a high value on an open culture to enable collaboration. This includes a high level of transparency so that the current work status, progress, and future plans are clear to everyone.

In principle, everything that is covered by an open-source license falls under open source. You can find a complete list at *https://opensource.org/licenses/*.

The key factor here is, of course, that the code is open. Not only can you view the code of these projects, but you can also adapt and redistribute it. This means that outsiders can easily suggest and incorporate new ideas and that the project is constantly evolving. It is precisely this aspect that is often missing within companies.

For open-source projects, it is vital that they are located on DevOps collaboration platforms such as GitHub or GitLab, where developers can easily view the code with the complete history. The documentation for the project is usually also available there, explaining how it is built, how to get it running locally, how to deploy it on production systems, and how to use it. This is the only way to make it easy for interested parties to contribute new features and bug fixes to the project.

> **Note: Code of Conduct**
>
> In addition to this technical support, most open-source projects have also published rules for collaboration, the *code of conduct*. The aim is to make it easy for as many people as possible to work together. Further information can be found at *https://opensource.guide/code-of-conduct/* and *https://contributor-covenant.org/*.

The result is that users and developers can quickly download, build, and get the project up and running without having to request permissions. And then it is easy for developers to get involved and invest their own knowledge and time into the project.

So don't be afraid to work on open-source projects. Of course, you can't submit code for security-critical parts of the Linux kernel right from the start, but almost all projects have smaller tasks that you can take on to get started.

You can then submit suggestions for improvement, which are checked by experienced project members—a code review, so to speak. These members decide whether the change is accepted, whether further adjustments are necessary, or whether the change needs to be rejected completely.

You likely have open-source software as a dependency in your own projects, so it is not unlikely that you'll come across a bug or a missing feature. Report such issues to the project. Of course, it would be even better if you could correct the error directly and contribute the solution. Ultimately, everyone benefits when you contribute directly to an open-source project you're using as a dependency, and countless open-source projects prove that this is an extremely productive way to develop software.

These practices need to be established within a company looking to transition to DevOps.

5.5.2 The Path to Inner Sourcing

Now that we've discussed open-source practices, we can discuss inner sourcing. Inner sourcing is about using the culture and practices of the open-source movement within your own organization. In Figure 5.12, for example, you can see that developers from different projects have access to the shared API, which in turn is managed by a separate group of maintainers.

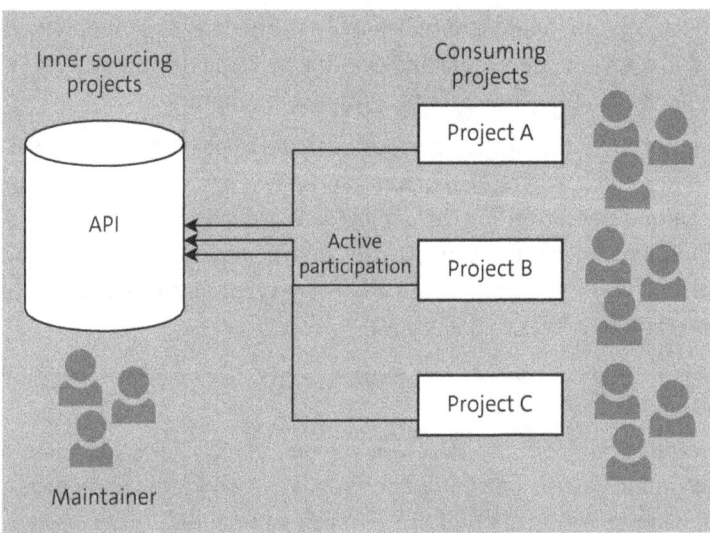

Figure 5.12 Inner Sourcing

Although inner sourcing is a time-consuming process, it has many advantages. However, there may be some barriers to implementing this process. For example, there must be a commitment from the management level; otherwise, this process cannot be implemented. In addition, such barriers are often organizational if there are policies that prohibit the opening of projects. So you can't enforce inner source with a sledgehammer; instead, you have to check where more openness and transparency make sense.

Step 1: Disclosing the Projects

The path to more openness and transparency should be taken in several steps. The first and simplest step is to disclose the company's projects so that everyone can at least read the source code. But here, too, you need to proceed with caution, because it's not that simple and does not come without risks. If passwords or other access data were previously stored in the source code, these must first be removed and rotated through, which requires further changes.

And it becomes even more complicated if the projects of other departments are also affected. If, for example, the operations team also uses repositories to store and version configuration files, you must ensure that introducing greater transparency does not lead to security problems. Information on certificates and other security-critical areas must of course continue to be protected. Regardless, informing the development team of what the setup of their own software looks like on the production systems is always helpful.

Step 2: Centralization of the Repositories

Although releasing the source code is an important step, the repositories will still be distributed across different systems. In addition, self-managed shadow IT by individual teams is not uncommon. In most cases, not all users will have access to all repositories, so releasing the source code is not enough.

Therefore, you need to set up a central collaboration platform that will host more than just the code in the future. The projects have to be relocated for this, which requires a lot of work, as interfaces to other tools have to be reconfigured. This is not a project for an afternoon!

Step 3: Writing Project Documentation

Once you have a standardized platform for your repositories, you can write the project documentation. The documentation should not simply be source code comments, but should explain how to build the project, how to get it running locally, and how to deploy it on a production system.

This information will make it possible for people from other teams to use the software to the extent that they can contribute changes. The barrier to entry must be as low as possible; nobody wants to start by reconstructing a build environment that is not documented.

The faster a team member can get a functioning development environment for the project up and running, the faster work can begin, and the lower the threshold for becoming active will be. And this is completely independent of whether this team member is new to the team or comes from another team.

Step 4: Enable Collaboration via Code Review

By using a central collaboration platform such as GitHub or GitLab, changes can be easily contributed to projects. However, introducing changes does not replace the original code, so you don't have to worry about other teams suddenly contributing changes without you noticing. Instead, changes are submitted via merge requests and pull requests.

The teams behind the project being contributed to are informed of changes in this way and can perform a code review, as described in Section 5.4. The process can be the same regardless of whether a change comes from inside or outside the team. It is only important that accepted changes are of good quality and that reviews are carried out properly.

5.5.3 Advantages of Inner Sourcing

But what are the benefits of inner sourcing, and what exactly does it have to do with DevOps? Strictly speaking, inner sourcing can of course also be practiced without DevOps. However, DevOps is primarily about breaking down silos between teams and increasing productivity through efficiency, which is achieved through inner sourcing.

More Efficient and Effective Development

So the first advantage is, unsurprisingly, more efficient and effective development. Imagine, for example, a team that manages firewall rules between the source and target systems. This is preceded by an approval process: The source system team has to create a ticket in the ticket system to access the target system in the test and production environments. The ticket has to be approved accordingly by the target system team. Only then can the operations team start working on the ticket.

This process means that a change that would normally take only a few minutes takes much longer. First, the ticket has to be approved by the target system team. The target system team then has to wait for free time from the operations team. And finally, the source system team has to check that everything really works and that nothing was forgotten or done incorrectly. Such a process could take several weeks for an intervention that is not much more than a change to one line.

Inner sourcing makes such processes simpler and faster; with an open repository and a culture in which easy collaboration is possible, there would have been no need for a ticket in that scenario. The source system team could have proposed a change directly via a merge or pull request. The target system team and the operations team would only have had to approve the change, and the change would have been rolled out directly.

This approach can help shorten the time to product rollouts, as some steps can be combined and thus reduced. Not only is this process more efficient, but it also overcomes the organizational limitations of multiple different teams, and everyone is happy because changes are pushed faster. There is no bottleneck, and the process is less error prone.

New Product Ideas from Inner Sourcing

Inner souring also has advantages from a software development perspective. In principle, open-source projects are structured in such a way that they can be reused. This type of structure is also useful within organizations, as it helps avoid duplicate developments. In some cases, such developments from projects even result in other products that can be used elsewhere in the company.

> **Note: Open-Source Projects from In-House Developments**
>
> If inner sourcing is implemented well, you can look into releasing one or two projects under an open-source license to give something back to the community. But that's not all: You will also benefit if other teams use your software and possibly develop it further.

In addition to fostering simple processes, inner sourcing also ensures better code quality in a gentle way. As the code can be seen by many eyes, there are fewer crude workarounds, because other teams that depend on these parts of the code would otherwise immediately sound the alarm.

Higher Employee Satisfaction

Last but not least, this way of working leads to better employee motivation. Nobody likes to deal with additional hurdles when working on difficult and complex problems.

The higher the hurdles, the more likely it is that people will instead try to solve problems with workarounds rather than correcting them at the source. But nobody is really satisfied with this. It is better reduce the barricades, and then everyone will work happier and more efficiently.

> **Note: How Not to Do It**
>
> An acquaintance of mine is a big fan of inner sourcing and getting to the root of problems. As a GitLab admin in his organization, he has access into all projects on the instance. One day, he wanted to contribute a change directly to a project to avoid

> workarounds. He opened a merge request with the desired change. However, his company was not a fan of inner sourcing, so, even though the change was useful and helpful, he got into trouble rather than being praised for his help: "Please don't interfere in other people's projects!"
>
> In the end, it all depends on the mentality and the culture: Only when this way of working is desired can it be implemented.

Even if the effort involved in such a migration is high, the advantages outweigh the disadvantages. Unfortunately, the changeover phase will also require a lot of effort if teams need to change their working methods and habits without seeing any immediate benefits.

After all, the new structures first have to be created before collaboration can function smoothly. And it can be frustrating if you put in the effort and create good documentation and transparency around these new processes, but other teams on your project are less interested and give limited input.

Nevertheless, inner sourcing is an important and indispensable point in the transformation to DevOps. It lays the foundation for an efficient software development lifecycle, and if the foundation is already shaky, the next steps such as the build process, testing, and deployment will also be difficult. I will discuss these steps in the next chapters.

5.5.4 Monorepositories

One question I hear again and again is whether monorepositories (monorepos) would make sense from a DevOps perspective. The main reason given is that it would be easier to refactor the source code, as changes could be made to several projects in one merge request.

But first, let's take a step back and define what a monorepo is. A monorepo is one repository that contains all of the company's projects, including both active and inactive projects (see Figure 5.13).

The best-known case of a monorepo in productive use is Google, which has built its own source code management tool to manage a single repository. Meta Platforms, the company that develops Facebook and Instagram, also essentially relies on a monorepo, which uses the Git-compatible, in-house developed Sapling platform (*https://engineering.fb.com/2022/11/15/open-source/sapling-source-control-scalable/*), which was released as open source at the end of 2022.

A monorepo like this is huge and contains the history of (almost) all of the company's projects. However, a monorepo approach with Git quickly reaches the limits of the software. Git is not particularly suited to working with a large number of directories and files. This should not be a problem for companies that work with small codebases.

5.5 Inner Sourcing: Sharing Code within the Company

Nevertheless, the size of the repository can quickly reach several gigabytes, which slows down the entire development process; for example, the pipeline clones the repository almost every time. This therefore not only costs time, but also money, as a higher load is to be expected due to automation.

However, many people have a general misconception of the term "monorepo." For many people, a monorepo is a repository in which several projects are stored in one repository, such as a repository storing the source code of the frontend, the backend, and the mobile apps. Some companies even speak of several monorepos, which of course makes no sense, as the term "monorepo" implies exactly *one* repository. For those scenarios, a more appropriate term would be "multiproject repository."

I would like to use monorepos as an example of how simply copying the approaches of very large companies is not always a good idea. It is also an illustrative example of not seeing the big picture.

As I have mentioned, integration into many phases of the DevOps lifecycle is immensely important to increase visibility everywhere. There are various advantages and disadvantages to using a monorepo. I won't go into all of them here, but it's always worth taking a look at the tools you use to see to what extent they really support this approach.

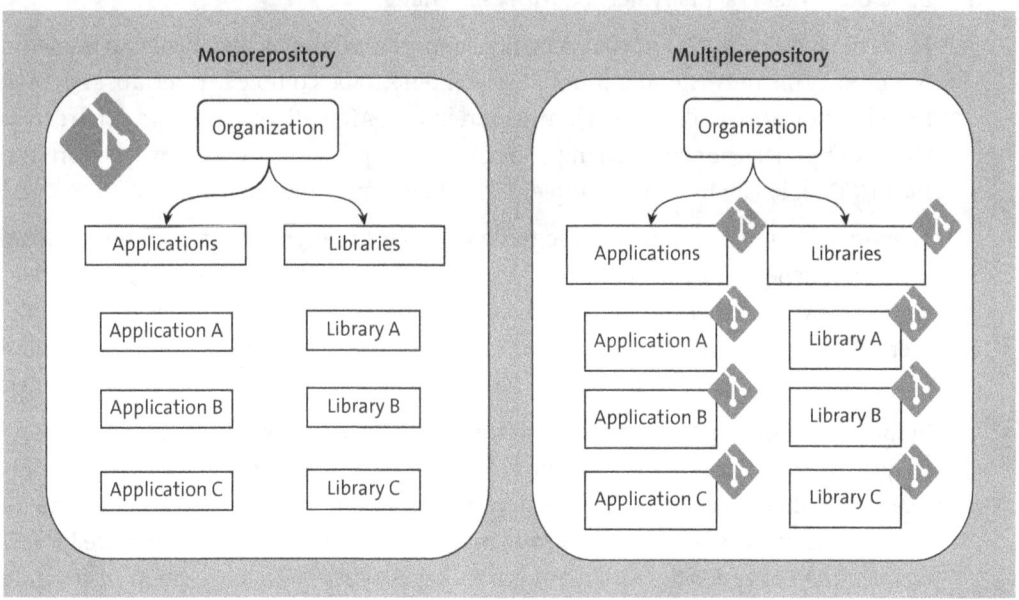

Figure 5.13 Monorepos versus Multiple Repositories

One of the main reasons for using a monorepo (or even a multiproject repository) is that it simplifies the management of internal dependencies. And this use case is not that rare. For example, it is often the case that there are several projects within a company that use the same dependency, and managing changes to that dependency can complicate development.

For companies that do not have a monorepo, changing a dependency requires a change in two repositories at the same time: one in the dependency itself, and the other in the main project. However, the change in the main project depends on the change in the dependency. Therefore, the change in the dependency has been incorporated before the change can be made in the main project.

The process necessarily leads to longer development times. In addition, the process takes longer if several projects use the dependency, as the change must be updated in each project, even if it is only a small adjustment.

This scenario already shows a significant advantage of monorepos and multiproject repositories: A change to a dependency and the projects that use it can be made in one go and would need only one review. There is no need to merge changes across repositories, and at the same time a review of the change can be carried out more transparently, as the change can be monitored across all projects in one view and with one review.

While there are these advantages to using a monorepo, there can be some disadvantages when it comes to the CI/CD setup for a monorepo. The reality is that the whole thing simply has to be made automatable via the pipeline, which should be able to build only what is actually needed with each change.

However, today's CI/CD toolchains do not support monorepos very easily, so the setup is tedious. This means that a monorepo will tend to be complex to set up, and long-term maintenance is also usually rather strenuous. After all, you don't want to rebuild all of your projects for every change. So you have to put a lot of engineering effort into the CI/CD toolchain in order to implement efficient pipelines.

Therefore, for most companies, the trade-off is far too high, and it is better to create smaller, independent projects and repositories in order to keep working with them simple. However, this has the disadvantage that some changes have to be implemented and tested in several repositories at the same time—though this is a problem mainly faced by larger companies, as the overhead is negligible for smaller companies.

In the end, it also depends on the software architecture: This works better with loosely coupled microservices than with a monolithic architecture, where every change means that everything has to be built once anyway. The more tightly coupled the dependencies are, the more worthwhile it is to manage them in a repository. The more loosely coupled the dependencies are, the more it makes sense to manage them in separate repositories.

> **Note: Resource on Monorepos**
> At *https://monorepo.tools/*, you will find a well-designed website that goes much deeper into the subject of monorepos.

The question here is to what extent the various tools on the market actually support your own ideas. Ultimately, the best idea is useless if it can be implemented only with a great deal of effort.

Giant tech companies like Google may be able to afford to do this, but most companies are neither that big nor that capable. Instead, my advice is to always take a better look at the different tools at your disposal to see how they can support you. You always have to make compromises.

> **Note: Reflection**
>
> The less authority you have in your organization, the more difficult it is to introduce inner sourcing. This applies both to opening your repositories and to changing the development processes in such a way that everyone in the organization can contribute.
>
> If you work according to DevOps principles, then the pure (technical) implementation is less of a problem. Simple steps in the right direction would be to grant internal people who regularly open bug reports access to the project. This way, these accesses can be used directly as a test balloon: Which guidelines would have to be changed as a result? Are we already seeing efficiency gains through closer collaboration?

5.6 Summary

In this chapter, I have discussed the management of the source code. At first, you may have expected this to be a very technical topic. By now, however, you have probably realized that it is more a question of culture and collaboration than a question of technology itself.

Code review and pair programming ensure that changes to a project are secure and of high quality before they are incorporated, and inner sourcing makes this approach available to people outside the team to enable the best possible collaboration and thus make development as efficient as possible.

Chapter 6
Continuous Integration and the Build Process

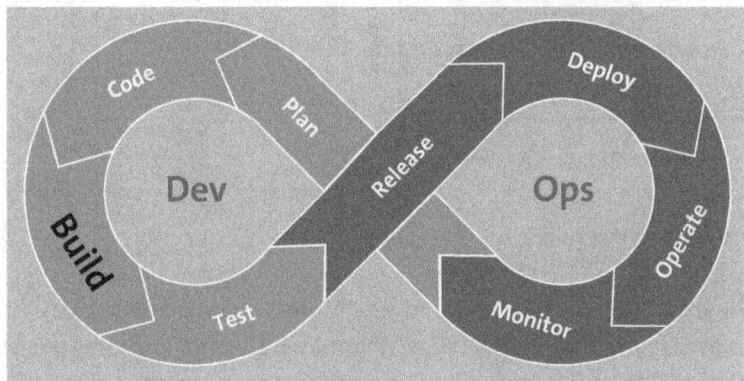

Figure 6.1 Build Phase of the DevOps Software Development Lifecycle

Now that we have looked at how development practices and direct collaboration between various teams can be improved, we can move on to the next phase in the DevOps software development lifecycle.

In fact, the last chapter served as an introduction to the topic of *continuous integration*, which in short involves integrating the code into the main development branch as quickly as possible. In order to achieve this, a clean build process is required. And that is exactly what this chapter is about.

6.1 Typical Problems in the Build Process

You already know that the *nicely-dressed.com* online store consists of various modules and applications that are developed by different teams. It also has various backend systems, such as a billing module for all invoices and the module where the latest offers are prepared and presented. There are also two teams that take care of the Android and iOS mobile app versions of the store.

As we have discussed, the teams all work separately from each other and in separate repositories. The processes for the respective project have developed differently over the years, and because each team member builds their own part during the six-month development cycles, the build processes are not very efficient. Only at the end of the cycle do the various teams put all the parts put together and integrate them on a grand scale. Conflicts are naturally preprogrammed because there are always interface changes between the modules that have not been communicated.

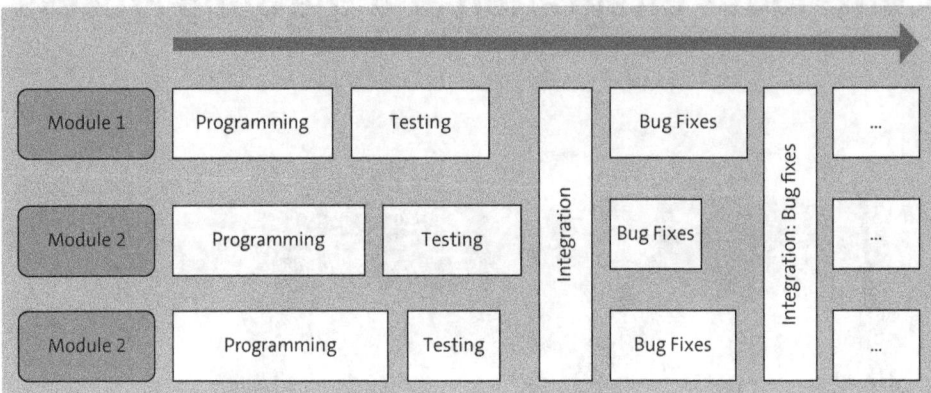

Figure 6.2 Typical Problems in the Build Process

As you can see in Figure 6.2, the development phases for the individual modules never run in parallel. The case illustrated here is relatively harmless: The programming phase and the test phase are of different lengths, and there are also waiting times until integration. After integration, only bug fixes are made, and then the bug fixes are integrated.

That sounds complicated, and it is. There are often several errors and conflicts in the integration phase, which means that integration cannot take place properly, which delays the whole process.

Ideally, such problems should be tackled at an early stage so that it is clear early on whether changes made during the development phase can be integrated seamlessly and will not lead to problems in the other modules.

In the rest of this chapter, we will look at how such a big bang integration can be mitigated.

6.1.1 Onboarding with Stumbling Blocks

While it is important to try and tackle problems early on, difficulties with the build much more often involve downstream problems that arise later in the development process. The problems with the individual builds from the various teams start much earlier—namely, when a new developer joins the team and needs to be trained.

One of the first tasks for a newcomer at nicely-dressed.com, immediately after installing the operating system, is to set up the development environment for the project. Depending on the team the new developer is joining, this process may be better or worse, but regardless, each team lacks documentation for it. Many of the tasks for setting up the environment have to be done manually. Every development environment is different, so it is not even possible to switch from one team to the next without having to change things manually.

One specific problem across all teams is that the dependencies needed to build the project are not documented, nor are there any automated mechanisms to speed up the setup.

Let's look at an example of problems a new employee at *nicely-dressed.com* may encounter during onboarding. A new employee downloads a working copy of the project from the repository. To get a feeling for the development of the software, she first wants to build the project herself. Once that's done, she can start working on small bug fixes (rather than complex features, which would take longer for her to do given her lack of experience with the codebase). Such *low-hanging fruits*, tasks that require little work but go through the entire integration process, are ideal for a new employee getting started with a new codebase, as they do not require too much time and effort.

However, the new employee immediately encounters a problem when trying to build the project: The project cannot be built without documentation or the usual programming language automation. Also, it needs many dependencies that have to be copied manually to the correct directories on the development computer, and it isn't clear which versions of those dependencies are needed.

Without a good build process, the new employee runs the build script, and chases after the dependencies each time the script aborts. Because the dependencies are not stored the project repository, the new employee has to download them from the Internet and store them in the correct directory—and hope that she is downloading the correct versions. She goes through this process with each error message that pops up during the build.

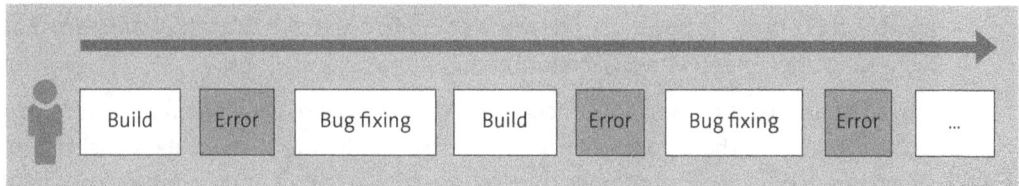

Figure 6.3 Build Errors That Constantly Occur and Have to Be Corrected Manually

She repeats this cycle (see Figure 6.3) again and again until the project builds without any problems. Not only is this process obviously time-consuming, but it also has little added value and is quite laborious.

Other teams have the same problem, but some of them tried solving the problem by creating a virtual machine with all the development tools and dependencies that new hires could copy to get started more quickly.

But the problem is that this process is done once, manually. Dependencies are not updated if they are no longer needed. So what seems like a reasonably good idea only slows down the development process even more in practice, as all team members have to work with this virtual machine, which is significantly slower than the actual workstation.

The problems with the build are repeated every time a dependency is updated or replaced because others on the team have not noticed. In addition, the problems then also run to all downstream teams, as the QA team and the operations team also have to go through the same process, only with a significant time delay, so several of these problems occur at the same time.

Such a strenuous build process leads to few and infrequent builds. Developers end up testing only their own code and avoid interacting with other parts of the project; some even avoid interacting with the online store at all because they don't want to deal with these build processes.

All in all, no one at *nicely-dressed.com* has an overview of the entire build process; this is a classic silo formation. Employees' knowledge is completely limited to their own projects, meaning that no one understands how the online store is programmed as a whole.

6.1.2 Build Difficulties Due to Infrequent Integrations

Each sub-team of the online store works with its own cumbersome build steps. The tickets are designed accordingly: cumbersome and too long. Several change requests and feature requests are often packed into a single ticket, so it takes a very long time until the work for a ticket can be finished.

In practice, this means that developer work on features over a period of several weeks before reporting to the team that they are finished. Before finally checking the code into the repository, however, they download the changes from the rest of the team that have been implemented in the meantime.

This naturally leads to merge conflicts, as several people are working on the same code at the same time. These conflicts need to be resolved before in the code is checked in, which requires a lot of communication. No one can keep track of the conflicts that arise due to the sheer size of the changes that have accumulated in the meantime (see Figure 6.4).

This is always frustrating for the people involved, as adding the changes to the repository is very error prone and laborious. In addition to the technical merge conflicts, there is also interpersonal conflict because many drastic changes have not been clearly communicated, even though they were known in principle.

Sometimes it takes days to resolve such conflicts, and reimplementing changes takes just as much time and energy as the original work. And that's not all: Once conflicts in the source code are resolved, the code is pushed to the repository and is thus available to the others in the team, who then encounter similar problems when they push the changes. It is a vicious circle.

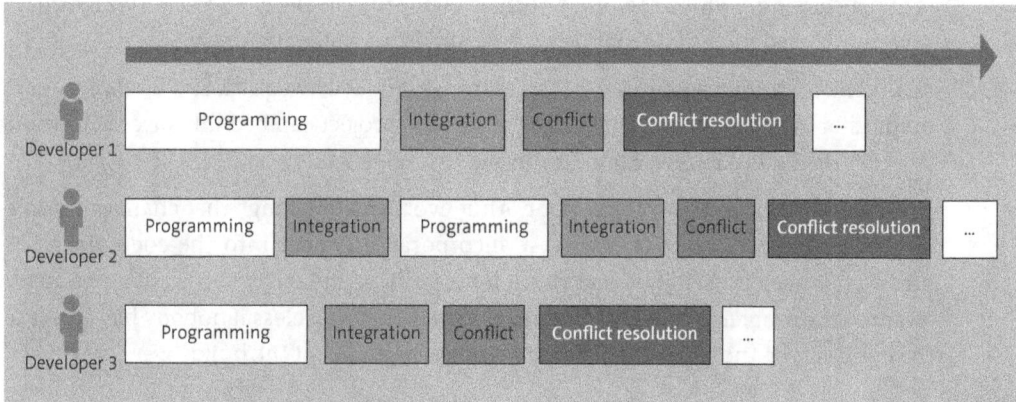

Figure 6.4 Not Continuous Integration

With this process, everyone is always waiting for the integration of the latest changes. And if developers hear that a major change is to going to be pushed soon, they stop their work to wait until the integrations go through, which results in even more waiting and inefficiencies. Developers avoid making even small corrections because the integration is time-consuming.

And these are just the problems of a single team! The problems continue to scale up and become more complex when more changes are involved, such as when individual modules have to be integrated. Collaboration is necessary for the entire project team so that the entire store can be built successfully. We will deal with this problem in Chapter 8, which is about release management.

6.1.3 Only a Few Tests

The role of the development team at *nicely-dressed.com* also involves the development of tests. Developers write tests for parts of their implementations when they see a need for testing, particularly when they implement a feature that doesn't work right away. Debugging is easier with small test cases.

However, the process of developing tests was never implemented stringently. Although the team has developed some unit tests for class methods, these are never executed in full. As a result, customers keep encountering errors. These bugs could have been found if the tests had been maintained.

Why is testing so neglected at *nicely-dressed.com*? Once again, the problem is the high walls between teams. The development team is convinced that testing is the task of the QA team: "After all, they were hired to test the projects." This view and the problems and challenges that arise from it are examined in more detail in Chapter 7.

6.1.4 A Build Server behind Closed Doors

To check whether the project is actually being built, each team at *nicely-dressed.com* has set up a server on which the build of their own projects should run every night. This server is located under an employee's desk and is managed manually.

Again, the basic idea makes sense: The teams should receive quick and direct feedback on the status of the project—that is, whether the project can be built successfully and whether the defined unit tests are running.

However, the implementation is poor: After every major change that changes the version of a dependency or is otherwise incorporated deeper into the code, the build breaks. It's not necessarily a bad thing for a build to break, as you want to be made aware of the problems causing it to break. However, it is useless if nobody has access to the build server and, therefore, nobody checks the status of the build.

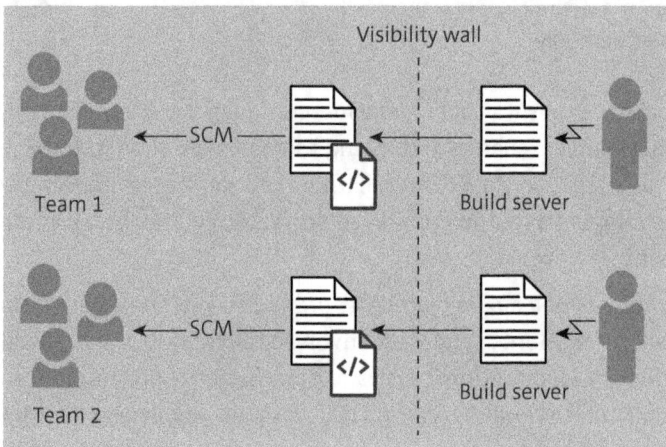

Figure 6.5 No Insight into the Build Server

Additionally, because the server is under the desk of a certain employee, who is also responsible for setting up and maintaining the build server, even fixes for problems like a missing dependency are left unaddressed (see Figure 6.5).

In organizational terms, there are several reasons why the build for *nicely-dressed.com* regularly fails: It is too cumbersome to handle, and there is no easy way for team members to see and review the failures. It has therefore become common practice to simply ignore these problems. And this applies not only to the pure build, but also to unit tests, which are never retraced and adapted.

As a result, the other developers also leave their work undone, as these problems also occur for them when they later try to integrate their work into the repository.

In addition to these organizational problems, there are also technical shortcomings. The dependencies are installed manually and only by the few people who have access to the build server. The rest of the team is not actually unhappy about this because it means they bear less responsibility and therefore cannot be blamed for failed builds—after all, they do not have access to the system, and someone else could have installed the updated dependency there.

Last but not least, a separate script was written to build the project. Many dependencies are copied there from hardwired directories whose contents were also managed manually.

But that's not all, because these are only the challenges of a team building a single module. The development team needed to set up another server to receive the results of the individual sub-teams in order to then build the entire online store; this server was set up in a similar way to the aforementioned team servers, but it never became functional in practice. As the individual modules were largely never able to build, the entire project could not be built. And because the individual teams already have difficulties integrating their own code promptly and with as few errors as possible, the server works even more rarely with changing interfaces in the overall project.

The result is that deployments are even more difficult because all integrations have to be straightened out again. Bottlenecks occur at a number of points during implementation, primarily when a release of the online store is due. Errors early on in the process lead to exponential problems in the subsequent stages.

> **Note: Reflection**
>
> One of the problems mentioned here is often seen in companies but can be fixed without a major change in culture: straightening out the build process, especially for the onboarding process.
>
> Check whether your projects can be built without stumbling blocks and whether there are any other aspects that are disruptive and do not add any real value.

6.2 Modern Build Management

As you can see from the rather long description of the current situation at *nicely-dressed.com* provided in the previous section, the teams' approach build management is full of pitfalls and problems. How can it be improved with DevOps?

The biggest benefit of transitioning a company's build management processes to those based on DevOps is automation. You may have noticed that most problems at *nicely-dressed.com* can be traced back to one point: too much manual work taken in too many individual steps that require coordination.

Too often, the teams at *nicely-dressed.com* face complications when merging major changes, which means that the projects can no longer be built. Too often, various team members wait for someone else's feature to be completed.

In transitioning build management processes to DevOps, the goal is to streamline the entire build and standardize and then automate as many steps as possible. Consider the First Way of the Three Ways: The entire system must be considered. With the current build management system at *nicely-dressed.com*, understanding the entire system is not possible, as all build processes are different among the various teams and are poorly documented.

And the overall goal of development should not be programming a feature that is then manually integrated into the team's code. The real goal is that improvements of and corrections to the code should reach end users as quickly as possible. This is possible only by optimizing the process so that changes end up in the build faster and with fewer errors.

But let's first deal with the cause of the problems, namely understanding what exactly a change is. To do this, you need to take another step back and look at what the project planning for a new feature or a bug fix actually looks like.

At *nicely-dressed.com*, entire requests are packed into one ticket, making tasks inevitably very, very large. In many cases, implementation takes weeks, if not months, and produces hundreds or thousands of lines of code that can be integrated into the development branch only with a lot of difficulty. This applies both to the developers themselves and to the other colleagues on the team who depend on these changes. We already touched on this topic in the previous chapter, and we are now taking it one step further.

These tickets are more like *epics*, in which several different steps are recorded in new tickets that have to be processed until the entire feature is ready. Epics are useful for basic orientation and help in understanding dependencies within a new feature, but they are poison for the actual implementation.

It is important to instead keep tickets as small as possible and to clearly define when the work is considered complete. Ideally, both developers and project managers who read a ticket should be able to understand what is to be solved without further questions. The smaller the individual steps of development are, the more likely planning errors become apparent at an early stage, as gaps become apparent only when the required steps are written down in detail (see Figure 6.6).

When developers start working on tickets, they should plan to incorporate the changes into the main development branch as soon as possible. Ideally, changes should be incorporated at least once a day, even if the work on the given feature is not completely finished. However, it is important that the project always remains buildable, both for the developer incorporating changes and for the other members of the team. This also includes the successful execution of automated tests.

This process also makes code reviews, which were discussed in Chapter 5, Section 5.4, much easier; smaller, manageable changes are easier to track and quicker to review. This makes *merge hell*, in which a lot of effort needs to be expended to integrate many different huge changes, much less frequent.

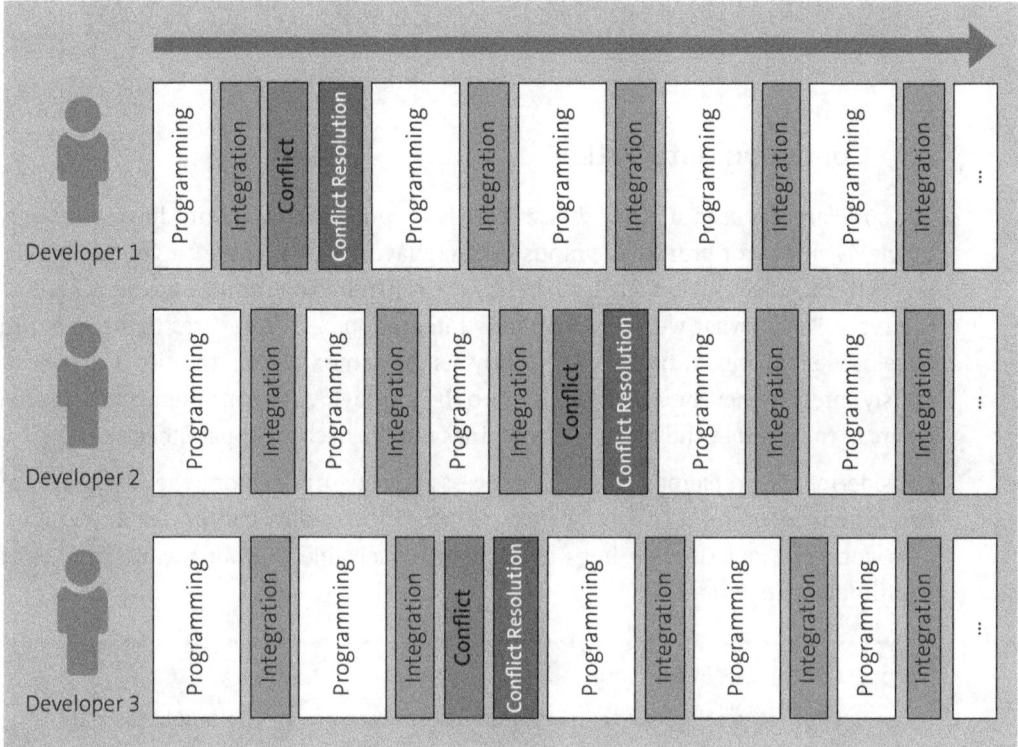

Figure 6.6 Workflow with Smaller Planning Steps

> **Note: Breaking Changes**
> Breaking changes are always a major challenge. They require a high degree of coordination and collaboration, which becomes more complex the more dependencies have to be considered. One option for communicating such changes is a versioning scheme such as semantic versioning, which I present in Chapter 8, Section 8.3.

Never forget the goal of it all: the rapid delivery of improvements to end users. The aspects that I have just briefly explained are all aspects that fall under the term continuous integration, which we discuss in more detail in the next section.

> **Note: Reflection**
> I would venture to say that for the majority of my readers, "modern" build management is no longer so new and modern. Though some companies still do not use such modern approaches, it is fair to say that modern build management is now common practice.

6.3 Continuous Integration

The build process used at *nicely-dressed.com* is so cumbersome mainly due to the large, unwieldy, and error-prone code monsters that have to be integrated at long intervals. It is better to have smaller changes that can be integrated continuously and at shorter intervals. This is what we call continuous integration, and it ensures that the entire development process runs more efficiently. As the name suggests, the aim is to continuously integrate new or changed code into the product. With continuous integration, errors in the product and other shortcomings tend to become apparent earlier.

Considering the problems of the build process at nicely-dressed.com, the advantages of continuous integration should be clear to you. Incidentally, continuous integration also works without DevOps, but DevOps is inconceivable without the advantages of continuous integration.

> **Note: Continuous Everything**
> Continuous integration is often discussed in connection with continuous delivery, so the term CI/CD is often used: continuous integration, followed by continuous delivery. The term continuous deployment also exists, but it does not mean the same thing as continuous delivery (but more on that later). For now, we will focus on pure integration.

The implementation of continuous integration consists of a technical and an organizational part. On the one hand, the team must work with small, agile bites, but the necessary automation must also be in place to ensure that changes are processed without any problems.

The problems with *nicely-dressed.com* primarily stem from the fact that code is integrated far too late, resulting in major conflicts during integration. As a result, the project has become almost impossible to build. Because it was built on a server that only a few people have access to and that is ignored most of the time, the project can rarely be built without problems.

As part of the *nicely-dressed.com* team's transition to DevOps, the aim is to remove as many of these build-related restrictions as possible. Certain technology can help with this goal, improving the culture surrounding build management.

The first rule to remember is that the build of the main development branch should always be "green"—the last build should have been completed successfully, which means that it should be possible to build the project again. Two questions should be considered:

1. How do you enable visibility of the build process?
2. How do you ensure that only code that can be built is integrated?

The most important point when switching to continuous integration is to increase the visibility of potential problems. One way in which the team at *nicely-dressed.com* can do so is to automatically create access to the previously build server as soon as new people join the team. That way, everyone will be able to see the build status. This is a good start, but it is helpful only if the underlying problems are taken care of.

The ideal solution would be to create maximum transparency within the organization. However, the minimum requirement is maximum transparency within the team: All team members must be made aware that a failed build will lead to further problems in the near future and must be rectified immediately.

Therefore, integrating changes and building the software must not be an appendage that follows the actual work; instead, it must be a fixed part of the daily routine.

These processes can become part of daily development work through the use of a continuous integration server that is built precisely for this purpose. Nowadays, GitLab CI and GitHub Actions are the most widely used tools for this purpose, as they integrate seamlessly into the source code management of the respective platforms. Jenkins is also widely used. (We cover in more detail the advantages and disadvantages of the individual continuous integration server solutions, and factors to consider in order to work as efficiently as possible with these solutions, in Section 6.4).

Put simply, the continuous integration server executes defined actions when they are triggered. A regular build that is executed daily at a specific time, as is done at *nicely-dressed.com*, is an example of a trigger. This is often called a *daily build* or a *nightly build*.

The problem with this approach, however, is that only the changes that were integrated throughout the day are built together. If errors that did not exist in the previous build occur during this daily build, the specific changes leading to those errors are not clear.

The concept of nightly builds is generally outdated. Instead, the pipeline should be configured so that a build is executed for each commit on each branch. The scripts that are executed for the build (but not exclusively for it) are part of the pipeline.

The *pipeline* is the composition of several actions that are executed in sequence to build the project, execute the various tests, and deploy the project to an environment (see Figure 6.7). A green pipeline is a good pipeline because all steps are successful. If a step fails, subsequent jobs from the pipeline are no longer executed.

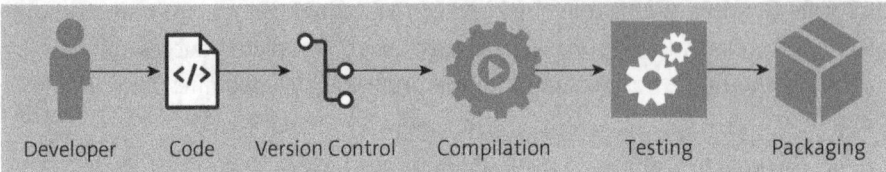

Figure 6.7 The Continuous Integration Pipeline

Pipelines should be defined in such a way that they are reproducible. This is not a problem with modern continuous integration servers, as they have built-in mechanisms that simplify the continuous integration process, including clean environments in which projects can be built without local dependencies and in which all data stored in the project's repository is used.

Of the build-related problems faced by the *nicely-dressed.com* team, first and foremost is the management of dependencies. Dependencies are manually processed on the server. This is a general problem if the project cannot be built directly from the source code.

But what exactly does that mean? Everything that is defined in the repository should be sufficient to build the project. A good practice is to use the build tools and package managers of the programming languages in use to ensure stable and reproducible builds.

For example, Maven and Gradle are build tools available for Java. Both can be used to define all dependencies needed to build and run the service in configuration files, allowing them to manage the dependencies. This should not be news to the majority the readers of this book, but it is a common practice worth noting.

The package management system automatically pulls the required dependencies from a central location, including both public dependencies, which are usually under an open-source license, and internal dependencies, which should be provided via a separate package repository.

If a new version of a dependency is used, then in addition to the code changes needed to reflect this new version, the configuration file also needs to be changed so that the newest version is pulled in the next build. Use of configuration files to define dependencies means that manual login to a central server is not necessary and helps with the setup for the local development environment.

> **Note: Not an Old Hat**
> If you have experience working professionally in software development, you will know that good and automated dependency management is a key factor for good builds; this is an established best practice. Unfortunately, however, this best practice has natural enemies: complexity and exceptions.

> When a build becomes so complex that management of dependencies and configurations becomes too complicated, it can be easy to fall back on past workarounds that have already been replaced. Counteract this tendency as quickly and decisively as possible! It should always be the standard practice that all dependencies are located directly in the same repository or that they are pulled through the build using automation. Any other practices will only slow down development and cause developers to be less productive.

It is also important that the complete automation of the build process is not only carried out but also made visible. And of course, errors in the build process should be corrected as soon as they occur.

And this brings us to our next point: How do you ensure that only buildable code is integrated?

A common practice is to use short-lived feature branches to ensure that changes are not integrated into the main development branch in an uncontrolled manner. These branches are useful not only for simplifying the code review process, but also for ensuring that only buildable code is integrated.

Continuous integration servers can also help with this goal. These servers make it very easy to automatically trigger the build process for each branch. After pushing a feature branch and opening a merge or pull request, the pipeline should start automatically, build the project, and execute the defined tests to provide feedback (see Figure 6.8).

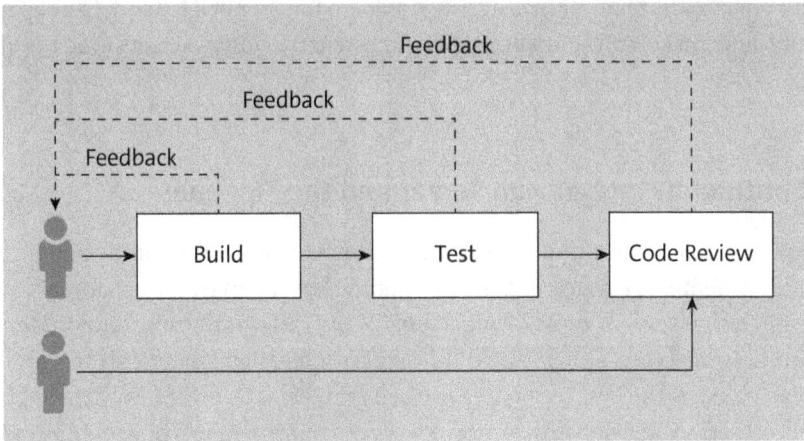

Figure 6.8 Feedback in Each Step

Feature branches are the key to continuous feedback, as they help ensure that only buildable code is integrated. They also make it much easier to contribute smaller changes without long waiting times. So the first step of improving the build process is to reduce the size of the changes; then, the tools in use help to support the process instead of getting in the way.

> **Note: Do Not Overload the Pipeline**
>
> Testing and QA are discussed in more detail in Chapter 7 and are also fundamental parts of the CI/CD pipeline. Projects should be built and tested as quickly as possible so that developers are informed within a few minutes as to whether their changes led to serious errors. Testing code changes is the only way to shorten the lead time.
>
> However, don't test without any limits or boundaries. I've seen a few pipelines over the course of my career (and had to work with them myself) in which everything that could be automated was put into the builds and tests. This led to a pipeline runtime of four hours. That is clearly too long.

The introduction of automation around the build process ultimately also noticeably improves collaboration. With more automation, errors are visible immediately and can be resolved straight away, rather than ending up on the plates of other team members who may encounter them long after they have been integrated.

Automation also helps eliminate finger-pointing in the event of problems, as the development team then becomes responsible for both the code and the build process. However, it's not just the developers in your own team who benefit from this, but also the operations team. Frequent and regular builds help deployments to the production environment run smoothly. (Basically, anything that simplifies deployment and makes it more secure will be met with great pleasure during operation).

And the QA team should not be forgotten either; in fact, a standardized, automated build process benefits this team the most, as it would no longer need to work in a separate repository and make time-consuming changes that involve workarounds and exceptions.

6.4 The Continuous Integration Server and the Pipelines

Now that you know that a continuous integration server with a corresponding pipeline can make development work much easier, we will now look at an overview of various continuous integration servers, most of which enable not only continuous integration but also continuous delivery. However, do not expect a detailed analysis of the pros and cons of each tool, as the best tool for your needs can be determined only through an individual evaluation. Instead, this section covers the more general concepts behind each tool so that you can consider whether they may fit your needs.

6.4.1 The Basic Structure of a Pipeline

First, we'll take a closer look at the core concepts of a modern software pipeline. In the next section, we'll look at how pipelines can be written and which continuous integration servers are commonly used.

Pipelines in every continuous integration tool have the same basic structure, even if some tools refer to their pipelines differently. Pipelines have *stages* and *jobs*. Stages, which are similar to typical DevOps stages, group together various activities (jobs) that are executed in the pipeline. Jobs are always part of stages.

There are usually at least these three stages: *build*, *test*, and *deploy* (see Figure 6.9). Each of these stages contains at least one job. For example, the build stage includes a job that builds the project. The build job drops build artifacts that are then required in the next stage, as the compilation must be tested in that stage, depending on the programming language and environment.

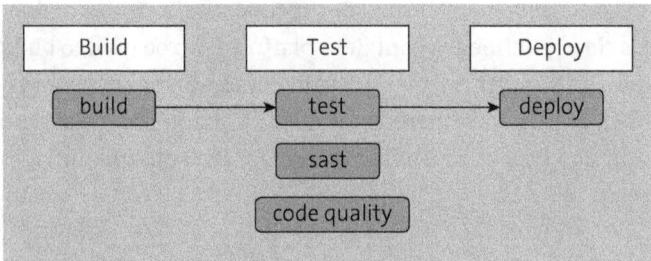

Figure 6.9 Conceptual Structure of a Simple Pipeline

The use of stages and jobs optimizes the speed of the entire pipeline. Jobs should always run as quickly as possible, so they should be written and configured accordingly. This means, for example, that the build artifacts from the build job are reused in the test job. However, if the build job fails, the entire pipeline terminates and provides early feedback.

Depending on the complexity and requirements of a given pipeline, the test stage includes several jobs that execute various tests, such as unit tests, integration tests, or various security tests, as further discussed in Chapter 11, Section 11.4. Jobs in the test stage may also include tests for code quality in order to automatically check compliance with the expected code layout.

The deploy stage includes a deploy job, which carries out the deployment. This job can be automated or rolled out after manual confirmation by authorized persons.

Each job requires a runtime environment that contains all the tools needed to complete the job. For example, a Java project that is built with Maven requires an environment in which Maven is installed. In the past, the required dependencies were simply installed on a system. This meant that the job could run repeatedly without requiring any major changes. One of the problems with this approach was that the build environment for each job was not always completely clean and the dependencies always needed to be installed manually.

Nowadays, software containers are used for this task, and container images are used to contain all dependencies. A number of container images already exist for many tasks and requirements, such as for Maven. In the publicly accessible DockerHub, for example, you will find numerous ready-made images, some of which are maintained by the teams behind the projects and some of which come from other sources. If the desired container images are not available in the right version and with the right dependencies, it is relatively easy to supplement existing images.

The advantage of this is that all jobs are reproducible because they are based on fixed container images. On many current development teams, debugging problems that occur in environments that have not been properly cleaned up beforehand is a thing of the past. Anything else is pretty inefficient. (There are exceptions to the rule, of course.)

The use of containers also makes it much easier to scale pipeline jobs, as the container that is needed at each stage is clearly defined. A container platform can be used to build and test jobs in a distributed and parallel manner. By using ready-made images, systems are also much easier to automate, as there is no need to manually install and maintain environments in order to use a dependency. Instead, the container images have to be managed.

6.4.2 Scaling and Reproducibility

Modern continuous integration pipelines ensure scalability and reproducibility, which are the two biggest advantages and should always be the goal of implementing a pipeline.

Scalability

First of all, pipelines ensure *scalability*. The task of building the source code into a usable artifact is organized in such a way that it makes no difference whether the build is a small "Hello World" project or the entire production environment. You can achieve scalability by fully automating the individual steps required for building and by reducing the management of the infrastructure around the CI/CD pipeline to a minimum.

While it used to be common for organizations to have a large team to manage the CI/CD infrastructure, today this is the case only in organizations that are not keeping up with modern build practices. Such a team is typically responsible not only for building the pipeline itself, but also for managing the infrastructure, which does not always scale well and therefore requires a great deal of engineering effort.

The larger a company is, the more likely it is to have teams whose only responsibility is to manage the CI/CD infrastructure. Take a close look at the extent to which these teams are involved in the entire process and the number of people working on them. The mere existence of such a team isn't necessarily a problem per se; on the contrary, with the right knowledge, they can be very helpful. Ideally, however, each team in an organization would use ready-made container images that can be created directly from the official images with little effort and that are stored in an internal registry for their projects. Some tools may need to be installed later, but those tools can be part of the pipeline job directly, or you can provide a separate container image that is always rebuilt.

Container images can be easily scaled by using an orchestration tool such as Kubernetes. Such a tool not only ensures stability and reliability of the production environment but is also practical for build environments if automatic scaling up and down can be configured depending on the current workload. Nothing is more annoying for a developer than a pipeline that does not run because the infrastructure is not ready for it. But it is also not necessary for a pipeline to constantly eat up resources. Make sure that your orchestration tool is as fully automated as possible; the current modern continuous integration tools already provide functionality for automated orchestration tools, or greatly simplify the task.

In some cases, container technology cannot be used. For example, this is the case when working closely with hardware, such as when software has to be installed directly on embedded devices in order to test them.

But even in such cases, as much as possible should be automated, even if only with bash scripts or similar tools, as long as they are easy to maintain. The goal of using CI/CD infrastructure as a tool is to add value to the teams using it, not frustration. Part of that value comes from the ability to automate. The aim is to work *with* the tool, not *on* the tool.

Static systems, in which dependencies have to be installed manually, have the major disadvantage that they are not very flexible for use in new or different projects. So if a team starts with a new technology and needs different build tools for it, it can take care of the container image itself, which can also be rebuilt regularly and automatically if necessary.

Reproducibility

The second important point is *reproducibility*, meaning that a project can always be built reproducibly.

First of all, it is important that the same results are always delivered for every build and during testing, which is why clean environments using containers are the state of the art today. They ensure that you always start from the same point and that the conditions are the same for every build. This requirement can be extended: If the source code and the build environment are known, it should be possible for someone else to build your application.

This may sound obvious at first, but builds are not always easily reproducible. If you ever try to build a huge open-source project like Firefox or LibreOffice from the sources, you will realize what a huge effort this task is. And it is usually not even certain that the result (i.e., your Firefox build) really corresponds bit for bit to what you can download via the installation files.

> **Note: Reproducible Builds**
>
> The *https://reproducible-builds.org/* project is committed to ensuring that open software can be built reproducibly, as reproducibility creates trust. After all, it is one thing to be able to check the source code of an application like the Firefox browser to ensure that there are no hidden backdoors, but if you can't also be sure that this source code is built in exactly the same way as the program used, you haven't gained much.
>
> Take a good look at the project's notes and considerations; they are a goldmine for anyone involved in building and deploying software.

The reproducibility of builds is particularly important in regulated environments to ensure that the pipeline always delivers the same output. If a bug fix has to be delivered for an older version, the build infrastructure should be able to do this relatively painlessly through the use of container images, as no systems from which build files no longer exist are usually required. In other words, it should be just as easy to build older versions as it is to build the very latest version.

6.4.3 Declarative Pipelines versus Scripted Pipelines

Before we discuss the most common continuous integration servers currently on the market, we need to look at two different types of pipelines, based on how they are written: *scripted* and *declarative* pipelines.

Scripted pipelines are written in a programming language. The advantage of using a programming language to write a pipeline is that the full range of the language's functions can usually be used. Depending on the continuous integration server, different programming languages are supported. The continuous integration server Jenkins, for example, supports Groovy, which runs in a Java virtual machine. In addition to the regular features of the programming language, Groovy comes with other continuous integration server–dependent features, such as the ability to define the individual stages and jobs of a pipeline.

Scripted Pipelines

With the power of a complete programming or scripting language, almost any solution you can think of can be implemented in the pipeline in various ways. Therefore, the use of a programming language makes writing a pipeline very flexible. For example, loop constructs can be used to generate jobs for a pipeline.

However, the high level of flexibility can also be a disadvantage. Pipelines can quickly become complex, which hinders the pipelines' overall maintainability. In addition, anyone who wants to write or adapt a pipeline must be proficient in the language in which it was written. With a high level of flexibility, there is also a risk that the implementation will be different for every project.

It is important that all automation in the DevOps environment is easily maintainable and customizable for all team members. But this is not always possible for scripted pipelines, as specialized knowledge is required. The result is often that there are central teams that are always called upon for support or that even write the pipelines for the various teams. This is not only time-consuming but also inefficient, as it unnecessarily reduces the autonomy of the teams. The respective teams should be able to write the pipelines themselves or add their own jobs to the pipelines if required.

The problem is that, in many cases, too much effort goes into working on the DevOps tools rather than with them; that is, there is a tendency to over-engineer. The core objective of a cross-functional team should ultimately be the development and operation of the application, not the large-scale management of automation.

> **Note: The Role of Programming Languages**
>
> The role of programming languages in the DevOps environment is not as exciting as articles with titles such as "DevOps engineers should know these languages!" want you to believe.
>
> No specific programming language is better or worse suited to DevOps pipelines per se. And that's a good thing!
>
> Nevertheless, there are a few languages that are used particularly often and that you should master: Bash and Python. If your environment has already arrived in the cloud and uses many modern tools, Go is also important. I go into this in more detail in Chapter 14, Section 14.3.3, which is about the technical DevOps learning path.

Declarative Pipelines

Declarative pipelines are described by configuration files written in markup language such as YAML (which, ironically, stands for "YAML Ain't Markup Language"). The logical syntax is ensured by correct indentation, which can be cumbersome with more complex YAML files. YAML files are actually very easy to read even without prior knowledge—unless they are overly complicated with excessive use of indentations and different levels.

However, this is often not a major problem with pipelines, as these are generally not particularly complex. If you write a pipeline file that turns out to be too complicated, then you should make sure to simplify the setup accordingly. Admittedly, this is easier said than done.

Modern continuous integration servers come with their own keywords that can be used in pipelines to use the various functions of the continuous integration server. This makes using and maintaining pipelines convenient and simple in many cases, as no additional programming language needs to be learned. Everyone on the team can therefore make changes quickly, as the hurdle is relatively low.

> **Note: Don't Reinvent the Wheel**
>
> A common mistake in practice is expending unnecessary effort to reinvent the wheel. With the right choice of tools, you don't need to! Nowadays, many problems have already been solved well enough that you don't need to come up with your own solutions. Regardless of the tools you use, it is better for everyone involved to stick as close as possible to the standard of the tools. This makes it easier for your team to familiarize themselves with the tools, keeping this hurdle as low as possible.

6.5 Efficient Pipeline Authoring

This section discusses key points to managing automations as efficiently as possible, regardless of the specific CI/CD pipeline solution in use.

6.5.1 Avoid Central Pipelines

You often see attempts to use a large central pipeline to manage all projects for an entire company. Curiously, the larger the company, the more common it is to use a central pipeline. The basic idea behind this decision is to relieve the individual development teams of the effort of developing their own pipelines, which is understandable, given that teams should work on the value stream and not with the tools around it.

In companies with a central pipeline, the development teams use only this pipeline and are no longer responsible for managing the build process. Such companies usually have a separate team whose purpose is to write this pipeline. Unfortunately, this team is often called the DevOps team.

This approach has major disadvantages that lead to the opposite of the intended results.

The problem is that, especially in large companies, there are often many different builds that work completely differently. An iOS app is built very differently compared to a server application written in Java. If all builds managed by a company are to be mapped in a central pipeline, numerous switches and branches have to be created. The result is an incredible level of complexity that can quickly become unmanageable and impenetrable.

But the use of a central pipeline is also the wrong approach from an organizational point of view. The teams responsible for a company's products know the builds best; they know which configurations are necessary and which dependencies need to be in place. Therefore, these teams should also be responsible for managing the pipelines used to build their code. And if development teams are not dependent on a central pipeline, they can better work autonomously on their projects.

However, this does not mean that every team should use their own tools and processes as they see fit. If you follow the basic tips presented in this section, your teams will develop their own pipelines, which colleagues from other teams should be able to easily familiarize themselves with. This is a better approach than creating a centralized monster pipeline that needs its own team to make it work.

> **Note: Compliance**
> While you should avoid using central pipelines, various compliance requirements may make them necessary. For example, it may be useful or necessary to carry out a final check to ensure that these requirements apply to all applications that the company builds. Compliance checks are intended to incorporate and adhere to specific quality guidelines. These checks can run during a build, but they are not a compelling argument for a large central pipeline. This topic is covered in greater detail in Chapter 11, Section 11.6.

6.5.2 Provide Pipeline Building Blocks

Instead of providing a central pipeline and imposing it on all teams (or at least trying to do so), provide a central library with individual pipeline components. Regardless of the specific CI/CD tooling you use, a central library can contain many smaller components that your teams can use.

This approach kills several birds with one stone. First, it simplifies administration: Not all teams need all parts of a pipeline. Second, it allows your teams to learn how to use the CI/CD tooling, which is absolutely essential.

Not everyone needs to be an expert in using the CI/CD tooling, but the teams should be able to adapt the pipeline themselves, by both writing their own pipeline components and using the components of the central pipeline library.

When building the central pipeline library, following the *KISS*, *DRY*, and *YAGNI* principles:

- KISS is an acronym for "Keep it simple, stupid." The idea is to keep the solution as simple as possible. In the context of a DevOps pipeline, this means using as little nesting as possible and keeping the pipeline components as simple as possible; that way, the components can be easier to maintain and to use individually.
- DRY is an acronym for "Don't repeat yourself." Instead of writing the same code over and over, try to abstract as much as possible (but do not forget the KISS principle).
- YAGNI is an acronym for "You ain't gonna need it." Avoid adding components for special use cases that are not currently needed; they make the code complex and ensure that development of the library takes much longer.

These principles apply not only to the build process, but also to testing and deployment. Therefore, it makes sense to write a separate library that provides default values used during testing and deployment. The teams can then import these components and have the option of making adjustments for their own projects.

But don't start mapping your own edge cases in the central library. This increases complexity, takes longer, and makes it more difficult for other teams to use. Settings that affect special cases for individual teams must then be overwritten in each case. Only when configurations are needed by other teams should they be moved to the central pipeline library.

Finally, when developing your central pipeline library, avoiding building in any dependencies between the individual components; otherwise, you'll end up with a large web of dependencies that you'll have to manage and debug somehow, which is tedious.

6.5.3 Create Visibility

It is important to provide a good overview for your teams so that they can do the following:

1. Inspect the central components themselves
2. Ideally be able to understand how the components work
3. Test the components in their projects and then use them productively

The better your teams understand the components, the easier the whole process of building a pipeline will be.

Visibility and understanding are important because the teams can simply choose to remove individual modules from the pipeline and replace them themselves if they do not work for their particular needs. They can then also be transferred centrally to the library to contribute to reusability there.

In general, a central library is useful, and all the more so for larger companies. Note, however, that the various CI/CD tools often come with their own templates and solutions, which should also be used so as not to reinvent the wheel, as already mentioned. After all, this applies not only within the company, but also outside it.

6.6 Overview of Continuous Integration Servers

There are many different continuous integration servers. Before you choose one, you should consider the effort involved in setting up and operating it, the effort required to write the pipeline, and its scalability.

Some solutions are introduced early on in companies and support the development workflow very well over time, but those responsible for managing the solution realize that they wouldn't choose that same solution today.

The chosen continuous integration system should be as easy to use as possible for those who have to use it on a daily basis; it can have few restrictions, but give enough autonomy. Ideally, you don't have to start from scratch every time, but you still have the opportunity to make adjustments.

The following are the most important criteria to consider when selecting a CI/CD system. I will revisit these criteria throughout the following sections:

- **System maintenance**
  How much effort is required to maintain the system? This criterion also relates to scalability, as scaling often goes hand in hand with maintenance work.

- **Complexity of pipeline construction**
  How easy or difficult is it to build pipelines with this system?

- **Permissions management**
  What capabilities does the system have for permissions management (that is, who is or is not allowed to take certain actions)?

- **Security vulnerabilities**
  Unfortunately, this aspect is often ignored, but it is important. What impact does the continuous integration system's software architecture have on security? This criterion also goes hand in hand with maintenance, because security patches that are delivered must also be installed. Unfortunately, this aspect of continuous integration systems is often ignored; it is important to take security into consideration when using any outside system.

- **Extendibility**
  How can the system's capabilities be extended? Each system offers different ways to extend the capabilities, each with their own advantages and disadvantages.

- **Clarity**
  Does the system allow you to quickly see whether a project has been successfully completed?

- **Role in the overall DevOps concept**
  What role does the tool plays in the overall DevOps toolchain concept?

> **Note: Not a Complete Analysis!**
> Before we get into the specific CI/CD systems outlined in this section, note that my aim here is not to provide a complete analysis of all available systems. Rather, my aim is to describe the concepts behind them, the problems they may present, and the solutions they offer so that you already have a solid basic knowledge when an analysis is required.

The following sections focus on the three most common CI/CD systems: Jenkins, GitLab CI/CD, and GitHub Actions. Of course, there are other options; however, describing them all would not only go beyond the scope of this book, but it would also be unhelpful, as they are generally are of little relevance today and will probably remain so in the future.

6.6.1 Jenkins

Jenkins is the oldest and probably still the most widely used continuous integration server. But even though Jenkins is widely used, it has recently been falling out of favor. However, Jenkins is still frequently used, and it will take years before it disappears from the scene. In this section, I explain why Jenkins is so widespread, and yet why we will hopefully not see much use of it in the future.

Jenkins can be used very flexibly. However, what used to be seen as a strength of the tool is now seen as a weakness. There are many ways to solve a problem in Jenkins, which means that every Jenkins installation and every setup can end up looking completely different.

Jenkins can be installed quickly. Jenkins is written in Java and therefore runs on many different kinds of host systems. In most cases, Linux servers are used, or Jenkins itself is started in a container. But while it can be quickly and easily installed, maintaining the infrastructure and creating the pipelines is a much more difficult task.

Architecture and Scaling

Jenkins consists of two components: The *controller* manages the pipeline, and the *agents* connect to it to execute the pipeline jobs.

> **Note: Other Terminology for Controllers and Agents**
> The controller and the agents used to be called the "master" and the "slaves." This terminology is no longer officially used due to its racist history. However, these terms can still be found in many places, so it is important to recognize their meaning.

With this architecture, it can be tempting to create a central controller for all teams, on which the different pipelines are then executed. As you should already know after reading Section 6.5.1, doing so would not be a good idea because it would mean that all dependencies, plug-ins, and so on would have to be installed on this controller. First, this would result in a very large Jenkins controller with many pipelines and plug-ins, which cannot be easily maintained without a separate team. If several projects have to share a Jenkins controller, the settings and plug-ins almost always conflict with each other, and it could cause problems if many people need to access the controller at the same. Second, many complicated pipelines would place a heavy load on the Jenkins controller. This heavy load could cause even simple pipelines to be processed at a snail's pace, causing frustration.

These drawbacks are why companies usually set up many smaller Jenkins controllers, such as one for each team or department. However, this approach results in the problem that all the independent instances cannot be managed centrally. Administration becomes difficult and characterized by manual work, which defeats the DevOps goal of using automation wherever possible.

The Jenkins Configuration as Code plug-in provides a remedy. It allows the Jenkins controller to be defined and configured in YAML code, which previously required many clicks in the UI. Using this plug-in, there can be several YAML files for various instances within an organization that contain relevant configurations, such as the necessary passwords and the plug-ins used. Ideally, the Jenkins container should be containerized so that it is started up on a container host on which a new image is built and deployed with every update.

Defining and configuring the Jenkins controller in YAML also has the advantage that the description of the pipeline can be saved with the actual project, making it versioned and portable! This is an important point for the reproducibility of the build. If pipelines are configured manually instead, the definition would be saved in the Jenkins controller itself, which means the pipeline would *not* be defined as code, so there would be no versioning. (In practice, I often see that the definitions are saved and versioned separately from the actual project, which is also not helpful.)

> **Note: Manage Pipelines within Projects**
> If you are working with such outdated pipelines, you should take steps to modernize them. Doing so will increase visibility and improve maintainability, which will make subsequent DevOps-related moves much easier to make.

But even if so much can be automated, the engineering effort is still high if many small installations have to be maintained and managed. In the event of problems and updates, a maintenance window must be found and communicated for each instance, and each automation must also be executed in a controlled manner in order to detect and rectify problems.

In addition to the controller, the agents also need to be managed. As usual, Jenkins offers many different options for installing, managing, and scaling agents. Nowadays, you should rely on containerized agents that automatically scale up and down depending on how much workload is currently needed.

Credentials and Permissions Management

Another problem with the maintenance and administration of Jenkins is credentials and permissions management. Because Jenkins is the central hub of automation, credentials and permissions for the entire DevOps toolchain need to be managed here. All other tools must be connected to Jenkins: the Git server, various tools for QA, security tools, deployment tools, and so on.

Credentials, such as passwords, tokens, and other methods for access control, must be stored and configured for each tool. The Jenkins server is therefore the central target of your infrastructure and must be secured accordingly. You absolutely need a well-thought-out system of restrictions and should regulate access as strictly as possible to prevent misuse.

In addition to credentials, you also need to consider permissions management in Jenkins. Permissions are closely related to credentials, as they also restrict who can and cannot take certain actions. It is usually necessary (and sensible) to duplicate the configuration of the Git server: Whoever is allowed to check in and change code is also allowed to start and edit the build. This means that you have to maintain these permissions in two places. This also involves a certain amount of effort and increases the complexity of the setup. But this is how the permissions for individual users and user groups can be managed in Jenkins. Role-based strategies are also possible, through a plug-in.

> **Attention: Security Vulnerabilities**
>
> Like all software, Jenkins has security vulnerabilities, some of which are found from time to time. These security vulnerabilities are contained in Jenkins itself and in its plug-ins.
>
> Updates usually require Jenkins to be restarted, which can be quite disruptive, as it can be difficult to find a suitable time for a restart if all pipelines are located on a central server. And if there are a large number of independent instances, manual work is required to complete a restart. These difficulties mean that teams often postpone updating Jenkins, which is not good in terms of security.
>
> Secure and viable management of Jenkins for infrastructure alone is quite complex and requires a team, which needs to be quite large for larger companies.

Limited Range of Functions and Plug-in Hell

The basic scope of Jenkins is quite limited. Almost all functions can be used only if plug-ins are installed. Even important core functions are mapped to plug-ins; for example, the Git plug-in is needed to work with Git repositories. You can find a plug-in for almost any task, and if there isn't yet one available for a given task, you can also develop a plug-in yourself. This allows you to move complexities from the pipeline to a plug-in.

The problem, however, is that even more effort is required to ensure that the different plug-ins fit together. Jenkins is notorious for compatibility problems, because while the core plug-ins are developed and maintained by the Jenkins team, additional plug-ins are the responsibility of the community. Some plug-ins are actively maintained, and others are not.

This can mean that you may be dependent on plug-ins that are deeply wired into your workflow and DevOps toolchain but are no longer maintained and, therefore, slow down updates of the entire system. In short, all Jenkins plug-ins that you use become part of your software supply chain (as discussed further in Chapter 11, Section 11.5) and need to be maintained accordingly. And if you use many Jenkins plug-ins, it may not be possible with reasonable effort.

> **Attention: Marketplaces**
>
> The prevalence of plug-ins applies not only to Jenkins but also to many similar tools. A marketplace offering plug-ins and templates seems to have become almost a flagship of many platforms. Be very careful with such marketplaces! By using plug-ins, you are not only making yourself functionally dependent on third-party offerings, but, in the worst case, you could be also be introducing nasty security problems into your system.

In addition to security problems, using a large number of plug-ins can lead to a loss of overview. This is especially true if different plug-ins are used for the same tasks in different pipelines—developers cannot simply jump back and forth between projects, as they would first have to deal with the foundation on which the pipelines were built in.

Clarity

A continuous integration server should ensure that you can easily see important information. The visibility of the project status is important, as they say.

Specifically, it should be quick and easy to see in the server whether the project is building properly and whether all tests are running successfully. However, Jenkins provides too many different options for seeing this information. For example, the cleanest way to see this information would be if the hierarchical structure from the source code management platform were mapped in Jenkins. This would make it easy to find your way around. Unfortunately, Jenkins does not enforce such a mapping; it does not even offer the option as a default, as Jenkins exists separately from the source code repository. The connection between Jenkins and the repository must first be established, which can be done—you may have guessed it—only via a plug-in. (Previously, this connection had to be established completely manually, meaning that a separate pipeline had to be configured for each branch and for each project. As this was tedious, it was often not done, which is why pipelines were rarely executed for short-lived branches.)

To map the information from the source code management platform to Jenkins, you need the Branch Source plug-in for GitHub, GitLab, or Bitbucket. With this plug-in, projects are automatically created in Jenkins and correspond to the structure of the integrated repositories. To do this, a Jenkinsfile is searched for in which the pipeline is defined. (The standard is a single Jenkinsfile, but this is only *one* standard of theoretically many that can be implemented.) If a branch disappears, then the corresponding job also disappears. This uniform structure makes it easy to find your way around, both in your own project and in the projects of other teams.

The main sticking point is, of course, that many Jenkins instances are not set up according to best practices, and there is often a lot of uncontrolled growth.

If you use GitHub Actions or GitLab CI/CD instead (see Section 6.6.2 and Section 6.6.3, respectively), the functionality of Branch Source is a built-in feature that works without configuration, as the repositories and the continuous integration engine are directly linked.

Let's draw a conclusion: Is it better to be able to add all conceivable functions in a modular way via plug-ins, or to rely on a monolithic one-size-fits-all solution? Both approaches have their advantages; the answer probably lies somewhere in the middle. In my opinion, the CI/CD platform should offer a certain basic set of functions that you can absolutely rely on and that are maintained by the core team. Custom extensions for details and special use cases work fine, but processes for accessing repositories and building the software must be smooth and without a lot of manual work.

However, in my experience with Jenkins, the tool does not offer this basic set of functions; far too much is outsourced to plug-ins. Their quality is inconsistent; some are good while others are mediocre. In addition, the many customizations decrease user-friendliness and increase the learning curve; no one can really claim to have mastered Jenkins because no two setups are the same. You have to be careful not to use too many different plug-ins and functions, which can make administration even more complicated. This often means that there is a dedicated Jenkins team responsible for configuration, which is something you want to avoid.

Complexity in the Construction of Pipelines

In Jenkins, pipelines can be configured and managed in many different ways. This makes administration flexible but also complicated, as each team can choose its own approach.

The oldest variant is to click together the pipeline configuration in the web interface and store the shell scripts and other integrations there. This corresponds to a manually configured pipeline and should never be done, as the resulting build would not be reproducible in this way.

Instead, the pipelines must be created in a file, namely in the already mentioned Jenkinsfile. You already know that there are two different types of pipelines: scripted pipelines and declarative pipelines. Jenkins naturally supports both approaches, and the approaches can also be mixed, which makes pipeline construction very flexible. Though this flexibility is convenient, it is also the biggest problem with Jenkins when it comes to defining pipelines. Each pipeline can be as complex as you like.

For scripted pipelines, Jenkins uses the Groovy language. Scripted pipelines written in Groovy can be executed in a Java virtual machine. Typical programming language elements such as loops and functions can be used to write these pipelines.

Declarative pipelines can be used in Jenkins to avoid the creation of multiple pipelines that all look different. Declarative pipelines can look relatively similar across all instances, which simplifies maintenance and makes it easier for developers to familiarize themselves with a CI/CD system.

```
pipeline {
  agent {
    docker {
      image 'maven:3.8.1-openjdk-1'
    }
  }
  stages {
    stage('Build') {
      steps {
        sh 'mvn install -DskipTests -batch-mode
        -Dmaven.repo.local=./.m2/repository'
      }
    }
  }
}
```

Listing 6.1 Example Declarative Jenkins Pipeline

Listing 6.1 shows an example declarative Jenkins pipeline. The pipeline definition defines agents and stages. The agent configuration indicates that the specified Maven image should be used for all jobs. The build stage defines a step in which Maven is executed with a few parameters within the shell.

> **Note: Keep Pipelines Simple and Decide on a Style**
> If you use Jenkins and use a colorful bouquet of different types of pipeline definitions, you should migrate to a uniform declarative standard to reduce uncontrolled growth. Ideally, however, I would recommend migrating away from Jenkins altogether.

The Role of Jenkins in the Overall DevOps Concept

There are many reasons why more and more organizations are saying goodbye to Jenkins and turning to other solutions. The main reason is the high cost of maintaining the systems developed using Jenkins. By combining the use of many different plug-ins, it is not easy to keep the toolchain simple. As it is very difficult to administer different Jenkins installations in the same way, isolated solutions are often created, which makes it difficult to maintain an overview of different projects. The developer experience suffers as a result, as developers end up working more on the pipelines than on the actual development work.

Theoretically, it is possible to implement all desired functions in Jenkins in this way, but in practice it is quite difficult.

6.6.2 GitLab CI/CD

Build pipelines are implemented quite differently in GitLab compared to Jenkins. GitLab sees itself as a complete DevOps platform, where the CI/CD component is the oldest core element alongside source code management before GitLab evolved into an all-encompassing platform. Many aspects of Jenkins that I am critical of look different in GitLab, as it is a complete platform, which means that the basic structure is already knitted differently.

Architecture and Maintenance

Technically, GitLab is divided into two parts: the GitLab server and the GitLab runner. The GitLab server contains all the essential functions of the platform, including not only source code management and CI/CD management but also other features such as package registries, security offerings, and monitoring solutions.

> **Note: GitLab as a Service and Self-Managed GitLab**
> GitLab can be used in the cloud as a software as a service (SaaS) offering or as a self-managed instance. The self-managed instance is most commonly seen in companies outside of Europe. If you don't want to host GitLab yourself, use GitLab SaaS, which has more or less the same range of functions.

The GitLab runner collects the jobs of the pipelines from the GitLab server and executes them accordingly.

The GitLab server can be installed via the Omnibus package for all common Linux distributions. This package contains all the dependencies required to get GitLab up and running. It also includes the PostgreSQL and Redis database. A release is made every month, and the server can be updated via the package manager. There may be a short period of downtime, but this has no impact on the running pipelines. Zero-downtime upgrades are also possible if a highly available setup is installed and managed. However, the latter becomes necessary only when several thousand users are working on the system.

While the GitLab server runs exclusively on Linux, the GitLab runner can be installed on a wide range of systems. It can be installed with different executors so that jobs can be executed on different environments, such as directly on the system via shell executor, in Docker containers, or with automatic scaling via containers in Kubernetes clusters. Scalability is particularly important when working with containers, as automatically scalable infrastructures can then be operated with relatively little effort.

In contrast to other systems, like Jenkins, GitLab delivers only ready-made integrations and cannot be extended via a classic plug-in system. Though the lack of plug-ins certainly has disadvantages, in my opinion, the advantages outweigh them: The integrations are developed and tested directly by the manufacturer, which means that breaking changes are rare and are usually announced.

> **Note: Security Vulnerabilities**
>
> Every application has security vulnerabilities, including GitLab. However, with GitLab, usually only the server needs to be updated, which is relatively simple to do. Apart from a short period of downtime for common installations, updates to the server have no significant impact on the pipelines; they can continue to run as normal.
>
> Therefore, security problems can be dealt with quickly and efficiently, as they are fixed and made available via the package management platform. Plug-ins and other dependencies that can block updates do not exist. Basically, only general patch management is required here.

Expandability via Integrations

Although GitLab does not support plug-ins, it can be integrated into other tools, and other tools can be integrated into GitLab. These integrations are not the same as plug-ins in the classic sense, as only the existing interfaces are used to accomplish the integration. The tools talk to each other directly, and no functions are outsourced.

For example, take a look at how Jira can be integrated into GitLab and vice versa. There is a Jira integration within GitLab allowing you to jump to Jira tickets from anywhere in GitLab, and there is a GitLab integration allowing you to display code, branches, and merge requests in Jira.

As part of the CI/CD pipeline, classic tool integrations as plug-ins are not very relevant in practice, as GitLab tries to deliver everything from a single source. However, you can use certain some relevant functions through the paid Enterprise Edition of GitLab. And if you do want to use other tools in your pipeline, you can rely on open standards.

An example of a feature available directly in GitLab is its standard set of security scanners. The complete administration of these scanners, including updates to them, is carried out by GitLab itself. By integrating the scanners into the pipeline using predefined templates, you can use them without much effort.

However, you could also use your own security scanner. You just have to take care of the execution in the pipeline yourself. And if you're using the Enterprise Edition in GitLab Ultimate, you can integrate your own security scanner into GitLab's vulnerability management platform.

Overall, GitLab does not offer the ability to work with plug-ins in the area of CI/CD, but because many features can be executed in the pipeline and a strong base set of features is available, plug-ins are often not that necessary anyway. GitLab describes itself as a DevOps platform for good reasons.

And GitLab is also a great tool for ensuring the key DevOps principle of visibility of work and progress. GitLab is particularly well suited for ensuring this principle, as everything is collected in one platform, provided, of course, that all functions are used.

As this section is primarily concerned with the clarity of CI/CD, it is again important to look at the interaction between source code management and CI/CD. In principle, each project should have only one pipeline. By using only one pipeline per project, it is easy to see the current status of the pipeline for the corresponding commits, and the pipeline can be reached quickly.

Permissions Management

GitLab includes predefined user roles that are assigned to users in groups and projects. These roles do not have much use in the pure CI/CD component, except perhaps when it comes to deployments. What is more important in this context is that the permissions apply across the entire platform. For example, if a container image is built in the pipeline and pushed to the GitLab internal container registry, no additional configurations are required for access to the various systems. This means that you can manage the many different functions of the platform efficiently without having to constantly adjust the authorizations for individual users, as is the case with Jenkins, for example.

Complexity in the Construction of Pipelines

The pipelines in GitLab are defined in YAML files. For a given pipeline, a *.gitlab-ci.yml* file in which the stages and jobs of the pipeline are defined is created in each repository.

This setup is relatively simple, though it can be used to build some complex setups. As the YAML file is located directly in the repository, no configuration is required to get the pipeline running, as the repository is part of the system.

With a platform like Jenkins, in which the CI/CD process is mapped via an external tool, not only must each project's or group's access permissions be configured in Jenkins, but the pipelines must also be set up to be configured and managed automatically accordingly.

Permissions do not need to be set up in this way with GitLab, as permissions management is a part of the platform. Only the *.gitlab-ci.yml* file needs to be created—the rest happens completely automatically. This saves a lot of work and reduces potential sources of error. The prerequisite is, of course, that a usable runner is configured to execute the pipeline jobs.

As I have already briefly mentioned, there are various ways to execute the pipeline jobs, but I recommend using containers to do so. The use of containers results in clean build environments and makes the process easier to scale. To minimize the effort required to maintain the containers used, consider using container images provided through the programming languages or build tools you're using.

As they are defined in YAML files, GitLab pipelines are declarative; this makes them easy to read. The complexity of the YAML file defining a pipeline is low; although there are some available keywords, they are well documented.

```
stages:
- build
- test

[...]

build:
    stage: build
    image: maven:3.8.1-openjdk-15
    script:
        - mvn install -DskipTests –batch-mode
            -Dmaven.repo.local=./.m2/repository
    artifacts:
      paths:
        - .m2
        - target
```

Listing 6.2 Snippet of a .gitlab-ci.yml File

The snippet of the *.gitlab-ci.yml* file shown in Listing 6.2 defines the build and test stages and a job called build, which is executed in the build stage. A Maven container is used to run Maven in the script section of the file. To ensure that the resulting artifacts can be used for subsequent tests in the test job, the file defines two paths that are to be saved as artifacts so that they can be used in the next job. Apart from the artifacts, this is the same example as the declarative Jenkins pipeline definition shown in Listing 6.1.

As GitLab CI/CD has been an available product for some time, GitLab has incorporated a number of features that deviate from the standard use cases, as well as keywords for many more specialized functions, which are available in the Enterprise Edition. As a result, you don't usually need to reinvent the wheel, as there are many features available.

Of course, there are many more details about GitLab that to consider, but that would go beyond the scope of this chapter. The key takeaway is that most pipelines created through GitLab are relatively similar, which is why it is not always a steep learning curve to read the definition of a pipeline.

Of course, you can also create complex setups through GitLab, using hierarchical templates to increase reusability. But note that in 2024, GitLab published the first release of CI/CD components to further simplify the process and thus increase reusability.

The Role of GitLab CI/CD in the Overall DevOps Concept

The role of GitLab CI/CD in the overall DevOps concept is pretty clear: GitLab's focus is on providing a complete DevOps platform. The proximity of the source code to CI/CD without the need for much configuration and without the need to retrofit missing functions with plug-ins is the core advantage of GitLab.

Its core advantage is also its main difference to Jenkins: While everything is configurable and extensible in Jenkins, allowing what feels like infinite implementation possibilities, GitLab offers only a few options for implementing functions. As a result, it is easy to implement functions in GitLab; no major training is required when switching between GitLab instances from different organizations. Of course, there are a number of exceptions depending on the use case, but this is generally the standard.

While this chapter mainly concerns continuous integration, GitLab functions related to the continuous delivery environment are also relevant to this discussion. Continuous integration and continuous delivery go hand in hand. For example, the relevant deployment environments can be addressed from the pipeline definition so that they are also displayed in the GitLab user interface to ensure seamless integration.

6.6.3 GitHub Actions

GitHub Actions is the automation component of GitHub. Its core area of responsibility is the setup of CI/CD pipelines, but it can also be used to configure and execute other automations.

Compared to GitLab CI/CD and Jenkins, this offering is still relatively new; it was introduced only in 2018. Since then, GitHub has been following a similar path to GitLab by transforming itself into a central platform allowing users to map the entire software development lifecycle.

Installation and Maintenance

As with GitLab CI/CD, GitHub Actions has two components: the server and the runners. However, there are some differences between GitLab's and GitHub's server component, especially in terms of the SaaS and the self-managed server instance.

GitHub, owned by Microsoft, is primarily known for *GitHub.com*, where many developers host open-source projects. GitHub adds new functions to the website quickly and regularly. In addition to the free version, paid subscriptions with more functions for teams and companies are available.

GitHub Enterprise Server and GitHub Enterprise Cloud are available for enterprise customers. While the latter is a dedicated GitHub instance hosted by GitHub, the former is aimed at customers who want to host an instance themselves. The SaaS offering can also be used via *GitHub.com*.

Note, however, that the feature set available in GitHub Enterprise Cloud and GitHub Enterprise Server often lags some time behind that of the free version. These are different products that are not supplied with the same functions at the same time. And this is also true of the GitHub Actions feature. While GitHub Actions was already available for *GitHub.com* in 2018, it became available in GitHub Enterprise Server only in January 2021.

GitHub Actions runners are available in the free version and can be used by everyone. These can be downloaded and installed on Linux, macOS, and Windows. There is no real package for this, as you have to download a *.zip* or *tar.gz* package in order to start them. Jobs can then be executed. As an alternative to the runners available on the platform, you could also add your own self-hosted runners.

The runners can be scaled automatically, but GitHub significantly reduced the options for doing so, compared to the options for doing so in GitLab, in the winter of 2023. GitHub officially recommends two automated scaling options: One is a controller for Kubernetes, with which actions can be executed on Kubernetes clusters. The other is a Terraform module for scaling on Amazon Web Services (AWS). Although other scaling options also exist, they are—just like the recommended Terraform module for AWS—available only from third-party providers and therefore require a great deal of maintenance and integration.

> **Note: Security Vulnerabilities**
>
> The handling of security vulnerabilities at GitHub is relatively similar to that at GitLab. GitHub regularly rolls out fixes for security vulnerabilities to the free version of the platform, so users hardly notice them. GitHub also regularly provides security updates for GitHub Enterprise Server; these updates must be installed by users.
>
> Security in the GitHub Actions solution is more important to consider. You can quickly and easily introduce third-party actions, which you must always keep up to date. The security concerns for third-party actions are similar to those for Jenkins plug-ins.

Permissions Management

Two aspects are relevant for permissions management in GitHub: management of users and authorization of the GitHub token, which is used by workflows to process other data.

Permissions can be set for individual users and user groups for the repositories and at the organizational level. In the free version, you can configure the read, triage, write, maintain, and admin roles, which come with a static set of permissions. With GitHub Enterprise, you can configure permissions in a more granular way.

This is because, as with GitLab, permissions are relevant across various functions. You therefore do not have to maintain permissions in several places in order to manage access to the source code, the CI/CD platform, and the package and container registry.

Complexity in the Construction of Pipelines

GitHub relies on YAML files for defining automations. What's special about GitHub is that automations are not limited to the execution of CI/CD pipelines—automations can be set to trigger other actions when certain events occur. The GitHub Actions documentation contains a long list of events that can be used for these automations (*https://docs.github.com/en/actions/using-workflows/events-that-trigger-workflows/*).

GitHub includes events relevant to CI/CD such as push and pull_request. A push event is triggered when new commits are pushed to the servers. A pull_request event is triggered, among other events, when a pull request is created. Outside of typical CI/CD use cases, automations can also be triggered when issues are created or when labels are set in issues. This makes the execution of workflows very flexible.

Workflows in GitHub are also written in YAML and are therefore declarative. They are defined in the corresponding repository in the .github/workflows/ directory. Several workflows can be stored in this directory, so various use cases can be defined in separate files.

A workflow can consist of several jobs, and each job can consist of several steps. Steps execute the actions required for the given job, such as building and testing the project. Listing 6.3 shows an example of a workflow for building a Java project with Maven:

```yaml
name: Java CI

on: [push]

jobs:
  build:
  runs-on: ubuntu-latest

    steps:
      - uses: actions/checkout@v3
      - name: Set up JDK 17
        uses: actions/setup-java@v3
        with:
          java-version: '17'
          distribution: 'temurin'
      - name: Build with Maven
        run: mvn install -DskipTests -batch-mode
                    -Dmaven.repo.local=./.m2/repository
```

Listing 6.3 Example Workflow in GitHub Actions

This example is basically the same as the example pipeline definitions provided in the sections on Jenkins and GitLab CI/CD: It builds a Java project using Maven. However, the syntax looks a little different here.

After the name for this workflow is set, the event is specified, as the workflow should run here for every push. The job definition in our example is more interesting. As in the previous examples, this file defines only one job, called build. However, this job consists of three stages.

Before the first step, the file specifies that this project is to be executed under Ubuntu. A containerized setup is not used in this example, but such a setup would also be possible if a Docker image were specified in the uses section. In the first step, the Git checkout is then executed so that a working copy of the repository is available for the workflow.

In the next two steps, the predefined actions that provide the runtime environment for the build are used. This is the installation of the required Java environment. Maven is then executed in the environment to build the project.

Expandability

One practical feature of GitHub Actions is that reusability simple. GitHub Actions has a marketplace (*https://github.com/marketplace?type=actions/*) that anyone can upload actions to. You can transfer predefined steps from existing implementations into your own workflows so that you don't have to start from scratch with every new workflow.

By using existing actions from the marketplace, you save yourself the work by using work from the community; you don't have to reinvent the wheel.

However, using actions from the marketplace also has an impact on the supply chain of your project. Think carefully about whether you really want to use existing actions and, if so, which ones you should use. At the very least, check where the action comes from. I would recommend using only official actions. The note **Verified creator: GitHub has verified that this action was created by actions** indicates that the action is an official action from GitHub Actions itself. There are also verified actions from third-party providers that you can use.

You should be careful with unverified actions created by normal users. It is best not to use such actions, but if you plan to use them, at least check the code to ensure that there are no obvious security vulnerabilities and that the action does (only) what it promises to. Always remember that these actions can gain access to your code and to connected systems.

For each action that you obtain from external providers, make sure to check each new version that the provider releases to ensure that the code still works as expected. Specifically, make sure that the changes are still compatible with your system and that you fully understand the functionality of all changes.

Ultimately, even with all the precautions I recommend taking, using actions means that you still have a layer of dependencies to manage, and there are many factors you need to consider in order for your system to be secure but also up to date. You should be aware that managing these actions requires a certain amount of extra work. Compared to Jenkins, it is easier to update these dependencies in GitHub Actions, but it has to be done for each project.

Clarity

Clarity is not such a big problem in GitHub, especially when you look at the interaction between source code management and CI/CD.

For each commit, you can quickly access the specified workflows and the actions they contain. Because workflows are linked not only to commits but also to events, you can quickly access them via the **Actions** menu item in the repository.

This means that newcomers to the code can get up to speed with a project relatively quickly and see a relatively standardized process, just as they can with GitLab. Only Jenkins is probably the exception here, in which there are a number of ways to find a solution.

The Role of GitHub (Actions) in the Overall DevOps Concept

At this point, it should come as no surprise that GitHub and GitHub Actions are just as valuable to the overall concept of a technical DevOps implementation as GitLab is.

GitHub involves fewer tools, fewer handoffs, less complexity, and a standardized permissions management approach. The arguments in favor of GitHub are almost the same as those for GitLab. Of course, the two platforms have some differences and can be easier or more difficult to implement depending on the various use cases.

We will take a closer look at this topic in Chapter 13, where we look at DevOps platforms as a whole and examine the factors outlined in greater depth.

6.6.4 Other Continuous Integration Servers and Tools

There are numerous other tools that you can use to implement the CI/CD process. In principle, you could probably meet your needs with any of them, but not every tool is convenient and easy to use.

In addition to GitLab and GitHub, Azure DevOps is another continuous integration tool to consider. Like GitHub, Azure DevOps is owned by Microsoft and is commonly used in companies. The closer you are to Microsoft technologies such as .NET and Azure Cloud, the more likely you are to find Azure DevOps.

It will be interesting to see how Microsoft shapes the future with its tool offerings. I think a long-term future for Azure DevOps is rather unlikely, as the internal competition with GitHub is fierce, but that is just speculation.

Atlassian also offers CI/CD tooling. The much older solution is Bamboo, which can be used only for on-premises installations. Because it is developed by Atlassian, it is closely integrated with Jira and Bitbucket. Although it is not a standardized platform, this Atlassian stack comes relatively close to a platform.

Bamboo is not available for the cloud, but Bitbucket Pipelines, another CI/CD solution from Atlassian, is. Bitbucket Pipelines also defines pipelines using YAML.

And there are numerous other tools, such as Teamcity, developed by Jetbrains, and Circle CI. The basic concept with a central server and the definition of the pipeline is available in almost every option. Other options include dagger.io and Tekton, a cloud-native CI/CD solution.

6.6.5 Continuous Integration at nicely-dressed.com

The longer those responsible for transitioning the teams at *nicely-dressed.com* to a DevOps environment looked at the various problems they had and tools at their disposal, the clearer it became that the complexity of continuous integration should be kept to a minimum so that each team can manage its own pipelines.

They also wanted the infrastructure to be as easy to manage as possible, which is why they decided against using Jenkins. They were also not interested in niche solutions that are not particularly widespread, as it was assumed that it would be more difficult to train people in these solutions or to quickly resolve problems with them.

They decided to choose between GitHub and GitLab. However, the final choice was hotly debated: Some knew GitHub primarily from open-source projects, while others appreciated the sophistication and open-source approach of GitLab.

> **Note: Reflection**
>
> Continuous integration is essential for a DevOps work environment. An important basis is clean management of the source code, as we discussed in Chapter 5.
>
> Regardless of the technology used, it is important that you keep integrations as small as possible so that they can be executed more quickly. Continuous integration pipelines ultimately support this aspect.
>
> Almost every time I see and talk about the setup of CI/CD environments in various companies, I notice that there is a very high level of technical debt. This has to do with the tools used, but also with the fact that many people feel the need to constantly reinvent the wheel.
>
> This is precisely why you should make sure that you keep pipelines and their definitions as lean as possible. Avoid "super smart" pipeline setups and try to stay as close to the standard version of the chosen CI/CD tool as possible.
>
> I've worked a lot on and with Jenkins in my career. I used to enjoy using this tool, but I recognize that it is becoming less relevant and less desired today, as GitLab and GitHub have are becoming fierce competitors. The administrative weaknesses that Jenkins has are ironed out in GitLab and GitHub.
>
> The less you work *on* a tool, but *with* it, the better. And that's exactly why many people tend to migrate away from Jenkins and why, despite my years of experience with Jenkins, I recommend not using it.

6.7 Summary

The build process is one of the first steps in a CI/CD pipeline. Although it can be considered technical, many DevOps principles play a role here as well. The most important and obvious principle is Automation from the CALMS model: The build process must be properly automated through the use of a continuous integration server and the package manager of the respective programming language.

CI/CD alone creates many advantages in terms of developer productivity for current team members, as well as for onboarding—and almost independently of DevOps.

Chapter 7
Quality Assurance

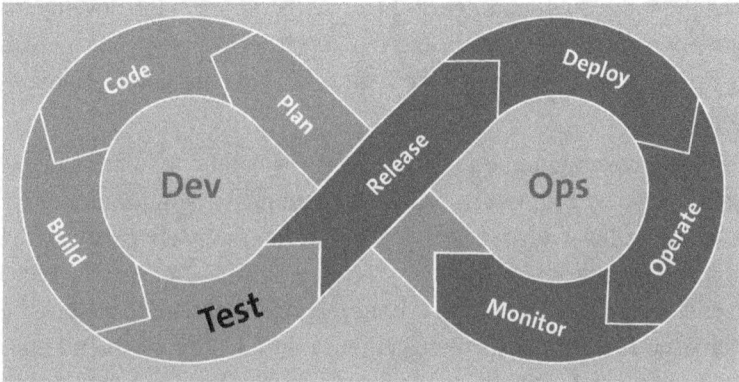

Figure 7.1 Test Phase of the DevOps Software Development Lifecycle

The last phase of the development side of the DevOps lifecycle concerns testing and QA. Two questions arise here: How does the QA team fit into the cross-functional matrix? How does the DevOps paradigm affect the testing activities of teams with different areas of focus?

DevOps is not just about bringing development and operations together; the QA team *also* needs to be integrated into the DevOps environment.

7.1 Typical Problems with Testing

The *nicely-dressed.com* team previously conducted testing of the software separately from the actual development work. Both the team structure and the processes in the software development lifecycle itself ensured that the development work and the subsequent QA work had no direct touchpoints.

This structure and approach to development work had many disadvantages and hardly any advantages. The teams were simply set up according to their separate tasks, which seemed to work over the years without any major changes—after all, it had always been done that way.

7 Quality Assurance

7.1.1 Teams in Their Silos

When the developers finished work on a certain component or ticket, they would throw the project over the wall to the QA team, because that team was responsible for testing. The QA team then would then work in its own repository and in its own test infrastructure. The code for the tests and the corresponding configurations and builds were therefore separated from the developers' environment.

The main problem with this way of working was that the QA team always had to chase down the changes to the individual modules of the online store from the developers. During the development work, both the code and the user interface changed so much that the previously written tests no longer worked; the effort that had gone into maintenance was wasted.

As it was almost impossible to update the automated tests by hand, a lot of testing was done manually instead, often with checklists and tables containing definitions and the sequence for manual testing. But even this documentation was rarely up to date, as the menu navigation in the online store was constantly changing. The QA team therefore focused on manual end-to-end user tests, in which the testers simply had to re-click the menus. This was also what the company's management wanted; after all, the menu is the part that reaches the user and is therefore the most important component to test.

Let's take a step back and take another look at the current situation: How did the *nicely-dressed.com* QA team test the online shop to try to ensure the quality?

The main work of the QA team always started after the implementation phase of the development team, as shown in Figure 7.2. The team consistently had periods of time when it had no work to do because the implementation phase was in progress.

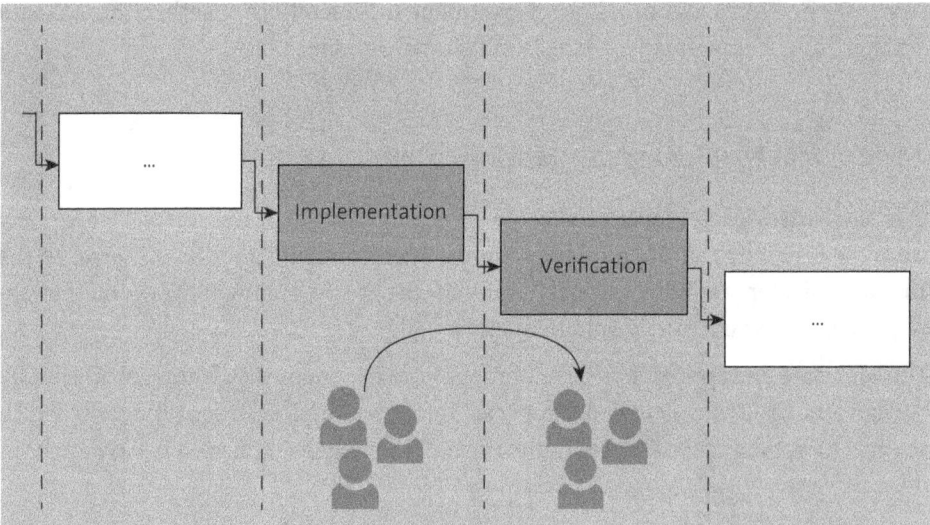

Figure 7.2 Workflow from Development to QA

As the store consists of many different modules that were developed independently of each other, there were also several QA teams responsible for one module. In some cases, there were even several QA teams responsible for several modules, which made testing even more complex. No two modules of the store could be tested in the same way, so the different QA teams also had to work differently.

If the QA team found a bug, it could not be corrected directly. Instead, the team had to open tickets describing the bugs for the development team. The development team then had to fix those issues. Of course, this did not make the QA team popular, as it only created more work for the development team. At the same time, the QA team was happy to submit tickets to the development team because the team's success was measured by how many bugs it found.

The fact that obvious bugs were found and corrected after some time was still the ideal case. The process was often further complicated because it was not always clear how exactly a feature should be implemented or how something should work. "It's not a bug—it's a feature!" was a phrase heard all too often. It was said jokingly, but the core of the problem remained unsolved: Who actually decides whether something works as planned? Only those on the business team who had originally formulated the requirements could answer this question. However, the path to the sales and marketing people was much longer, and even if the question made it to that side of the company, the team often couldn't say exactly what it wanted six months ago.

7.1.2 Different Understandings of Requirements

The communication problems between the teams became even more apparent as release dates approached. Before release could take place, a technical review had to be carried out by a separate team. To conduct the review, this team had to find out whether the defined business requirements had been met.

So this review concerned deployment of the release to the production environment. There was no real staging environment, only manually configured test environments for when really large features had to be tested. Of course, these environments never corresponded to the real production environment, so bugs were not noticed during testing and review, only in the actual production environment.

As they wanted to prevent bugs from popping up in the production environment at all costs, the testers concentrated almost exclusively on technical details and stumbling blocks that could hinder deployment and operation of the code. They clicked through the menus and tried to get the application to crash or make obvious bugs. (The team was particularly willing to take this approach because finding such undeniable bugs ensured that the testers met their metrics quickly.)

However, with this approach, the testers did not bother addressing the question of whether the business requirements were met which is the most important consideration for the end users of the system. Any requirements that were not implemented correctly were noticed only at the very last stage—namely, when the changes were presented to the sales and marketing people who had commissioned them. It would then become clear that the feature that the development team had worked on for several months did not look the way the business team had imagined, and the business team would have to deliver this feedback to the development team.

But this feedback would go to the development team several months after the team worked on these changes. The development process would then have to start all over again: The development team would start working again on the code that it had actually completed several months ago, would correct the issues described in the feedback they received, and then would build another version for the QA team, who would then check it again (and would possibly find some bugs again) so that the technical review could then take place again.

The packages for the individual modules could be handed over to the operations team to start the deployment only after the changes were approved.

7.1.3 The Number of Bugs as a Metric

I have already mentioned that the success of the QA team at *nicely-dressed.com* is measured by how many bugs the team finds. Unsurprisingly, this metric is not a good one, because even if the development team programs software that has no bugs at all—which is highly unlikely, if not impossible—the QA team still has to invest a lot of time and effort into verifying the whole thing. And that means that if the QA team does not find any bugs, it would receive a correspondingly negative rating.

Admittedly, everyone should be aware that all software has bugs. But even with almost perfect software, the QA team will have to invest a lot of time and effort into checking the software. So if the team doesn't find any bugs, that doesn't mean that the team members failed at their job.

However, this is not how management saw the role of the QA team at *nicely-dressed.com*: The individual QA teams and people on them were measured by how many bugs they found. This metric was set at management level, which meant that it was the employees lower than the managers on the hierarchical structure who suffered. This shows once again that the culture of DevOps must be practiced at all levels: The wrong targets for measuring success lead to the wrong ways of working within and between teams.

> **Note: Metrics and Targets**
>
> The big problem with metrics is that they are easily confused with targets. Or they gradually become goals, which is almost worse. Of course, striving for software that is as bug-free as possible is a good goal, and counting the number of bugs found is a useful metric. But if these two points are mixed up, both quality and statistics ultimately suffer.

> Consider this analogy: Developers have the goal of writing good code. And lines of code can easily be counted. But a large number of lines written doesn't that the code is good, and vice versa; the measurement should not be equated with the goal!

7.1.4 Is It Fixed Yet?

Let's go back to our discussion of testing: When the QA team at *nicely-dressed.com* finds a bug, lengthy processes that lead to further waiting times and context switches occur: When a person from the QA team opens a ticket to inform the development team of a bug, that person is currently in the context of the bug. Once a ticket has been opened but not processed, this bug and its context also remain unprocessed until the fix is available, usually within the following few weeks. In the meantime, the QA employee then moves on to their next task and gradually forgets the context of the bug situation because it is impossible to document everything. Once the correction is ready, the QA employee has often already lost this context.

This loss of context is also a problem not only for the member of the QA team but also for the developer: To address the bug, the developer has to revisit an implementation they worked on some time ago. Both parties therefore have to rethink the problem and the context each time they encounter the bug. The necessary frequency of these context switches cause both parties to waste a lot of time.

As these processes are running so slowly at *nicely-dressed.com*, it has become necessary to check whether reported bugs had been fixed, as there is simply no visibility into the processes, which you should know from reading about the build process in Chapter 6. In terms of the build process, the lack of visibility simply makes it more difficult to know whether the build was successful. In testing, however, the lack of visibility is even more serious. As the QA and development team work separately, the testers don't know what is being worked on until the development process is complete, and the development team has no insight into how the QA team carry out the tests.

This in turn often means that the development team is unable to reproduce certain bugs, as not all information from the teams' various systems is accessible to everyone. Some QA teams work with test management tools, for example, in which a number of manual tests are documented with each intermediate step. However, many of the development sub-teams do not have access to these tools, so there is, again, no way around manually copying the tests if they want to ensure that all information had been handed over, which, of course, never worked properly in practice.

> **Note: Reflection**
> Teams working in their own silos is a tiresome topic, even in QA. Incorrect incentives and poor communication ensure that problems linger for longer than they should.

> You should look at whether there are metrics or disincentives in your organization that need to be avoided. Also, reinforce the idea that finding bugs is a good thing and is not just the testers' responsibility.

7.2 Testing as Part of the DevOps Process

The fundamental aim of tests in the DevOps concept is to find bugs at an early stage before they cause problems on production systems. Correcting the errors early on ensures quality, which enables frequent deployments with significantly lower risk.

As explained earlier, DevOps is not, as the name suggests, a pure merger of the development and operations teams, but a general philosophy for collaboration that involves other teams as well, including QA. In the eight stages of the DevOps pipeline, the QA team is integrated into the software development lifecycle before the operations team (see Figure 7.3). Depending on the size of the company, there are various ways of integrating the team. But regardless of what the role designation ends up being, as a rule, testing should take place as early as possible.

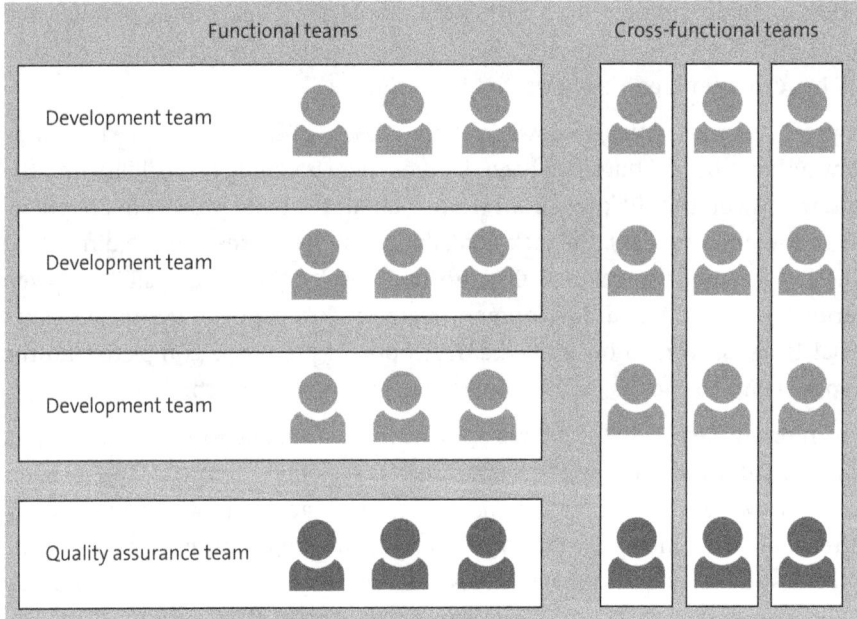

Figure 7.3 Functional Teams versus Cross-Functional Teams

In smaller companies, integration of the QA team into the software development lifecycle is often not necessary, as such companies often have no individual QA teams and their development team takes care of the tests. However, even companies that do not have QA teams need to consider certain technical and cultural aspects of QA, particularly with regard to automation.

In larger companies with an independent QA team (such as *nicely-dressed.com*), making the transition to a DevOps environment can lead to major changes on the QA side. This is because integrating DevOps requires splitting up the QA team. Each development team is assigned at least one person from the original QA team to serve as a quality expert and take on the relevant testing tasks within the team.

However, the original QA team does not disappear completely. Though in a DevOps environment there is overlap between QA and development, the QA team remains as a separate team within the company. The QA team may continue to work with the development team and take on tasks as required, or its members see themselves as part of the development team and also work on the pure QA team on joint tasks that are required beyond pure development work. They are generally responsible for all QA issues and, in this role, are responsible for developing extensive test concepts for the rest of the team and for optimizing the test process itself.

However, the way the QA employees are organized into teams is really only a question for the organization chart. It is more important to know how the role and therefore the tasks of QA employees will change, as they will now be part of two teams: one that continues to be the QA team and the other that becomes part of the existing development team.

In many companies, several similar problems occur during development, only with a time delay. Instead of leaving each team to evaluate these problems and introduce its own solutions, companies should have teams work in a more or less uniform way and use the findings and ideas of the other teams to collaborate on a single solution. For example, a developer may want to try out different test frameworks to see which one works best. Instead of trying them out on their own codebase alone, they should work on it together with the experts from the QA department.

The team structure according to DevOps also changes the roles, responsibilities, and tasks of the previous standalone QA team. By setting up cross-functional teams, in which a QA employee sits on the development team during day-to-day work, but is in close contact with other QA employees sitting on other teams, knowledge sharing can be carried out quickly (see Figure 7.4). The original team can therefore learn directly from the mistakes and findings of the other teams, facilitated by the open exchange across team boundaries.

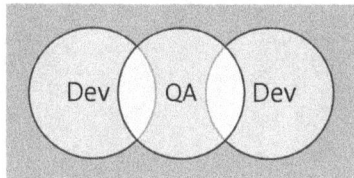

Figure 7.4 Development and QA Team Overlap

If we look at this from the perspective of DevOps principles and the CALMS model, this change in team structure alone brings two CALMS value directly to light: culture and sharing.

7.2.1 Tests in the Build Pipeline

This change in team structure strengthens not only the culture and the sharing values of the CALMS model, but also the Automation value; more automation reduces the amount of manual work that needs to be done during testing.

The integration of QA employees into the previously standalone development team opens up a number of advantages. The main problem with the former QA team at *nicely-dressed.com* was that it could start its work only at the end of the development phase; until then had no insight into the source code. During the test phase, the members of the QA team always had to correct their test scripts in line with the changes and therefore had to do a lot of manual testing.

Of course, it doesn't make much sense to continue working like this. In the DevOps world, proper testing is the standard. Bugs should be found and corrected as early as possible in order to drastically shorten the such long cycles for correcting bugs. Clean software with as few bugs as possible and, above all, with comprehensible automated tests is the basic prerequisite the successful rollout of the software to production systems later on. After all, nobody would feel comfortable rolling out software to production systems if they were unsure as to whether bugs are present.

We have already discussed how the build stage of the CI/CD pipeline works, in Chapter 6. In the build stage of the pipeline, the project is built reproducibly. After a successful build is the execution of tests; this is an elementary component of the pipeline. The tests run immediately if the build process is successful before the deployment job is triggered (see Figure 7.5). This ensures that those who submitted the changes receive feedback at an early stage.

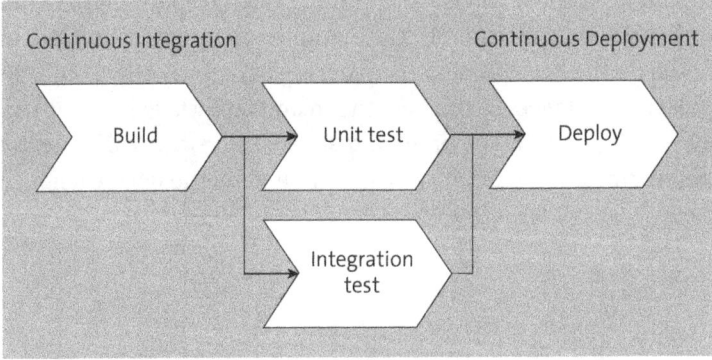

Figure 7.5 CI/CD Pipeline: Build, Test, and Deploy

But let's first take a step back: The first problem with the QA process at *nicely-dressed.com* was that the QA team had no access to the repository. By opening up the repository and adding QA specialists to the team, this problem was addressed quickly at the start.

The relevant people can now see directly which changes are being made and can directly adjust the automated and manual test processes accordingly. In addition, in principle, the closer to the source you can run tests, the easier it is to detect dependencies between the application and the tests at an early stage. If tests are executed only periodically or rarely, you still have not created an advantage. We take a closer look at this in Section 7.2.3, specifically at how tests can be integrated into the continuous integration pipeline in such a way that they are executed automatically and cannot be forgotten.

However, the measures taken so far have not really solved the problem; although QA employees have been integrated into the development team, no processes have been adapted and no cultural onboarding has been carried out.

After this change in team structure, the developers at *nicely-dressed.com* initially continue to work as they always have. The presence of QA employees on the team is initially irrelevant for the developers, as they still believe that those from the QA team want to find bugs only to make the developers' work harder.

However, the focus QA employees should change as a result of being included on the development team. They should no longer focus on finding as many bugs as possible before deployment; rather, they should focus on helping the team to identify problems as early as possible. In addition, the QA specialists must show the developers how the processes and software should be adapted to avoid obvious bugs as early on in the process as possible. The QA team member therefore becomes the enabler of the test concepts and the QA subject matter expert on the team.

Adapting processes in this way includes, for example, directly considering and adapting the tests when refactoring code. By doing so, the team can gradually move toward test-driven development (discussed further in Section 7.2.4). However, this can work only if the developers involve the QA experts directly in their development work.

The next point is that the test setup must be designed in such a way that tests can be executed as easily as possible. This applies not only to the workflow of individual developers but also to the continuous integration pipeline—namely, the addition of automation. After all, the best tests are useless if they are never executed.

> **Note: The Duration of a Test Run**
>
> Ideally, the run through a continuous integration pipeline with build and test processes is roughly equivalent to a coffee break or a trip to the toilet. However, this is not possible for every programming language and every project, so you always have to decide what is the greater evil: wait longer and find bugs early on in the process, or tackle the next developments more quickly because serious bugs will not become apparent so quickly anyway?

7.2.2 Different Tests for Different Tasks

We discuss automation further in the following section, but first, let's take a look at the different types of tests. These tests should be fully automated if possible so that as many bugs as possible can be found in good time:

- Unit tests
- Integration tests
- System tests
- Acceptance tests

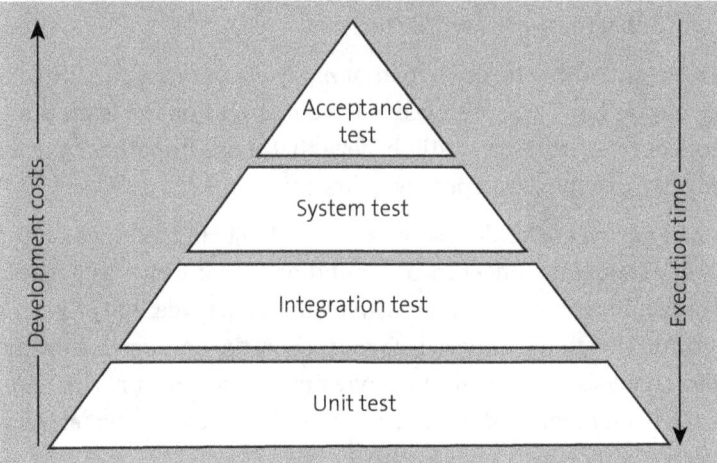

Figure 7.6 Test Pyramid

The test pyramid in Figure 7.6 shows how these testing types build on each other with regard to their development costs and execution time.

Unit Tests

Unit tests are the smallest tests; they are generally quick to write and to execute. A unit test tests only one function or method of a class. The goal of a unit test is to quickly find elementary bugs in the implementation of a method. The idea is that if unnoticeable bugs creep in at this low level, debugging at a later stage becomes time-consuming and, therefore, laborious. Ideally, there should be a suitable unit test for every method that is written.

If you have ever had to write tests for an older codebase that had no or only a few unit tests, you probably noticed that the comprehensibility of the code correlates directly with how easy it is to write the tests. Methods that are too long and too complicated to understand straight away are difficult to test—this is especially true for methods that are supposed to perform several tasks at once. For code like this, it is usually necessary to refactor the code to increase its readability and testability. (For brownfield projects that have existed for some time, some refactoring is usually necessary in order to be able to write proper unit tests.)

Therefore, writing unit tests can help developers find ways to improve the readability and quality of the source code. If you have problems testing individual methods, that is a good indicator that those methods need to be broken up into smaller methods.

The problem with unit tests is that they really only test one method. This means that dependencies on other objects that are required must be abstracted as mock objects so that, if possible, only the target method is tested.

Integration Tests

Integration tests go one step further than unit tests, as they test several components of an application. The effort required to write these tests is higher than for unit tests, as they test code more systematically. Dependencies within individual components and modules must first be established in the source code before it can be tested with integration tests.

Integration tests allow the combined functionality of the various classes of a system to become visible. The components being tested have to be mapped and tested in a matrix, so the more components there are, the more complex the complete testing is. This is the main reason why writing integration tests is more developmentally expensive than unit tests and why their execution usually takes longer.

System Tests

System tests are the next level of testing. Both unit and integration tests can run only if the application is buildable, so they are executed against the built software. System tests run when the application is running.

For these tests to run, it is important to ensure that the application can be started and deployed. For the *nicely-dressed.com* online store, all modules from the various teams must be executable so that the deployment can be carried out and the system tests can then be performed.

System tests can also be automated—for example, by running tests via the user interface in order to carry out a complete end-to-end test. In the specific example of *nicely-dressed.com*, not only does setting up the test environment and starting the online store take a long time, but the test runs take even longer due to the loading times.

Accordingly, the development costs are also high because these areas have to be adapted more frequently. This is the case whenever a change is made to the interface, for example, which may not change the functionality but ensures that the mapping for the automated user interface tests no longer works.

Acceptance Tests

Acceptance tests check whether the implementation behaves as intended and meets the requirements. These are the only tests that cannot be performed automatically. They are often called user acceptance tests, or UAT for short. Bugs can occur if the implementation does not meet the requirements, even if the functions are implemented technically correctly. Accordingly, unit, integration, and system tests are not useful for this type of testing.

Acceptance tests are usually carried out directly in the graphical user interface by automatically clicking through the application to see if everything works as intended. The challenge is that you also have to check whether acceptance tests are working correctly. It doesn't matter whether this is done manually or automatically. The question is rather, How do you involve the client, and how do you collect feedback? This takes us into the area of requirements engineering, which is out of the scope of this book.

> **Note: Different Programming Languages for Testing and Development**
>
> A common problem for efficient and effective collaboration between the development and QA teams is that these teams often use different programming languages. For example, an application may be written in Java, but acceptance tests are typically written in scripting languages such as Python and JavaScript. Even if APIs are to be tested, it is actually better if the language of the application is used for testing; however, it is often easier and faster to use scripting languages for tests.
>
> This is unfavorable for collaboration, as not all developers are proficient in every programming language, which makes it difficult for many developers to write tests independently. In many companies, there are already too many different programming languages in use, which limits collaboration. Switching between languages within a team helps even less.
>
> Ideally, the same programming languages should be used wherever possible so that you can concentrate on writing tests without necessarily having to learn a new language.

7.2.3 Automate Tests

Writing and executing tests is all well and good. However, it is also important to optimize the test duration. Tests that run for too long should be avoided if you want to make changes faster and more frequently.

Analysis of the code—including code quality as well as performance and load testing—can also be integrated into the pipeline. Manual tests should be completely automated wherever possible; otherwise, the lead time cannot be sufficiently reduced. However, it is not possible to completely avoid manual tests in every environment. In the financial sector in particular, you often see a lot of manual testing—partly because there is no other option and because the focus is typically on end-to-end testing this this sector, but also partly because of regulatory requirements to conduct manual testing.

Test Automation in the Pipeline

Tests incorporated in the CI/CD pipeline are automated. Failures of testing jobs in the pipeline indicate that the corresponding tests have not completed.

The common test frameworks usually generate an XML file in JUnit format. These XML files contain information about which test cases were executed, which lines they belong to, and whether tests have failed. The various CI/CD tools can read these XML files, process them graphically, and evaluate them. For the big picture, it is important that everyone who is involved can quickly see how many tests were successful and if they have returned any errors. Tests that have returned errors must then be marked as failed in the pipeline and be clearly visible.

Executing tests is useless if they are also not evaluated—that is, if nobody actively looks at the logs of the executed tests in the pipeline jobs.

Finally, with most CI/CD tools, you can set not only a *green* mode and a *red* mode, but also a *yellow* mode to indicate that only a few tests have failed or that only warnings have occurred.

Although this feature is certainly justified, I am not a fan of it. Tests should always pass correctly, without exception. If you define a number of tests that can fail, or even a percentage below which tests can fail, you are taking the concept ad absurdum. Some tests are more important than others, but whether they are serious failures or not is almost impossible to assess across the board.

> **Note: Flaky Tests**
>
> Flaky tests are those that sometimes run successfully and sometimes not, depending on their mood. This can be quite annoying for the developer, as failing pipelines due to individual failing tests tend not to bring any added value. I can only recommend that you either fix such tests for good or expand them. However, fixing them may be tricky because the presence of flaky tests may mean that there is a problem in the environment in which the tests are executed.

Test Coverage

Test coverage is a common metric for measuring testing. This metric indicates how many instructions, branches, and paths in the code are covered by tests during test execution. It is generally assumed that high test coverage is better. However, I would not want to make such a sweeping statement.

A major advantage of trying to achieve high test coverage is that it ensures a large number of paths in the code are tested and that more bugs are detected, especially if the code is changed regularly.

Another advantage is that stability and robustness of the code are increased. This is particularly important regarding the continuous rollout of changes.

However, focusing only test coverage is not recommended, as it indicates how *much* testing is being done but not how *good* the tests are. So don't set a goal of 100% coverage. This might be just as pointless as asking about the absolute number of bugs found.

Also, the *trend* of changes in test coverage is often more meaningful than the amount of test coverage itself. Test coverage that deteriorates over a period of time could indicate that there may be problems. To ensure that testing is as comprehensive as possible, the tests should not only be calculated in the pipeline but also made visible. Ideally, this should be done directly in the development environment. Some integrated development environments (IDEs) support this by default.

In addition, the test coverage should also be made visible in the review process. If you use GitLab, you can display the test coverage in the merge request during a code review. Any problems will then become visible in the review.

Optimize Test Duration

The more tests you write and execute, the longer the total runtime of the pipeline will be. However, tests can often be executed in parallel. And if setting up parallel execution of tests is difficult, then the test setup should be modified to make it possible.

For example, instead of running 10,000 tests one after the other, which takes a total of 20 minutes, try running four sets of 2,500 tests to reduce the runtime to six or seven minutes. (The reason it would be six or seven minutes and not five minutes is that provisioning the test infrastructure, downloading the repository, and preparing for the execution of the tests also take some time.

A scalable container infrastructure can help you to run more test processes simultaneously. Various caching mechanisms can also help to ensure that tests do not take too long to start.

Considering observability is also helpful for optimizing pipelines and tests, as it can help identify any bottlenecks more quickly. This topic is discussed in Chapter 10.

Ensure Code Quality with Static Code Analysis

Static code analysis is relevant for general code quality and especially for security. It involves analyzing the code without executing it—in other words, statically.

As with test coverage, static code analysis helps identify inconsistencies and syntax errors before the code ends up in the production environment. It is therefore also an important part of the continuous integration pipeline in order to detect such errors at an early stage and thus minimize any risks.

Because it can identify code quality issues, static code analysis is also helpful for improving collaboration; a standardized code style and easy-to-read code are helpful during code reviews and are generally relevant for rapid and ideally error-free further development.

As with test coverage, the trend is important here. Results from static code analysis should always be processed promptly, but not all results are important. You should ensure that the overall balance of the static code analysis tends to develop positively.

Simple static code analysis scans that are run directly when writing the code are called *linters*. Static analysis is performed by a program. Ideally, the static analysis run is triggered both locally in the IDE and via the continuous integration server. Both runs are relevant and take different approaches. Early scanning in the IDE is practical so that you can see when unclean code is being written while programming. This is the same idea as the spell checker in a word processing program.

To ensure that these rules are applied uniformly and apply to all participants, linters should also be triggered via the continuous integration server and the results should be evaluated.

A common tool for static analysis is SonarQube, which is offered as both a free open-source version and a commercial version with more features. For local scanning in the IDE, there is a separate tool called SonarLint.

SonarQube supports many different programming languages, making it relatively easy to get started using the tool. It also supports plug-ins and integration options for the various continuous integration servers so that the results are immediately visible.

Code Climate is another, though less commonly used, static analysis tool. As with SonarQube, it is available under an both open-source license and a commercial license with more features. It supports a wide range of programming languages.

Code quality is relevant not only for pure source code. Linters are also useful for analysis of configuration files and files for infrastructure as code (IaC), as well as for Terraform code and Dockerfiles, as they can provide you with direct information on recommended procedures that you otherwise might have had to find manually.

Performance Testing

Performance tests are another testing option. There are different types of performance tests. For example, a typical use case is a load test, which tests the effects of loading an application. One tool for running load tests is k6, developed by Grafana Labs. Other performance tests include stress tests and scalability tests.

Performance tests should not be executed on every feature branch during every pipeline run; they may take far too long. Instead, they should be executed only on the main development branch. But again, restrictions apply, because a stubborn pipeline on the main development branch should also be avoided. Instead, it is better to run performance tests on a nightly build so that you can still get feedback on significant performance losses relatively quickly.

As always, it depends very much on the given setup. If you run a proper CI/CD pipeline anyway and deploy to production environments several times a day, then performance tests are likely to play a lesser role. However, if you only deploy infrequently, such tests are very important to determine the quality of the software.

Tools such as Apache JMeter (*https://jmeter.apache.org/*) are available for testing various types of applications, servers, and protocols, as well as the HTTP protocol for web applications and a number of other interfaces such as REST, SOAP, JDBC, and LDAP.

Regardless of which tool you use, it is important that you carry out *meaningful* tests and evaluate them regularly. Comprehensive performance tests take a lot of time. And don't forget the feedback loop; otherwise, even the best implementation of load and performance tests won't help.

A typical mistake made when implementing performance testing is placing too much focus on purely read or purely write requests. For typical applications, requests both read and write data, so healthy mix should be tested.

Sometimes testing every change isn't useful, but it is better to run the test in parallel once a week or even at night. The only problem is that these errors must also be made visible, which can be more difficult with a separate pipeline.

Problems with performance tests can always have different characteristics. For example, the application basically works, but the load or performance has just deteriorated. Therefore, the long-term trend is primarily important in performance testing.

> **Note: Security Tests**
>
> I have not covered security tests in this chapter. These tests are also important and are becoming increasingly implemented in pipelines. As is the case with other types of testing, the basic idea is to find issues as early as possible. The whole topic of security in DevOps is covered in Chapter 11, Section 11.3.

Testing at nicely-dressed.com

With regard to the tools, there were no major discussions at *nicely-dressed.com*: The various continuous integration tools support the execution of tests. The parallel execution of tests using a matrix configuration in order to reduce the overall runtime is also a fairly common feature for all tools, so there were no major differences here.

Note: Reflection

A high number of automated tests with high test coverage is important and necessary. Make sure that your tests can be executed quickly by running as many as possible in parallel. The faster the tests run, the better, because nobody likes to wait for the pipeline to finish.

Use other tools in the pipeline to keep track of test coverage and code quality at all times. Don't be too strict about adhering to these two aspects: It is more important to look at the trends.

7.2.4 Test-Driven Development

Test-driven development is often used in agile software development and, therefore, in DevOps environments. The idea behind test-driven development is that, instead of developing code first and then testing that code, you think about the test first and then look at how best to implement the requirement (see Figure 7.7.)

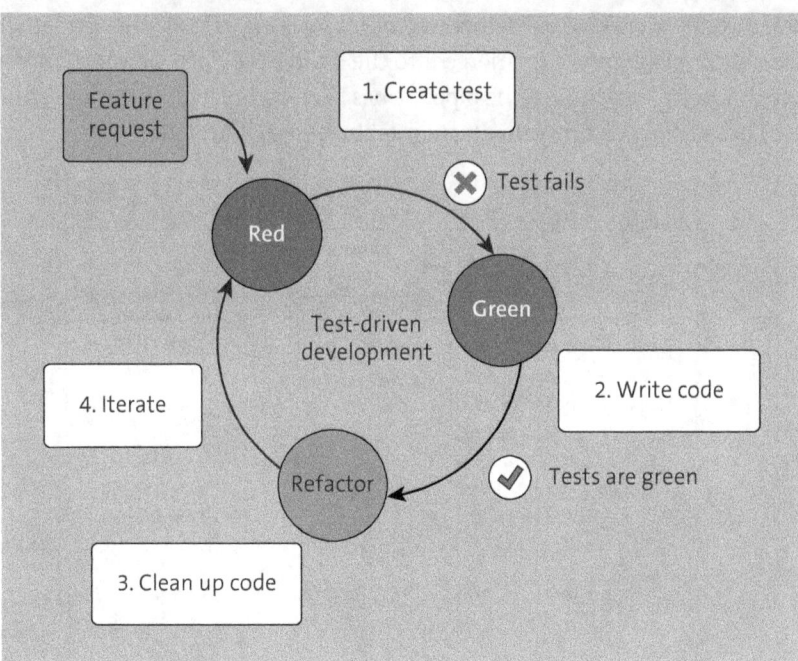

Figure 7.7 Test-Driven Development

Test-driven development can encourage more careful development practices. Consider these steps:

- A developer writes a test case first directly. It fails immediately because the implementation it should be testing is not yet available.

- The developer then writes the corresponding implementation of the function much more carefully than if they wrote the implementation first.

So developers automatically think about what a good implementation might look like after they write the corresponding tests. The quickly executable tests that result from this process make it easy to see whether the implementation work has been completed without having to do much testing or thinking.

Almost all tests developed through test-driven development should run automatically in the continuous integration pipeline. The main goal should be to run tests not only after certain milestones, but regularly and as early as possible.

As a developer, you take care of the implementation of the test case as well as the actual implementation. If tests are executed locally, there may be a lot of waiting time, or you may have to first chase after dependencies that are not available locally.

For this reason, tests should always be executed in the continuous integration pipeline before the given branch is merged into the main development branch. As part of the code review in a merge or pull request, both the build and the tests should be executed in order to obtain a comprehensive overview of the status of the feature. This provides developers with direct feedback as to whether the code is working as it should or whether there are any errors in other tests. The merge into the main development branch takes place only once the entire run through the pipeline has been completed. That way, errors can be corrected before they disturb other team members (see Figure 7.8).

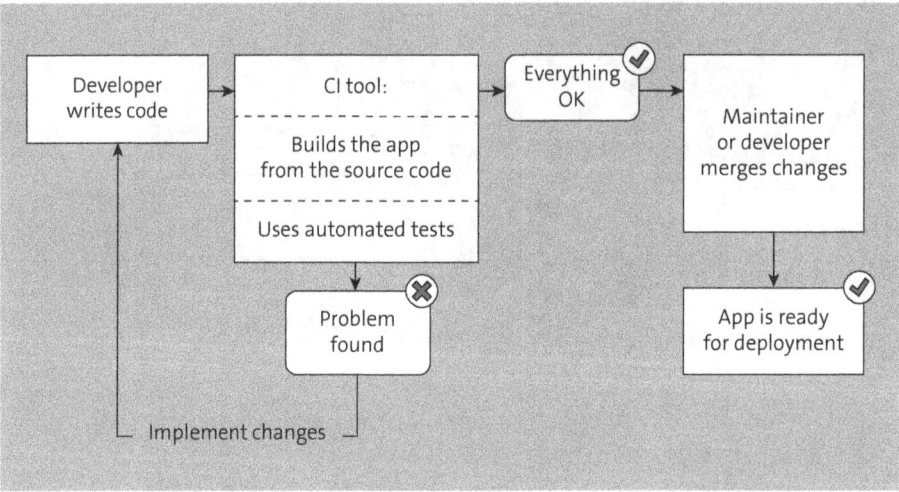

Figure 7.8 Continuous Integration Pipeline, Test and Review

As part of the code review, the reviewer should therefore also check whether the solution and the test cases have been implemented correctly and comprehensibly. However, reviews should occur only when all tests have completed. For larger projects in which a large number of developments are running in parallel, it is also necessary to carry out a dry merge against the target branch in order to run the tests.

Unit, integration, and system tests can obviously be executed automatically. In the case of acceptance tests, despite the fact that some may think that they cannot be put into the continuous integration pipeline at all because they are executed manually, that's not true—they can! At this point, the smooth transition of the continuous integration pipeline into the CI/CD pipeline is crucial—that is, the continuous integration pipeline should be supplemented by continuous delivery functions.

7.3 Summary

Early automated testing is a core component of DevOps and the associated continuous integration pipeline. The team structure changes by including QA team members in the development team. By merging the source codes of the project and the tests, a single source of truth is created for the entire project.

By opening up the development team and allowing new team members with a QA focus to form a link to the cross-team QA team, silos in the development process are broken down. Information on problems reaches the relevant developers much more quickly from the sister teams.

Early automated testing is an essential cornerstone for tackling and implementing the continuous delivery aspects in the next step. We will deal with these in the next chapter.

Chapter 8
Continuous Delivery and Deployment

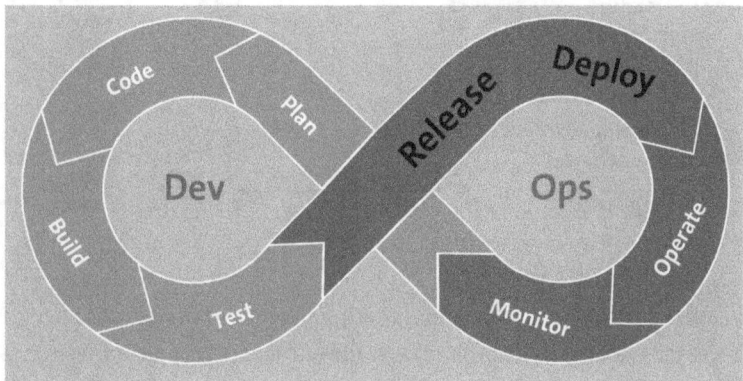

Figure 8.1 Release and Deploy Phase of the DevOps Software Development Lifecycle

In the previous chapters of this book, we have looked only at the development process, which involved planning, programming, building, and testing the application. The operations team and its activities have been left out so far. Moving forward, we'll now cover the operations team, the problems that exist when development and operations are separate teams in an organization, and the advantages of merging the two teams.

8.1 Typical Release Management Problems

All modules of the *nicely-dressed.com* online store—and, therefore, all involved in developing and operating it—are fundamentally dependent on each other. For a release, all work had to be completed at the same time so that everything could be published in one go; an example can be seen in Figure 8.2. For this reason, a new version was rolled out only roughly twice a year. Coordination between the teams took a very long time, and there were always delays somewhere.

The main problem is that all modules had to be put together individually before the release. The teams worked in separate repositories behind their large walls. Coordination between the teams was necessary but was only half-heartedly achieved via a ticket system, and the teams coordinated only when someone thought that a change to their own project would affect the other teams.

8 Continuous Delivery and Deployment

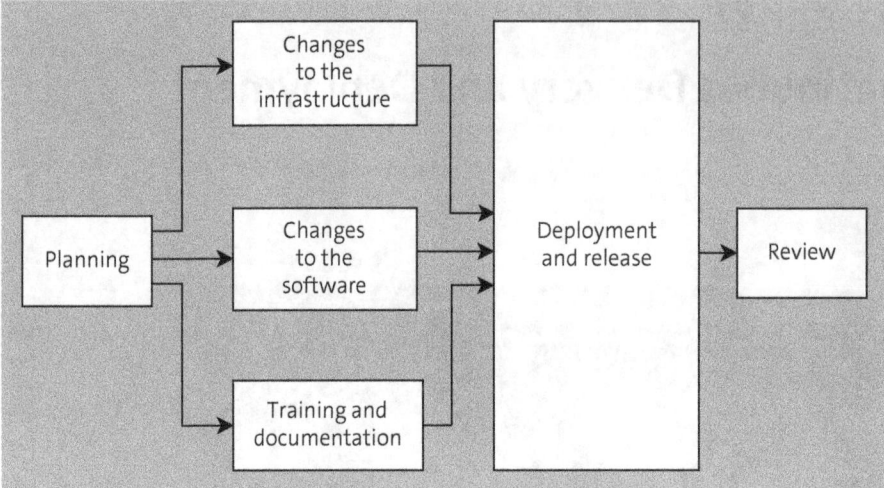

Figure 8.2 Waterfall Release Management Process

For a new release, adapting the source code and updating the software were just two of the relevant and necessary changes. In most cases, the infrastructure also had to be changed or even rebuilt to enable deployment before the new version could be tested and then pushed.

8.1.1 Separate Handling of Changes and Documentation

The team that took care of API changes in the backend documented the changes poorly during their development work. The reason for this was because the team made changes to the API as required based on requests from various other teams, and they would update the API documentation only after they completed their development work and the code entered the test phase. At this point, the developers would review their changes again to record and document them (see Figure 8.3).

The time that passed between implementation and documentation meant that valuable context was lost; even worse, the person who wrote the documentation was not always the person who had originally made these changes to the API.

Because the teams developing the iOS and Android app versions of the online store were the most dependent on the changes to the API, they often faced problems. They typically had to reimplement changes at the end of the development phase because changes to the API had not yet been documented or had been only partially documented.

This process repeatedly delayed releases by a few days or even weeks because new adjustments and corrections to the mobile apps were necessary, as they were no longer fully compatible with the upcoming deployment.

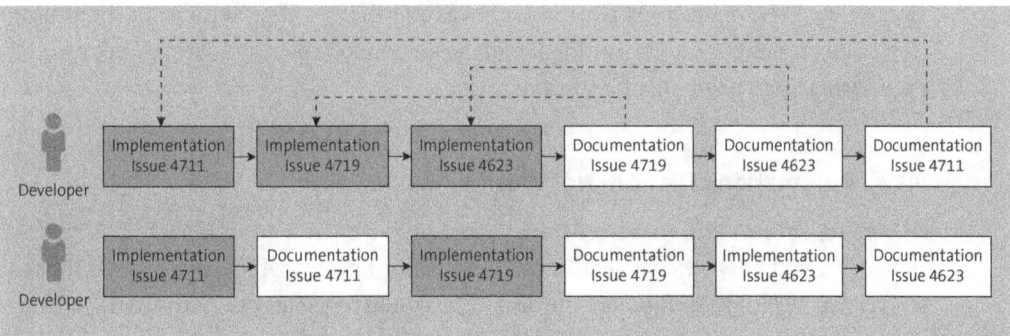

Figure 8.3 Decoupling of the Implementation from the Documentation

Again, the lack of coordination between the various teams was the main cause of the problems. Developing in secret not only led to problems during the implementation phase, but it also contributed to a more difficult release process.

Once the teams had finally been able to coordinate on the changes and the code was in freeze mode for all teams—meaning that nothing significant could be changed for the time being—the first task was to build the release in order to deploy it to a *staging environment*. This refers to an environment that is similar to the production environment, the preliminary stage for the actual production operation, so to speak.

Once the final tests passed, the store would be deployed to the production environment to make it available to customers with the latest changes.

8.1.2 Lengthy Release Process

This release process had to be carried out almost entirely manually. As described in Chapter 6, Section 6.1, most teams had only one build server each, on which the projects were periodically built—if the build was successful at all. Access to these servers was available only to the team leaders, so the complete release process required all team leaders to be available at roughly the same time to ensure a successful release.

A release manager managed the release process itself. This manager was responsible for communicating with the various team leaders, who in turn spoke to their teams if something was not working.

The process followed a long, predefined release plan and was, of course, incompatible with the changes made in the meantime.

The team leads were responsible for ensuring that the teams provided the code so that a local build could be created on the build server. Once this was done, the team leaders manually collected the various files required for the deployment.

This repeatedly led to errors, as files were sometimes missing or the leaders ended up collecting unnecessary artifacts. Such errors were noticed only when the data was already with the operations team, which was preparing a deployment to the production environment. A programming language–dependent package format, a commonly used format, was used only by a few teams.

8.1.3 Automations That Are Not Worthwhile

Of course, automating these processes would have been a good idea, but as releases were deployed only twice a year, it was deemed unnecessary; it was assumed that the effort required for to implement automations would be greater than the benefit. Under these circumstances, this is actually understandable.

Instead, a release manager was responsible for transferring the compiled artifacts of the projects to the operations team. These artifacts were simply stored on a file server so that the operations team could start deploying them to the staging environment. Of course, the time frame for this step was always tight. The date for the release of the new version was already fixed, and any delays in the development and testing process were noticeable at his stage.

However, not much time could be gained at this stage, as the process for deploying the artifacts to the staging environment also consisted of manual work: Someone from the operations team simply copied the transferred artifacts by hand to the (hopefully) correct location.

The operations team relied on instructions for deploying the respective packages. The development teams created and delivered these instructions in the form of comprehensive documentation. In particular, the documentation listed all dependencies. In theory, this documentation seemed comprehensive; in practice, however, it was full of information gaps, was outdated, and did not match the configuration of the staging and, ultimately, the production environments. And how could it? The developers had no insight into (or influence on) the exact version status and settings of the server. And nobody knew exactly what the operations team was doing in its production environment.

8.1.4 Hostilities between the Teams

The result of all these problems was sometimes venomous hostility between the teams: The operations team assumed the developers were lazy, while the developers accused the operations team of simply being too incompetent to put a few modules in the right place.

Deployment to the staging environment always took time, and nobody enjoyed the process. To reduce these conflicts, it was decided that the operations team should always contact the release manager in the event of deployment errors.

The release manager in turn spoke to the respective team leaders, who then discussed the deployment problems with their teams. The usual result of these internal team meetings was unsurprising: The team would assert that the errors were somewhere else and that someone else had to deal with the problems. Because there was no coordination and no insight into the mistakes made by the other teams, the chaos would start all over again with every deployment.

After the operations team got the online store's staging environment up and running—with a lot of manual work, digital duct tape, and other undocumented changes—the QA team would enter the final spurt. The QA team still had to run some end-to-end tests to ensure that there really were no errors. As usual, the team would discover a few more bugs, so the individual teams had to rework their code. The changes were then copied back to the staging environment—with all the same conflicts over again.

Once the QA team gave its approval, the release manager would also have to approve the deployment, and then the managing director would have to give a final approval.

8.1.5 Deployment on Production Systems with Obstacles

With the final approval from the managing director, the operations team was finally given the task of rolling out the deployment to the production environment via the ticket system. It was important that the team rolled out the update at a specific time on a specific day. For example, a major change was planned to be rolled out at the same time as the Black Friday offers, which were advertised both online and on television at the same time.

Successful deployment at the right time was important for several reasons. First, if the deployment were rolled out too late, sales could be lost if customers were to the online store and not see the offers that were advertised. Second, the deployment were rolled out too early, the related products may sell out before the offers are supposed to begin, which could to resentment among customers, who may decide not to return to the store.

A maintenance window was announced internally at the time when the deployment to the production environment was due. The operations team started copying the module files to the production environment again. Once again, the team did this task manually. Even though the team tried to eliminate the missing dependencies and other stumbling blocks that were already hindering the process during the deployment to the staging environment, errors that should have been fixed long ago occurred again.

But that's not all! The operations team faced new problems, which increased the tension among the team even more before the tight deadline. One reason for this was that the staging environment was too different from the production environment. To cut costs, a lightweight environment with fewer machines was set up, so it was not set up exactly like the production environment.

On the other hand, the staging environment was often used for experiments and tests that could not be replicated in the production environment. Every move the team made in the staging area would have had to be repeated in the production environment, which was completely unthinkable with the number of necessary manual changes.

And finally, in addition to the problems in the development process, there were also problems with the tests. It turned out that the end-to-end tests only ever concerned the purchasing process, as this process was how the company earned money.

Nobody had considered updating the workflow for integrating new products, so the new goods for the Black Friday offers could not even be entered into the system. And finally, the staging area worked only with a significantly smaller database, so performance problems were not noticeable.

The developers had to rework their code outside their normal working hours, and they then threw their changes over the fence again to the operations team. With such short-term, unchecked changes, there were errors in the reworked code. Too much information was mistakenly being written to a log file, as debug outputs were not removed, causing the filesystem to fill up regularly.

Instead of solving the problem at the source, a resourceful admin wrote a script that regularly deletes the log file. Because this seemed to solve the problem, there was no need to raise a ticket with the development team. After all, the admin found communicating with another team to be laborious and assumed that the developers wouldn't mind that the logs were being deleted every few hours.

At the time of the planned deployment, the services were restarted so that the deployment could be completed and the new software could be run. However, the services did not restart properly and errors that could have been prevented by better tests in the staging area occurred.

The only solution was open-heart surgery: interventions in the running production system to get the store back online and functioning as quickly as possible. This was a horror for the operations team and a huge amount of work for the developers, who had to recreate the changes in their development stages if they didn't want to repeat the process next time.

8.1.6 Conclusion

In my experience, these types of problems occur frequently in the real world—perhaps not all of them together, but every deployment has surprises in store. One big reason for these problems was the lack of automation; management assumed that implementing automation wasn't worth the effort for a project that was deployed only a few times a year. And since deployments had always been ultimately successful in the past, management did not see any need for additional automation solutions and other processes. Releases were always very stressful for everyone involved in the project, as a lot had to be coordinated and various errors always occurred.

> **Note: Reflection**
>
> Release management, paired with deployment, is the first look at collaboration between the development and operations teams. Before you dedicate yourself to automating this process, it is important that you recognize the existing problems with this process on your specific teams.
>
> The problems given as examples in this book are common in the real world to a certain extent. The first step here is also to investigate whether there is any resentment between the development and operations teams. This should then be eliminated as far as possible.
>
> Invest in team-building measures for your team at an early stage to strengthen cooperation.

8.2 Implementing Continuous Delivery and Deployment

As part of a company's transition to a DevOps environment, the development team and operations team need to be brought together into a joint team with a single goal. In the previous chapter, we discussed how to integrate the QA team into the development team. In this chapter, we'll discuss how to integrate the operations team.

8.2.1 Bringing Development and Operations Together

A common myth surrounding DevOps is that the development team simply takes over the operation of the application. This assumption suggests that the work of the operations team is easy enough that it could simply absorbed by another team. After all, there is also the cloud, which provides lots of automation and where infrastructure is available on demand.

Of course, that is not how the operations team is handled when a company transitions to DevOps! What is required instead is closer cooperation, goal-oriented communication, and greater shared responsibility.

In order to better understand what this means, it is worth taking a closer look at the various tasks of the operations team in order to see how they can be implemented in a cross-functional team.

So let's take a step back. What problems are we actually trying to solve here? The aim is to provide a service for users. This requires not only developing and testing the software but also operating it. In a company structure with separate teams, these tasks are also separate, and that separation cannot easily be reconciled. In Section 8.1, you saw what these problems can look like in practice at *nicely-dressed.com*.

The rigid division of tasks and employees leads to a work culture in which teams, at best, work side by side, but far too often end up working against each other. This "unculture" must be broken up in order to create symbiosis. This is what the DevOps concept stands for.

Figure 8.4, shows a depiction of the cross-functional team structure. This depiction is, of course, very simplified. You can't expect to put developers and operations specialists in a room and magically create a DevOps team. Instead, people need to be properly trained on how to work in such a team structure. It is also necessary to select employees for each cross-functional team based on their knowledge and experience and their individual strengths and weaknesses.

The roles and tasks of all those involved have to change: The goal of a single functional development team is to further develop the software, while the operations team is responsible for the stable operation of the application. The fact that these objectives are to a certain extent contradictory to each other is a result of the requirements.

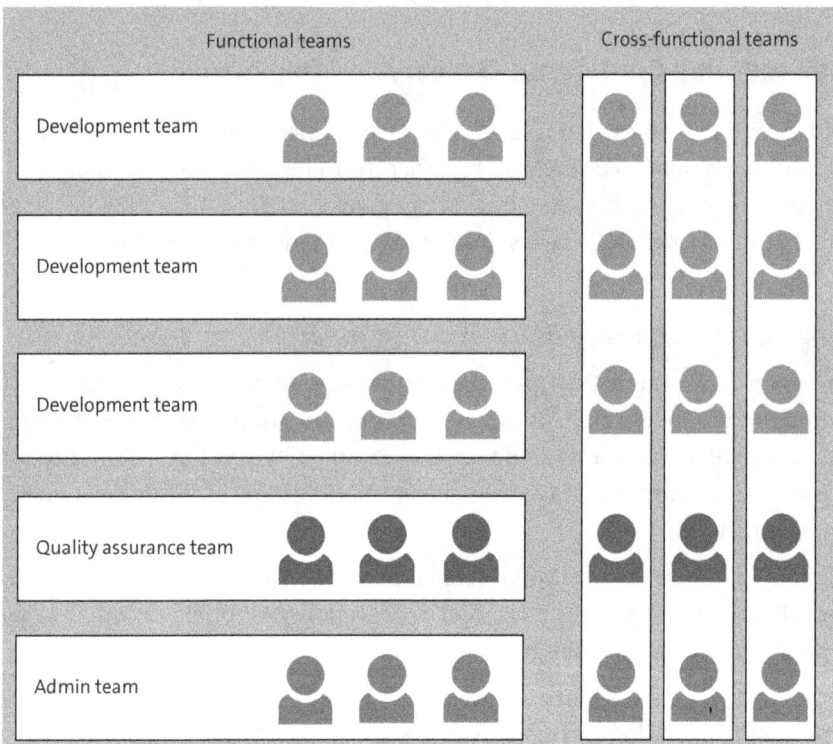

Figure 8.4 Cross-Functional Team, Including Development, QA, and Operations

This is one of the most important reasons why simply merging the roles or even eliminating the operating team is not helpful, as the different objectives remain the same.

Instead, both roles with both goals need to move closer together in cross-functional teams that pull in the same direction. We will look at how cross-functional teams can be implemented in detail in Chapter 12, which discusses the transformation to a modern DevOps culture.

On a cross-functional team, the administrator from the operations team will move much closer to the actual code and should have no objections about helping with packaging and preparing the deployment. On the other hand, the developers will have to take much more responsibility for deployment and operation. The development team has knowledge of the code, the development history, and the architecture of the project; its members therefore know how the application and the individual features work.

This is knowledge and experience that people from the operations team typically don't have. Instead, they know where the problems are in the operation of the application, where things get stuck during deployment, and when the application behaves differently than expected.

The interface between separate development and operations teams at the technical level is the package format: The development team knows in which format the application has to be packaged, and the operations team expects the defined package format so that it can receive the artifact to put it into operation.

However, just because the package format is known, that does not mean that new versions can be put into operation without any problems. Teams that are strictly separated are always passing the ball back and forth. This should not happen in a cross-functional team based on DevOps principles; instead of working against each other, the team tackles the problem that needs to be solved in a goal-oriented way! Everyone contributes their experience and knowledge by talking to each other directly and over short distances.

One place for this type of discussion can be a *daily stand-up*, which has been a common practice in agile teams for a long time. A daily stand-up is a short meeting in which the team addresses current challenges and problems. If a developer suspects problems with an upcoming deployment and wants to discuss them with the operations specialists, an uncomplicated meeting like this enables the exchange of experiences and perspectives.

There are also training sessions and workshops where people learn from each other. Ideally, these events are organized and carried out directly in the team; rather than tedious frontal lessons in which a developer tries to teach the admins how to program, for example, they are small, unbureaucratic events for the exchange of knowledge, preferably in a dialog format.

Let's imagine a team's task is to transform an old monolithic application into a microservice architecture and containerized so that it can be operated on a container orchestration platform.

The package format is essentially a container image. But before we get that far, there is a lot of work ahead of the development team, which has to rebuild the old application. Of course, they make life easy for themselves and use base images that include as many tools and dependencies as possible. This is practical but unnecessarily leads to an image that is several gigabytes in size.

What may not be a problem for the developers immediately raises the eyebrows of the admins: Several gigabytes for each new deployment? That will lead to rollout problems and more costs.

If teams are sitting at the same table, they can quickly identify and solve this potential problem before the base image used becomes too firmly anchored in the container. Everyone contributes their expertise, be it in the operation of containers or the packaging of software. It is important to work closely together to achieve the common goal of a cross-functional team: rapid and structured further development *and* secure and stable operation of the application. Both roles continuously support each other over time to ensure that both goals are better achieved.

New Tasks on the Team

With cross-functional teams, just as there will continue to be software developers, there will also continue to be a kind of operations team, albeit with different areas of responsibility. The focus will shift toward application operations, in close cooperation with the developers. Everyone works together on and with the CI/CD pipeline and ensures that changes reach customers as quickly as possible.

> **Note: Operation, Infrastructure, and Platform**
>
> Very large companies often have teams that deal more with the pure infrastructure. Its main task is managing the infrastructure—whether it exists on-premises or in the cloud doesn't really matter—while the other teams handle the deployment and operation of the individual applications.
>
> The terms *platform team* or *infrastructure team* may therefore be more appropriate, as they are not concerned with operating the application. Instead, the task of the infrastructure team is to provide a stable platform so that the other teams can work on it.

The DevOps idea is noticeable on cross-functional application teams but also on platform and infrastructure teams; since work is done more quickly and with more agility, it is necessary to automate manual work and standardize processes as much as possible. Even those who are still "only" responsible for the availability of the infrastructure must be able to react quickly to short-term deployments, which are possible only by automating these tasks as much as possible; having a new version ready for deployment in a few minutes isn't useful if it takes two weeks to install a new hard disk to meet the increased space requirements.

The key to this is infrastructure as code (IaC), or infrastructure that can be adapted quickly and easily using commands and configurations. (IaC is discussed in more detail in Chapter 9, Section 9.5.1.) A team looking to use IaC can use IaC tools that are very similar to programming languages or, depending on the scope, develop and maintain their own IaC software. Developers can draw on the skills of their colleagues to convert manual work into the best possible automation scripts.

In cross-functional teams, everyone benefits from these types of exchange with each other: Veteran operations specialists can learn how to solve tasks with a few lines of code, while developers can gain practical experience in operations. This ensures that the team can better work toward a secure deployment in the production environment, giving everyone much greater confidence in the continuous integration pipeline.

Everyone on the team must understand the basic structure of both the continuous integration pipeline and the actual application, but of course this does not mean that everyone knows every detail. There will always be specialists for different tasks, but on a cross-functional team, they can work more closely with each other.

Painless Deployment

Given the experience of all team members, the team can tackle problems in the release process directly at the root. The team can quickly get feedback on problems with the build or rollout. This feedback must go directly to the team member who can solve the given problem. Instead of communicating back and forth via tickets and pushing tasks away from each other, as is more typical in separate development and operations teams, the whole team sees directly that a problem needs to be solved.

Basically, there are three different areas of responsibility that all need to be covered by the team: the software development, the deployment, and the operation of the application. Both areas of responsibility have a smooth transition between the two sides of the team, connected by the CI/CD pipeline.

I don't discuss the various tasks and roles of a cross-functional team here, because they depend heavily on how the software and the company are structured. It is important that the originally separate tasks are also, but not only, linked with the help of automation. Ultimately, everyone on the team has to shake hands in order to work together efficiently.

Encourage Direct Cooperation

To improve the collaboration process, everyone should work together on problems via pull or merge requests and contribute changes instead of playing ping-pong with tickets. The same principle applies as to the code review process (see Chapter 5, Section 5.4), with the small difference that, in on a cross-functional team, code review no longer concerns only the source code but also configuration files and documentation.

The team members working on the changes can then simply test the changes themselves before they are incorporated into the actual development for everyone. This not only saves a lot of time but also prevents conflict and problems downstream.

It is crucial that the release process is as completely automated as possible. The First Way of the Three Ways plays an important role here, as it ensures that the changes flow from development to operations as early as possible and thus from the business to the customer. Teams working with more traditional six-month release schedules tend not to implement automation because it requires a lot of time and effort for a benefit that is used twice a year.

When deployments are done more frequently, then full automation is essential. Automation that is as complete as possible can prevent more errors than partial automation. Automations will certainly not all work right away without any problems. Instead, implementing automations takes a steady approach in which the processes are improved with every step.

Several factors come into play here. From a purely technical point of view, automating the deployment process is not incredibly complicated once is the automations are in place. However, you have to look at the big picture, where the question always arises: How much do you really trust your software and your team's ability to roll out changes, ideally multiple times a day?

Increase the Deployment Frequency Step by Step

To achieve the ultimate goal of rolling out multiple deployments every day, the intervals between deployments must be reduced step by step. With each deployment, the team gains experience and confidence in its own software and in the release process so that it can roll out deployments regularly without fear and with as few errors as possible.

As a reminder, as you can see in Figure 8.5, the more time there is between two deployments, the greater the risk of problems. Therefore, as mentioned before, it is important to keep the size of the changes as small as possible.

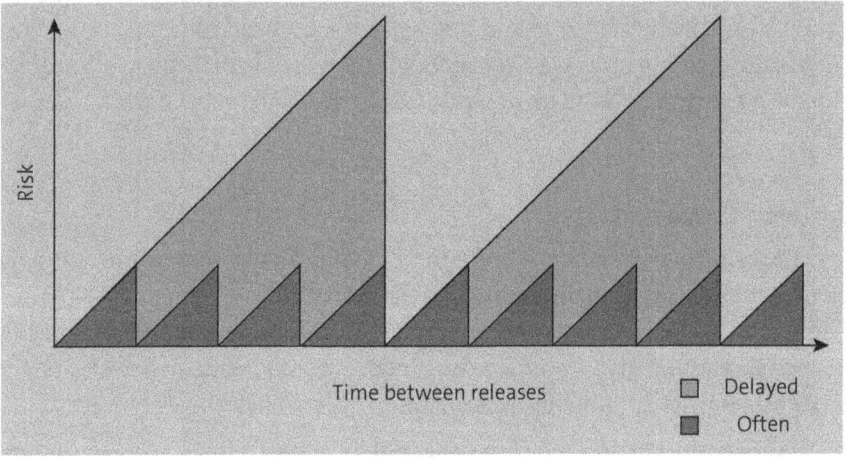

Figure 8.5 Relationship between Time between Releases and Risk

To achieve the goal of rolling out multiple deployments per day, the intervals between deployments to production environments must be gradually shortened in the long term. However, as you cannot start directly with the production environment, you should first concentrate on the other environments.

As part of the CI/CD pipeline, the project is built, the various tests are run, and the project is packaged.

The package is required to trigger the deployment in the next step. To trigger the deployment, the package is uploaded to a package registry so that there is a version history of the packages that were built. In the event of a rollback, an older version can be rolled out quickly.

The CI/CD pipeline should be accessible to all team members so that everyone can understand whether the project is being built and whether all tests have passed. Without an overview of what the processes and tools are doing, it is difficult to establish joint ownership of the software.

Before a deployment is rolled out on the production environment, automation on staging and review environments should be prioritized in order to increase confidence in tools, processes, and ultimately, of course, the people developing and operating the project.

8.2.2 QA, Staging, and Production Environments

The cross-functional team structure and improved team mindset we discussed Section 8.2.1 take the conflict and difficulty out of deployments. Development work is no longer put on hold for months but is delivered quickly and painlessly.

If CI/CD is fully implemented, changes are rolled out directly to the staging environment after each merge. A staging environment provides immediate feedback on whether the deployed changes are working and allows the team to check it technically.

Whether and when to deploy to the production environment depends on a few factors. If you implement CI/CD stringently, then you can actually deploy to the production environment immediately after a successful deployment to the staging environment. Though rare, some teams do not use a staging environment. It is more common for teams to trigger the deployment to the production environment manually in order to have more control over when and whether the deployment should be carried out.

The exact process of deploying to the production environment depends on various factors: trust in the software, the time on the relevant day of the week, and the specific business criteria. Feature flags and canary deployments are often used instead of a completely blind deployment (see Section 8.4).

In practice, work is usually carried out as depicted in Figure 8.6. In development, there are separate review environments in which the code to be deployed can be reviewed quickly. If the code passes review, it flows into the main branch and is deployed to a staging environment. This is where QA takes place to ensure the code works as expected. Only when this step has been completed is the deployment to production carried out.

8   Continuous Delivery and Deployment

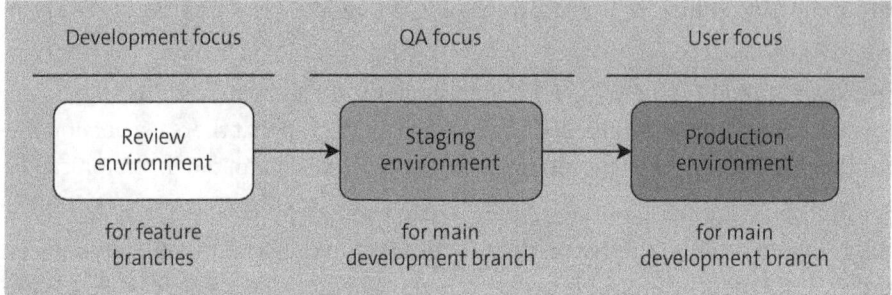

Figure 8.6 Focuses of Different Environments

Review Environments for Developers

The aim of a CI/CD pipeline and the review process is that the work on a feature or bug fix should be completed before the merge into the main development branch. To be able to assess a feature or bug fix, environments are needed to test the changes. These environments are called review, QA, or development environments.

As depicted in Figure 8.7, code changes are tested in review environments. However, this is the end of the line; the environments are pure playgrounds that have nothing to do with the production environment and can be structured completely differently. If the changes are approved, they flow back into the main branch and are then tested in a staging environment.

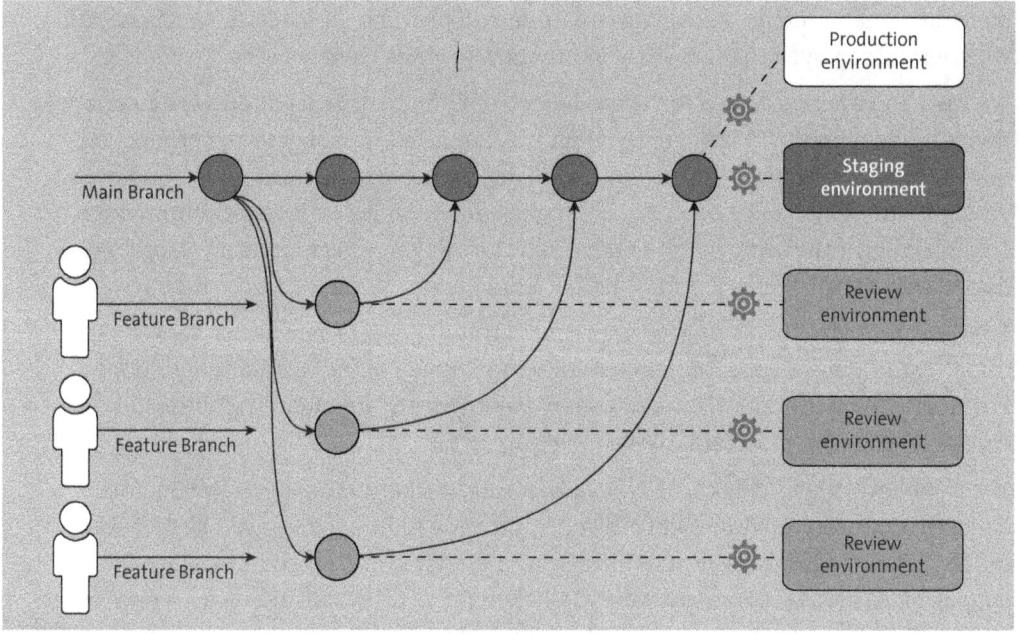

Figure 8.7 Productive, Staging, and Review Environments for Feature Branches

To ensure that the review happens quickly and without friction, the application must be deployed automatically. Every change is tested before it is merged into the main branch. The review covers several aspects at once:

- The project is being built.
- The project is being tested.
- The project is deployed.
- The changes are reviewed.

The review allows many potential sources of error to be identified at an early stage, much earlier than they otherwise would have been. Full automation ensures that problems are identified in good time, which significantly reduces the pressure of deployment to the production environment.

Deployment to the review environment also has the advantage that technical acceptance can take place at an early stage. This means that many tasks that would have been due much later in the overall development process can be moved to a much earlier point in time so that problems and errors can be corrected once those tasks are implemented. Ultimately, this is a seamless transition from QA, discussed in Chapter 7, and it leads into the topic of the operation of the product, discussed in Chapter 9.

Staging Environments as a Preliminary Stage for Production

Once the code review and deployment to the test environments have been successfully completed, the actual test happens: Does the application run as it should in practice?

The application must be tested under the same conditions under which it will later run; even if the changes to the code have passed the code review, that doesn't mean they will work in production.

This preliminary stage before the actual deployment is called *staging*. The main branch is deployed and tested under real conditions.

At *nicely-dressed.com*, there was also a staging environment in addition to the production environment, but there were a number of errors with each deployment for several reasons:

- The first major reason for the numerous problems during deployment was that the production environment differed far too much from the staging environment. This affected not only the infrastructure itself but also the data and configuration.

  The staging environment should be a scaled-down version of the production environment; it should be located in similar networks and it should have the same configuration, perhaps with less infrastructure required for high availability.

  The more different the configurations are, the more likely it is that the team will have to manage special conditions between the automation of the deployments for the staging and production environments, and that errors will creep into the production environment that were not noticed in staging.

- The next essential point is the automation of the infrastructure. It should be possible to populate all environments with new configurations with as little effort as possible. The team should be able to rebuild the entire infrastructure quickly and without much effort.

  The topic of *immutable* infrastructure and *mutable* infrastructure is relevant here. I address this topic in Chapter 9, Section 9.2, which deals with the question of on-premises versus cloud hosting.

- The third point is that the data on the staging and production environments should be as similar as possible. This has less to do with the pure infrastructure and more to do with the data records within the deployed software.

  For *nicely-dressed.com*, this means that as much data as possible from the store in production should also be included in the staging environment—specifically, data on all the products that are sold on the online store. Some data will inevitably have to be cut back, such as real customer data, which should not be found in staging environments to protect data privacy.

> **Note: Test Data in the Staging Area**
>
> In addition to differences in the technical setup and configurations, there is often another problem that cannot be easily resolved: The data on the systems is different. For *nicely-dressed.com*, this starts with the fact that not all products from the online store are included in the staging environment.
>
> For data protection reasons, however, there is no real customer data in the staging environment. It shouldn't *really* be a problem if you work with a few placeholders instead, but especially when it comes to scaling issues, it can't be ruled out that problems will occur in production even though everything worked in staging.

Automate Deployment Environments

You have already learned that an application must be deployed in different environments such as production, staging, and review for each feature branch that is to be included in the main branch.

Of course, there are several ways to deploy to these environments. Unsurprisingly, a manual setup of the environments is not advisable. Instead, you can automatically pull up fresh infrastructures again and again or create *golden images* as an alternative.

Newly Built Infrastructures

One option is to pull up the complete infrastructure required for the environment on demand. For this to be automated, a cloud must be available, regardless of whether it is a private or public environment. In addition, a CI/CD platform in which the individual deployment steps occur must be used. Every time a feature branch is pushed in the source code management system, the pipelines should start to run, build the environment, and deploy the new version of the application.

In practice, however, this process proves to be quite complex: A long chain of tools is required to get the infrastructure ready. First, the virtual infrastructure has to be set up with Terraform. Then, we need a tool such as Ansible, which is also controlled via the pipeline and installs all dependencies so that the application can be deployed.

Depending on the deployment concept, Ansible itself can also be used to deploy the build artifact from the pipeline to the new system. In any case, it must take over the system configuration, which also includes setting up the web server and certificates.

Although this method is certainly possible, it has one major disadvantage: It would be very slow. Setting up the infrastructure and loading the systems take time, and both actions generate high costs; creating a large number of virtual machines requires more power than getting containers up and running.

Another problem is that a large number of tools that need to be mastered by everyone who uses the pipeline are required.

Golden Images

An alternative method is to provide *golden images*, which are images of the entire base system preconfigured with all its dependencies.

A golden image only needs to be started in a cloud or a virtualization hypervisor and is then directly available for the artifact from the feature branch build.

Perhaps the concept of a golden image reminds you of the container principle!

The process is quite similar, even if the underlying technology is different. In both cases, an operating system is required and then the dependencies that are relevant for the operation of the application are installed.

In comparison to the previous example, there are two pipelines here: one pipeline is responsible for building the golden image, while the other uses the last built golden image to start it and deploy the application.

To build golden images, you can use Packer (*https://developer.hashicorp.com/hcp/docs/packer/*) from HashiCorp. The tool was initially under an open-source license until HashiCorp changed the licenses for its projects. Alternatively, you can also use the features of the hypervisor to create an image of a sample virtual machine.

This image must then be stored in a registry, also equivalent to a container image, so that it can be used as a basis when the application is deployed.

The main advantage of this approach is that the pipeline runs much faster. Golden images do not have to be built on demand with every deployment, but it is sufficient if the golden image is built once a day or on demand. The most important elements are in the build artifact anyway.

By regularly rebuilding golden images, errors are detected and security vulnerabilities in the operating system are patched more quickly.

This is particularly useful for production environments because it allows teams to implement the "cattle, not pets" paradigm (see Chapter 9, Section 9.2.1) even in a non-containerized world. This also reduces the need for manual intervention in the running systems, because after some time, everyone on the team realizes that they have to solve problems at the source if problems are to be solved sustainably.

A clean golden image is just as important as a clean container image, so a golden image can also be helpful if you want to implement canary deployments or A/B testing, even if you are not using containers. To do this, the golden images must be built as clean and reusable as possible.

8.2.3 Deployment on Fridays

Should deployments be carried out on Fridays? Developers always joke that Fridays are read-only: no changes, no deployments, nothing that could cause overtime!

This is actually completely understandable: You want to finish work early on Fridays, and you don't want to have to deal with the stress of possible downtime after a deployment to the production environment, which can also happen with minor changes, on the weekend. After all, you want to spend your weekend in peace and quiet, away from work.

But doesn't this logic also apply to every day of the week? You don't want to have to work after hours on any day of the week to solve deployment issues either.

You may have guessed it: The question is not about whether there should be changes in the production environment on Fridays necessarily, but simply about critical dates. For an online store like *nicely-dressed.com*, not only do the day of the week and the time of day affect when to deploy, but also whether a major marketing campaign is currently underway. Christmas business is very important for many companies in the retail sector, so there should be no downtime during this season. So should companies completely avoid deployments during this time?

No, because companies may need to do critical deployments during this season, or during any risky period of time. The goal of a well-coordinated DevOps environment is to ensure the following about deployments:

- Deployments are available in small, manageable parts so that the effects can be estimated.
- Deployments have passed all previous QA stages in the continuous integration pipeline, and the developers are sure that the build works.
- Deployments are accompanied by rollback mechanisms and canary and A/B tests (see Section 8.4).

It is also important to consider when a task can be considered "done" in a DevOps environment. While in a traditional, non-DevOps environment, work is considered finished for most team members once the code has been integrated; according to DevOps principles, everyone continues to work together to ensure that the product reaches the customer smoothly. In other words, everyone continues to take care of the deployment to production and the activation of functions so that they can intervene quickly in the event of any errors.

So in a DevOps environment, you have to internalize a different way of working. If you do continuous delivery and continuous deployment *right*, then Friday deployments are no problem. Nevertheless, you should schedule a lot of time on such critical days so that you merge and deploy changes only when you can monitor and observe the result accordingly. Importantly, only do as much as you and your team are comfortable with. This is a process that does not change overnight. And if you have been deploying on Fridays without issue, that doesn't mean that issues won't arise in the future. So it's better to always take a step back and see what can be improved if you face problems, regardless of which day of the week you deploy on.

> **Note: Reflection**
>
> Remember that the path to continuous delivery is not a purely technical matter. Changing the mindset on a cross-functional team is not particularly easy, at least if you want to achieve real continuous delivery.
>
> Real continuous delivery is still rarely done in practice today. I have observed many companies that claim to do continuous delivery actually deploy maybe every two weeks at the end of the sprint. Although this is much better than rolling out a deployment every six months, it is not continuous delivery!
>
> On both the technical and cultural level, I recommend steadily increasing the frequency of deployments. And if you encounter resistance higher up in the hierarchy, then you should make sure to at least set up a staging environment and work with review environments. It is also important that you accompany the changeover with measurable metrics. This makes it much easier to quantify whether something was successful. Far too often, you tend to see deployments based on the motto, "Hopefully it will work." But hope is not a strategy!
>
> If the deployment works smoothly, then this can be a cornerstone for subsequent discussions in order to continuously deploy not only to staging environments but also to production environments.

8.3 Build Management for Deployments

Applications should always be built from the source code. This means that all necessary dependencies are downloaded automatically when the repository is cloned and the build steps are executed.

8.3.1 The Question of Version Numbers

How and whether version numbers are assigned is a much-discussed topic, especially in the DevOps world. If you deploy often and quickly, the version numbers change quickly and, therefore, lose their significance. The days of Windows 95 and 98, when you could tell the year of the release from the version number, are over.

How a version number is updated and whether there is a version count at all depends on whether "real" continuous delivery is used and whether your software is used externally (i.e., outside your own project).

Unique version numbers are particularly important if your software is used externally as a dependency. So if you publish software that is obtained and installed by others, you should use a unique scheme that is based on semantic versioning (explained in the next section). After all, both you and the users of the software want to know which version offers which features and which version no longer contains certain bugs. Communicate this clearly and strictly adhere to the specifications of the scheme.

However, the fact that a project is used as an external dependency does not mean that it is used only outside your own organization. An organization may develop a project that is used both external and within the organization, such as on another team. If this is not the case, then classic version numbers are not required. This is because with genuine continuous delivery, with the use of A/B tests, canary deployments, and so on, there are numerous versions that can run simultaneously.

Semantic Versioning

Semantic versioning is probably the best-known way of defining version numbers. You can find a longer description of the concept on the website *https://semver.org/*. In summary, a version number using semantic versioning consists of three parts, following this pattern: Major.Minor.Patch.

1. Major is increased when API-incompatible changes are published.
2. Minor is increased when new functionalities that are compatible with the previous API are released.
3. Patch is increased if the changes include only API-compatible bug fixes.

In addition, identifiers for pre-releases and build metadata are available as extensions to the Major.Minor.Patch format.

In practice, this naming scheme does not always work smoothly: New versions are often hidden in patches, and in some cases, minor versions contain API-incompatible changes that do not actually belong there.

This is not so surprising; although the schema reads clearly, it is not always clear how a change should be evaluated. If you are working on a service, for example, an error cannot always be corrected without breaking compatibility. Strictly speaking, this update would then have to be published as a minor version and communicated accordingly—

that is, the error is contained in the existing version, on which users then have to build transitional solutions.

Internal Version Numbers

If your own software has no external dependencies, then you can theoretically manage completely without version numbers, or at least without classic, artificial version numbers. Some kind of versioning is still required, because when using true continuous delivery, you need a version designation to be able to track which version is running on the various environments.

One approach is to simply use the commit IDs from the Git repository. No separate tagging is required for this, as commit IDs can be used to determine the status of the project. Since ideally the project is deployed regularly anyway, the commit ID is actually sufficient in itself.

However, this is not ideal in every case. If necessary, the commit ID can be combined with a version number, implementing an approach similar to semantic versioning. This is particularly helpful when looking at dashboards where various current versions need to be identified using version numbers. The use of the commit ID ensures traceability.

8.3.2 Packaging

There is not much to consider when packaging the software. Of course, the version number plays a role here, but otherwise, only a deliverable package that contains the build artifacts needs to be created.

Different types of package formats must be supported, depending on whether the project is released internally or externally and on the programming language used.

A basic distinction is made between language-dependent and target-dependent packaging. The latter can be a container image as well as a package for deb- and RPM-based Linux distributions.

Note that you don't need to package the project as frequently as you may think. I often see that pipeline jobs run too often and packages are unnecessarily built repeatedly. For example, the pipeline runs for the first time when the release for a staging environment is built and deployed. If that process is successful, the pipeline is restarted, the project is built and tested again, and the artifacts are deployed to the production systems.

If you do it right, you don't need to rebuild the project. In fact, you actually shouldn't rebuild the project if you don't use the exact same package that ran through the staging environment. It is better to build the package once and then pass it through the various environments. This ensures that the package that was tested beforehand is actually rolled out.

A few more factors play a role in how often and to which environments you deploy. It is also important to use a package registry to build the project only once and to pull the artifact from the package registry during deployment. The package registry and the container registry are covered in later in this chapter, but first, it is worth taking a look at containerization in the build process.

8.3.3 Containerization

In container technology, continuous integration and continuous delivery are closely connected, because if a container is built in the continuous integration part of the CI/CD pipeline, it is usually also deployed.

This section does not discuss how and when containers are relevant and practical in operation and when you should use them over bare metal and virtual machines. For a discussion on those topics, jump to Chapter 9, Section 9.2.

Container Building for Applications

Whether you can port your application into one or more containers depends on the architecture of the software. Using containers is usually suitable for web applications such as the *nicely-dressed.com* online shop.

Basically, container construction works like this: All the information required for construction is declared in a construction description, the *Dockerfile*. This is where you first define the *base image* that forms the foundation. In the simplest case, these are very lean Linux operating systems that have minimal functions. This ensures that your containers carry as little extra weight as possible; however, it may be easier to use an image in which a web server such as Nginx or Apache is already preinstalled.

Further information is specified in the Dockerfile, such as the files that need to be copied, where they need to be copied to, the ports that must be opened and forwarded, and the environment variables that must be set. Listing 8.1 shows an example of what a Dockerfile may look like:

```
# Use the official Node.js base image from the Docker Hub
FROM node:14
# Set the working directory within the container
WORKDIR /app
# Copy the package.json and package-lock.json
# into the working directory
COPY package*.json ./
# Install dependencies of your application
RUN npm install ...
 # Copy the rest of the application code to the working directory
COPY . .
 # Expose the port that your application uses (e.g. 3000)
```

```
EXPOSE 3000
# Start application at container startup
CMD [ "node", "app.js" ]
```

Listing 8.1 A Dockerfile That Manages a Web Application

You can use this description to instruct a container engine such as Docker to build and start your application. This means you always have reproducible instructions for the build.

> **Note: Linting for Container Image Construction**
> There are various tools that can link the Dockerfile (i.e., check for errors and inconsistencies). One option is KICS (*https://kics.io/*).

> **Note: The Choice of Base Image**
> You have many options when choosing the basic operating system. You can use a normal Linux distribution such as Ubuntu or Debian; however, it is usually more sensible to use specialized distributions such as Alpine Linux. Red Hat offers specially certified base images at *https://catalog.redhat.com/software/containers/search/*.
>
> Always make sure that the base image really comes from the upstream project and is verified. Do not use container images whose origin you cannot verify for the build process, and remember the basic problems of the software supply chain (see Chapter 11, Section 11.5).
>
> If you use the programming language Go, you don't even need the base image; since compilations of Go projects include all dependencies in a single executable binary file, such a container image can consist of only one file.

In principle, container images should be kept as small and lean as possible, as they are constantly being rebuilt and deployed. The smaller they are, the faster building and deployment run, and the less likely you are to have failures and security problems. And you also save on storage costs for container images.

In addition to the appropriate base image, *multistage builds* should be used when building so that unnecessary build tools do not end up in the container image and bloat it. For this purpose, a clean-up is carried out in the build step, in which all unnecessary tools are removed. In addition, the caching of container image layers ensures that the build runs faster.

Container Engine

The container engine is responsible for building and executing the containers. For a long time, Docker was the dominant player on the market, which is why software containers are still often referred to as "Docker containers" and are always referred to as

"Dockerfiles." However, Docker, Inc., the company that develops Docker, has made a number of decisions in recent years that have been met with criticism, particularly from IT professionals. The Open Container Initiative (OCI) was formed (to which Docker, Inc. also belongs) to formulate standards for software containers.

This is why there are now several alternatives to Docker that can build and start OCI-compatible images. Podman and Buildah are the most widely used of these alternatives, but you can also find numerous other options.

While there are many alternatives to Docker, don't worry too much about which container engine you choose. Yes, there are very important differences between Docker and its alternatives, particularly in terms of architecture. In particular, there is the question of the permissions the container engine itself must have in order to be able to run containers. This is not trivial, but it does not play a big role in day-to-day development. It is more important that the container images are stored, versioned, and managed sensibly so that they can be used for deployment. This is the subject of the next section.

8.3.4 Container Registry and Package Registry

Both the packages and the container images must be saved in a registry so that they can be managed and used in the next steps of deployment.

In my opinion, the registry is the *most boring* tool in the DevOps toolstack. The core functionality is essentially the same for registry tool options. The most widely used tools are Nexus, developed by Sonatype (*https://www.sonatype.com/products/sonatype-nexus-repository/*), and Artifactory, developed by JFrog (*https://jfrog.com/artifactory/*).

GitHub and GitLab also offer their own container and package registries. The main advantage of these offerings is that no complex authorization setups need to be configured, as these registries fit seamlessly into the authorization concept within the projects and repositories. However, these built-in registries do not include all the functions that enterprise customers in particular need. They are certainly sufficient for smaller projects; larger companies are more likely to use Nexus or Artifactory.

Both solutions have everything you need to manage packages and containers. They support a wide range of programming languages and package formats, making administration easy for the development team.

> **Note: Package Managers and Registries**
> In addition to these registries, which can basically manage all artifacts, there are also package managers that are developed specifically for individual programming languages, such as pip for Python. In principle, you can also use individual tools, but then you'd have a hodgepodge of tools, increasing the effort needed for tool integration. This can work in a very small team with a very manageable number of programming

> languages and tools. In a corporate environment with different teams, numerous projects, and dozens of artifacts that are built and updated every day, however, use of such a package manager may make development and operations work more difficult. There is no getting around professional artifact management.
>
> There are also other standalone container registry tools. In addition to the Docker Hub, there is also Harbor (*https://goharbor.io/*), which can be easily hosted in your own infrastructure and focuses only on container images. The common cloud providers also offer their own container registry tools.

Although the topic of storage and administration is not particularly exciting, it is still important to consider; the registry manages all artifacts resulting from development work, so it is important to check which groups of people are allowed to upload, delete, and change which packages. You should also define retention policies that determine how long artifacts are retained. This is very important for the effective use of the CI/CD pipeline, as a number of packages and containers are built every day. It is not possible or necessary to keep everything—there would not be enough storage space to do so. However, you must also ensure that important artifacts are archived. Good versioning is important for this effort.

Dependency Proxy

Another important tool is the dependency proxy, which manages dependencies and keeps them available. The dependency proxy pulls the required dependencies from the primary sources. By using a dependency proxy, developers don't have to pull dependencies from the Internet every time the pipeline is run, saving them time and ensuring that less storage space and network bandwidth is required.

Application Security

Artifacts should be subjected to security scans before they are stored. Both Nexus and Artifactory have their own security scans and can also scan artifacts outside the pipeline. You can then specify that packages or images can be stored only if no (serious) security vulnerabilities have been found, and that dependencies can be pulled only if they do not contain any critical security vulnerabilities.

Although such security features are important for preventing insecure software from being rolled out, don't rely on them too much for application security. The registry becomes active relatively late in the development process, which is not good for visibility. If outdated dependencies with security vulnerabilities are already being used in the development phase, it is far too late to make quick changes, because by then the development work has long been completed. This contradicts the *shift left approach* (see Chapter 11, Section 11.3.2), which states that not only should security scans be carried out as early as possible, but also that visibility should be provided as early as possible.

Therefore, a dependency proxy should work as a transparent proxy so that it applies to all pulls during the development phase. that way, the pipelines can be checked during deployment, and the dependencies that the development team uses locally to build the project can be checked during development. This ensures that only clean packages and container images are used.

As is the case for the container engine, choosing the right security tool is not as important as ensuring that security scans are done in the right place and that their results are evaluated and corrected!

> **Note: Reflection**
>
> You are unlikely to have to make many adjustments during the build process. Packages that are built always end up in the package registry.
>
> The same also applies to container image construction. Make sure that the container images you build are as lean as possible and that they are constantly being rebuilt.
>
> Artifacts that are not stored correctly or are not up to date typically disrupt the course of the project. Make sure that this does not happen!

8.4 Rollbacks, Canaries, and Feature Flags

So far, this chapter has discussed the influence that DevOps principles can have on the technical implementation of CI/CD. A proper CI/CD pipeline primarily ensures trust in the software and processes. Many assume that deployments can be rolled out completely from one day to the next.

But should all changes be released to a deployment in one go, especially if deployments are carried out several times a day? I can reassure you that there are various ways to deploy often but not putting all changes into operation at the same time.

This approach to deployment concerns the differences between continuous delivery and continuous deployment. The two are closely related and are often thought to be the same concept, but there are a few differences:

- With *continuous delivery*, one or more software packages are delivered continuously. This is usually independent of the environment, so it is much more about being able to deliver *something* regardless of whether it goes live.
- With *continuous deployment*, you go one step further by putting the changes into operation. Continuous deployment is about mapping the entire process: from coding to testing to deployment. Continuous delivery is therefore part of continuous deployment.

Take the analogy of a washing machine manufacturer that constantly delivers washing machines to customers. The mere fact of the delivery does not mean that the machines go straight into operation (i.e., are immediately used), which would be the next step.

8.4.1 Rollbacks

If you roll out software several times a day, errors will inevitably occur. They won't necessarily be the types of huge errors that could occur with a six-month deployment schedule, but errors can never be completely avoided. That's why it is important to have a good error culture in the DevOps environment: What do you do when errors happen, and what conclusions do you draw from them?

Although error culture is primarily a human aspect, let's take a look at how we can start at a technical level to minimize errors and the resulting costs.

The use of complete release automation means that not only can new versions be published, but rollbacks can also be carried out just as quickly. All you have to do is run the pipeline to roll out an older version of the software. Of course, rollbacks aren't always the right decision:

- To avoid any downtime when an error is detected after deployment, then a complete rollback to an older version makes the most sense. To do this, you need to know which version is still running properly.
- However, if only isolated errors that do not take the system completely or partially out of operation occur, then it may make more sense to develop a fix quickly and roll it out immediately.

Deciding which path to take in the event of an error is not easy and always depends on the individual case. You absolutely need good monitoring that makes the deployment transparent so that you can assess whether a quick error correction or a retraction of the changes is the best way forward.

It's good that you don't have to make this decision alone in a DevOps team—you can draw on the expertise and experience of both the admins and the developers! Listen to everyone involved; the developer who worked on the feature is probably best suited to assess where the problem lies. It is important to work together constructively in the event of an error and not to get caught up in assigning blame (see Chapter 9, Section 9.4.2, which covers the importance of not assigning blame).

> **Note: Database Schemas**
>
> Probably the most important question with a rollback is how changes to the database schemas are handled. If the database schema is changed with a release and there is no downward compatibility, then a rollback will not work because it will lead to other errors in the application.
>
> In such cases in particular, it is absolutely essential that the development team accompanies the deployment; otherwise, data loss and inconsistency can result from an incorrect rollback.

8.4.2 Step-by-Step Activation Using Blue-Green and Canary Deployments

One way to reduce the number of faulty deployments is to use blue-green and canary deployments. Both types of deployment are technical options for not making all changes available to end users at the same time when rolling out deployments.

Blue-Green Deployments

With blue-green deployments, there are two systems on which the software is deployed in two different versions. The green deployment is active, while the blue deployment is inactive and is slated for the next deployment.

Tests should automated as much as possible and should be run as early as possible. However, not all code can be tested automatically in the review and staging environments, and manual testing can also reach its limits in these environments. Ultimately, you can only really make sure that everything is working after the latest version is deployed to the production environment.

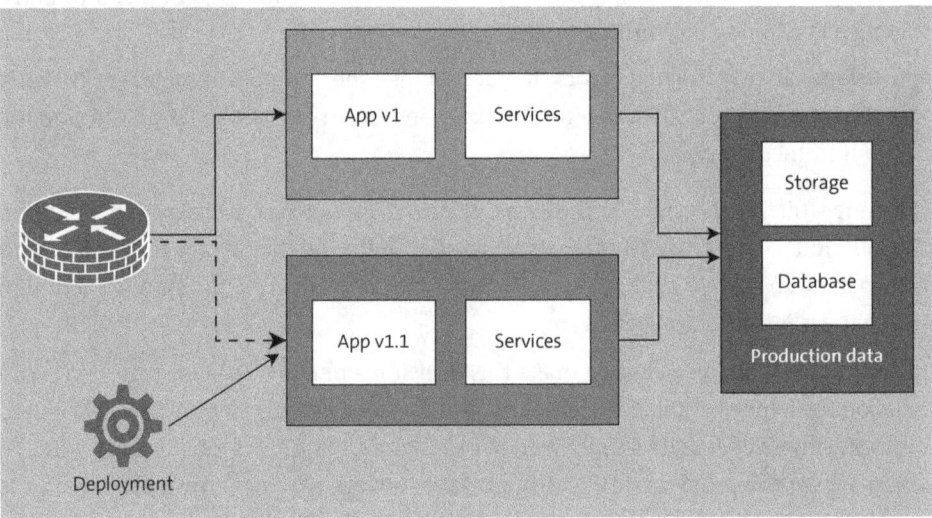

Figure 8.8 Blue-Green Deployments

With blue-green deployments, as shown in Figure 8.8, you can split the deployment of the application and commissioning into two steps. The previous version of the application initially continues to run normally on the production environment and becomes the green environment.

When a new version is deployed, the blue environment is used. The environment is inactive—that is, it can be deployed to—but no traffic is directed to it. This means that only one of the two deployments is active at any one time.

With this approach, there is no great hurry to correct errors, as only the inactive environment has been touched and the active environment continues to run with the old version.

> **Attention: Blue-Green Deployments Are Not Possible Everywhere**
> However, it is important that the software is written in such a way that it can run twice in different versions at the same time and that errors occur where there is no user traffic. As with rollbacks, database schema changes should be viewed with great caution, as such changes will also cause problems.

The advantage of blue-green deployments is that there is direct feedback as to whether the deployment works on a technical level. If not, further deployments can be added at leisure until the errors are corrected. If no errors are detected, the system can be converted so that users can access the newly rolled out version.

This ensures that a functional version is always in operation. If errors do occur, a rollback can be carried out quickly without even having to run the deployment pipeline with the older version. Instead, you simply flip the switch in the load balancer to send users back to the older version of the software.

So doesn't every company use blue-green deployments? It's because it is very time-consuming and expensive to maintain a completely redundant second environment. Especially if there is no scaling through cloud systems, a second environment is usually simply too expensive. Therefore, blue-green deployments are typically used for particularly difficult or large deployments.

In addition, blue-green deployments are not always possible—namely, when major changes to the database schema are unavoidable. For such changes, the architecture of the application must be adapted in such a way that changes to the code are decoupled from the actual changes to the data structure. These changes must be implemented and known and could increase complexity.

Canary Deployments

Canary deployments are an extension of the blue-green deployments. The basic idea and the name originate from the mining industry, where canaries were once used in coal mines. The birds served as an indicator: If too many toxic gases were emitted in the mine, the canaries died, which was a sign for the workers in the coal mine to leave the mine as quickly as possible.

Fortunately, no canaries are sacrificed in canary deployments. As with blue-green deployments, canary deployments have two production environments on which the deployments can be executed (see Figure 8.9):

- The environment on which the previous version runs
- A second, inactive environment on which the new version then runs

As with the blue-green deployments, the environments are switched on or off in binary form. In concrete terms, this means that the environments are either fully active or inactive.

However, end user traffic is handled differently with canary deployments; whereas traffic is directed to only the active environment with blue-green deployments, with canary deployments, a small percentage of traffic is constantly redirected to the new deployment. During this time, the system monitors whether errors occur with real, productive traffic.

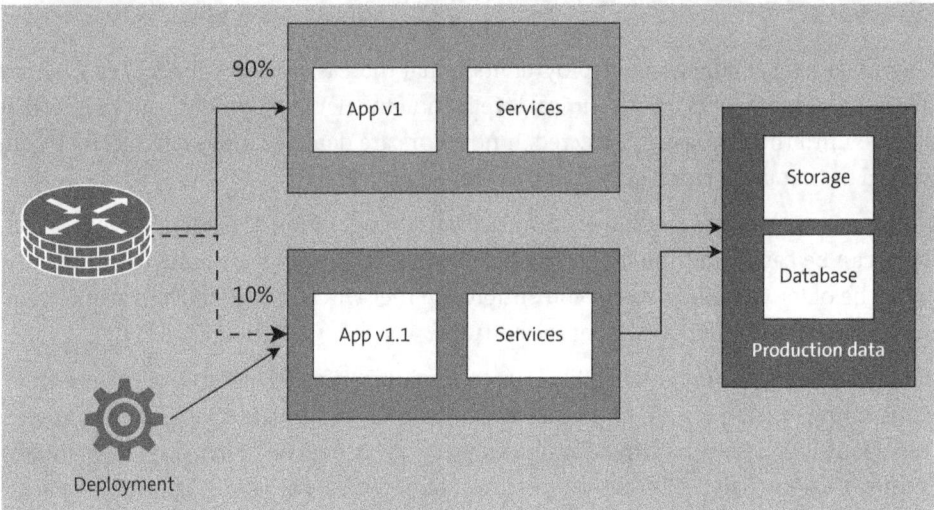

Figure 8.9 Canary Deployments

For a new deployment with a new version, the percentage of user traffic that flows to the new version is gradually increased. There are two options here:

- There is exactly one environment with several servers on which the artifacts are replaced step by step, thus increasing the percentage of new deployments.
- The deployment is rolled out on two environments of the same size, and the traffic is redirected. It is common to use a load balancer, which must be supported by the infrastructure accordingly.

The major cloud service providers offer various strategies for their deployment environments, and canary deployments are also possible via Kubernetes. The challenge is monitoring the system when deployments are rolled out on two environments. Everything can be automated to a certain extent, but that doesn't help much if you don't know whether the conditions for a smooth deployment are being met.

Canary deployments are much easier in theory than in practice, and they depend completely on good monitoring. If you cannot analyze in real time how well the latest deployment is running, you are better off with simple blue-green deployments.

Regardless of the technical implementation, the typical workflow is the same: While you ramp up the deployment, you monitor it to see if any errors occur (see Chapter 10, which covers monitoring and observability). The possibility of errors mainly depends on how many users are currently active and which functions they are using. If there are no errors, the new deployment can be scaled up further. However, if errors occur, the deployment can be turned down again to run the previous version.

> **Note: Keep It Simple**
> Of course, such an approach works only if a sufficient number of users are active in the system. For systems with few users, it doesn't make sense to use this approach.

In addition to being used to detect functional errors, canary deployments can also be used for other tests. One example is performance tests. Although these tests can be run in staging environments on which traffic is simulated, such environments will never be able to simulate real traffic exactly. Canary deployments can be helpful here. If a deployment includes performance improvements, traffic can be gradually redirected to the deployment to monitor whether the performance improvements are visible. If errors occur or the load is too high, the traffic can be quickly and easily reduced in a controlled manner without causing major restrictions for the system users.

Both blue-green deployments and canary deployments are technical implementations to reduce and ideally prevent sources of error. However, both options can be implemented only once there is trust in the release processes, and this cannot be done overnight. As stated in previous sections of this book, get the people on board, then establish the processes, and then implement the technology. It can also be done the other way around, but that only leads to a lot of frustration and little success.

Another option that can also be considered is feature flags, which we discuss in the next section.

8.4.3 Feature Flags

Feature flags, also known as *feature toggles*, allow individual features to be activated after deployment, as shown in Figure 8.10. With feature flags, new or changed functions run with a newly released version after deployment, but old code is still active.

Specifically, this means that there are two implementations in the source code: the previous implementation and the new implementation. A *feature flag service* is used in the source code to check which part of the code should be executed. This provides greater control over the deployment of changes and a stronger separation between the deployment of the application and the actual activation of the changes.

The following case is common for *nicely-dressed.com*: As part of the offer phases for various marketing campaigns, specially developed offer pages must be active at a certain time. This means that when the offer phase of a widely advertised campaign begins, the new version must be running and active, and the deployments had to be carried out at exactly the expected time; this has often not been the case in the past because deployments were rolled out far too rarely.

8 Continuous Delivery and Deployment

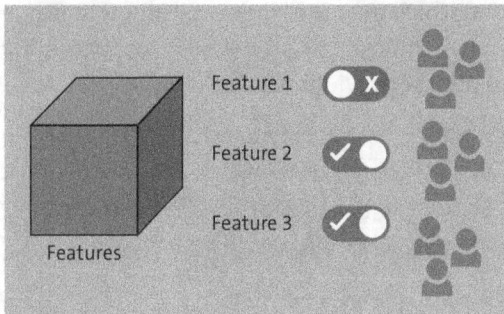

Figure 8.10 Feature Flags

Feature flags can be used to aid in this process. The deployment can be rolled out much earlier, reducing a stress factor. The offer page could be deactivated at the time of deployment, so it has no influence on the live system. At the start of the campaign, someone from the DevOps team would not necessarily have to be present to carry out the deployment. Instead, the switch could be flipped in the feature flag service at the desired time to activate the function.

Feature flags can also be used to enable functions on the production system for individual users for testing. Feature flags do not necessarily have to be binary—they can be used similarly to canary deployments to roll out features on a percentage basis rather than doing a full rollout. With a slightly more complex setup, feature flags can also be switched for individual users or groups so that, for example, internal users can already access a feature that external users cannot, even on production systems.

Of course, this does not mean that all changes should be deployed behind feature flags to production systems without going through QA. Testing in the continuous integration pipeline, deployment to review environments, and code review should still take place even with the use of feature flags. Feature flags should not be misused as a *quality gate*, but should simply be used to provide more flexibility when rolling out new features and help reduce the impact of errors as much as possible.

> **Attention: Deactivate Feature Flags**
>
> If feature flags have been activated for all users and are no longer needed, they should be promptly removed from the code. This is because the use of feature flags means that there are usually two or even more implementations of a feature in the source code. Anything that is no longer needed should be removed in the next deployment. A clean-up must therefore always take place in order to keep the quality of the software high and rule out side effects.

There are different ways of working with feature flags. One possibility is to store flags in configuration files that can be used to activate and deactivate individual functions. The flags would then be active for the entire deployment. However, activating such feature flags would require an additional deployment, so this approach takes comparatively longer. The main purpose here is the technical activation of various functions. However, this approach is not particularly convenient. In addition, this approach is intended only for the complete instance of the application, so it is not very flexible.

An alternative way of working with feature flags is to use a feature flag service. An application developed with feature flags must communicate with a feature flag service via an API to retrieve the data for each flag. Client libraries for common programming languages exist for this purpose, so such queries can be made conveniently. One commercial provider is LaunchDarkly (*https://launchdarkly.com/*), and an open source option is Unleash (*https://www.getunleash.io/*). What is particularly noteworthy about Unleash is that you can finely configure which user or user group should have a feature activated.

> **Note: The Dangers of Automation**
>
> Knight Capital, an investment firm, has lost hundreds of millions of dollars due to the incorrect configuration of feature flags. A deployment was rolled out on eight servers but was successful only on seven. However, the name of the feature flag was used twice: once in the new release and once in the old release.
>
> The old release was still active on one server, which is why seven out of eight servers were operating normally when the feature flag was activated. However, the eighth server then unexpectedly behaved completely differently, which is fatal for a high-frequency trading platform, as a number of sales were carried out that should not have been. As the deployment itself had also been rolled out manually and the monitoring failed, the error became apparent only when a lot of money had been lost.

8.5 Deployment Targets

Where does the deployment end up once the cross-functional development team has built the finished artifact? This section provides a basic introduction to continuous delivery.

8.5.1 Orchestrating Deployments with Kubernetes

If you have organized your deployments in such a way that you can deliver your software reliably and quickly, you have already achieved a great deal. The next step is to optimize the *orchestration* of your deployments, which nowadays always takes place with the help of containers.

The aim is to ensure that all services play together harmoniously, as in an orchestra. Or, to use an image that is a little closer to the IT value stream, imagine a software container (see Section 8.3.3) as a cardboard box that is filled with your software. Once you have a little practice with this, packaging is no longer a major task. Orchestration is the complete merchandise management of your company, which must ensure that there are always enough, but never too many, filled cardboard boxes in the right place at the right time.

If you have ever seen the logistics of a large company, you know how complex and demanding this activity can be. Unfortunately, this is also the case with software containers. The tool that helps with this—the conductor of the orchestra, so to speak—is Kubernetes.

Kubernetes is a big topic in the DevOps context. However, as I wrote in Chapter 1, many people call for Kubernetes in their workflows without first having understood the concepts or techniques surrounding it. So for you to understand the value of using Kubernetes, I'll explain some of the concepts behind it in more detail, particularly about how to deploy, scale, update, and perhaps even roll back applications, because that's the strength of Kubernetes.

> **Note: A Brief Look at the History of Kubernetes**
>
> Kubernetes was initially published in 2014 by Google. Google's proprietary tool Borg served as a model, as Google had been running containerized applications for its services for a long time. Since then, Kubernetes has long been developed independently of Google, by the Cloud Native Computing Foundation.

What Is Kubernetes?

Kubernetes is an entire system for managing container applications with many functions and incredible complexity. It's hard to describe Kubernetes in just a few sentences, but here are the basics: Kubernetes is open-source software that is used to orchestrate and scale containerized applications in a cluster. In principle, Kubernetes is there to run applications that run in containers; think of it as an operating system that is completely independent of the hardware.

What is exciting (and challenging) is the technical possibilities that Kubernetes gives us in the area of CI/CD.

> **Note: The Complexity of Kubernetes**
>
> Kubernetes has a very steep learning curve and requires quite a lot of effort to maintain. Even if the advantages are undisputed, you must first ask yourself whether you can use the features of Kubernetes sensibly.
>
> Kubernetes cannot be learned as a part-time job, and it requires fundamental changes to the software architecture and infrastructure. If you cannot say exactly which problems you want to solve with Kubernetes and how, you should first take a step back and check whether the effort is justified. After all, "Kubernetizing" your own environment as an end in itself will only lead to more work and trouble.

With Kubernetes, containerized applications can be deployed and operated *relatively* easily and in a standardized way. Although software containers are a good idea even without such an automated deployment, they only really show their advantages when they are properly managed.

One such advantage is that Kubernetes can automatically start up more containers with the correct configuration of the application in order to distribute high loads. When the load decreases, it can then scale down again. Imagine that *nicely-dressed.com* did not operate a web store but a traditional department store; Kubernetes would ensure that more checkouts were set up, staffed, and opened depending on when the rush starts. Customers would be directed straight to the new checkouts so that there would be no waiting times and everyone would be served quickly. And when the rush is over, it would reduce the operating area and the number of checkouts.

These deployment features allow possibilities that cannot be implemented in traditional applications, or only with great effort. For example, if you are planning a risky deployment and cannot completely rule out errors, Kubernetes could you perform an automatic rollback if something goes wrong; the techniques described in Section 8.4 can only really be implemented with Kubernetes. Self-healing mechanisms are also included both at container level and the node level. If a node fails, Kubernetes can ensure that the containers are distributed and started on the other nodes. This makes it easy to implement high availability, as Kubernetes can compensate for the failure of individual areas.

Kubernetes offers many other features; always keep in mind that Kubernetes was actually developed by Google to manage its data centers. If your IT infrastructure is smaller than Google's, you will probably never come into contact with many parts of Kubernetes. To get you started, let's take a look at how you can set up Kubernetes in a test environment so that you can take your first exploratory steps with it.

> **Note: Do You Need Kubernetes?**
>
> When you read this chapter, you may get the impression that there is no way around Kubernetes for the technical implementation of DevOps principles and CI/CD. Of course, this is not true.
>
> However, it is true that without Kubernetes, many solutions for deployments have to be built in-house. Deployments using Kubernetes follow a certain standard, but deployments with self-built solutions can look rather wild, as aspects of deployments such as configuration management, provisioning of the environment, deployment scripts, and monitoring.
>
> To deploy Java projects, `mvn deploy` may be executed in the pipeline. This may work, but it covers only a small aspect, because the target virtual machine must also exist and must be provisioned beforehand, such as with Terraform. Configuration management is then required to ensure that the runtime environment is available.

> Are these individual steps in your project so complex that errors and downtime occur frequently? Is the path to deployment so cumbersome that it takes new employees months to understand and eventually master the workflow? Do you need the scaling and deployment features of Kubernetes? Then you should think about using it in your application and infrastructure.
>
> Otherwise, YAGNI—You ain't gonna need it. Not every aspect that I deal with in this book necessarily has to be implemented.

Setting Up Kubernetes

Although Kubernetes can be used in its basic open-source form, commercial distributions are more suitable for productive use, as they do the work for you and provide support. You also need to decide whether you want to run Kubernetes in your own data center or in the public cloud.

> **Note: Documentation**
>
> Kubernetes is extensively documented. At *https://kubernetes.io/docs/home/*, you will find tutorials for getting started as well as more in-depth information. Even professionals cannot possibly know all the options and details about Kubernetes by heart, so the documentation should always be consulted.
>
> The big hurdle is that you have to find the right information for the version you are currently using. Kubernetes is developing very quickly, with features being introduced, changed, renamed, and discontinued on a monthly basis. Even if the basic concepts are now stable, there is still a lot of movement in the details.

Running Kubernetes on Your Own

There are various options for installing Kubernetes on self-managed infrastructure. As with Linux, there are separate distributions that bundle various Kubernetes functions and make them easy to use. Using distributions makes it particularly helpful to update, install, and manage Kubernetes clusters. Therefore, in practice, it rare to see clusters created with Kubernetes in-house resources; commercial distributions are usually used.

> **Note: Vanilla Kubernetes**
>
> The term vanilla Kubernetes refers to the open-source version of the tool. It is the standard version, without any special flavor. Of course, you can also add chocolate sprinkles or caramel sauce yourself; the distributions already include such additional features.
>
> If you only want to manage your clusters with Kubernetes in-house resources, you must use the `kubeadm` tool. This is not easy, especially if you don't have much experience in dealing with clusters—and this applies to essentially everything Kubernetes-related. The distributions offer their own administration tools and often also web interfaces that make your work a little easier.

The most important Kubernetes distribution is OpenShift, developed by Red Hat. OpenShift is a commercial solution that is offered with Red Hat's own support. Although it is based on Kubernetes, it implements some features differently. For example, while standard Kubernetes uses the command line tool `kubectl`, OpenShift uses the tool `oc`.

OpenShift comes with a lot of tools that you have to install and manage yourself in a standalone Kubernetes cluster. A container registry and web interface are already included. OpenShift differs significantly from standard Kubernetes in terms of user and role management, in that it uses role based access controls (RBACs), and in terms of network management. Anyone who is used to the OpenShift route concept, for example, will have problems getting used to the Kubernetes network stack again.

Another commercial distribution is Tanzu, developed by VMware, which offers its own tooling for the installation and management of Kubernetes clusters. A particular advantage of Tanzu is its integration with other VMware offerings; if you are already using vSphere for virtual machine management, Tanzu can be a good alternative to OpenShift.

Yet another important distribution is Rancher, developed by SUSE, which specializes in Linux. For small installations or for learning purposes, K3s (*https://k3s.io/*) is a good lean distribution that requires fewer resources than vanilla Kubernetes.

There are numerous other commercial solutions from other providers. You can find a good overview at *https://nubenetes.com/matrix-table*.

Kubernetes in the Cloud

In principle, you could simply use Kubernetes distributions on a rented server in the cloud. However, if you want to outsource the management of the hardware, you usually also want to outsource all the associated tasks directly. All major cloud services offer their own Kubernetes distributions for this purpose:

- Amazon Elastic Kubernetes Service (EKS)
- Microsoft Azure Kubernetes Service (AKS)
- Google Kubernetes Engine (GKE)

This approach is called *managed Kubernetes*, as the actual administration of Kubernetes is handed over to cloud services. You access the service via command line tools or the web interface to create containers and services and plan your infrastructure, asking questions such as the following:

- How many compute nodes should the cluster consist of?
- What happens in the event of a fault?

Even with a managed Kubernetes solution, the process is and requires good planning and time to familiarize yourself with it.

Finally, you have the option of fully managed Kubernetes clusters. GKE, for example, offers an autopilot mode in which clusters and the underlying compute infrastructure are completely managed by Google so that you don't have to worry about managing the cluster yourself and can concentrate fully on using it. With on-premises systems, such a service would hardly be conceivable. If there is a high load, for example, Autopilot not only scales up the application pods, but can even add new nodes automatically. When these additional nodes are no longer needed, they can simply be shut down again automatically.

This is incredibly practical for anyone who wants to develop and operate the applications without having to think about the underlying stack. The only important thing is that it works and that the automation doesn't have to do any unnecessary work.

Of course, if you go down this route, you will have less flexibility, and the service is more expensive. You need to carefully weigh these factors and determine whether a managed or self-managed solution is better suited to your requirements. With the first model presented here, you are usually billed by the number of compute nodes used; with the latter model, you are normally billed by the number of running pods used (i.e., the containers). Therefore, how your software is structured and what you intend to do with Kubernetes has a direct influence on the costs.

> **Note: Who Operates the Kubernetes Clusters?**
>
> Operating a production Kubernetes cluster is a demanding task, even when using distributions that provide additional tools and support. It is unrealistic for development teams to be able to do this on the side, and a Kubernetes cluster is also a major challenge for the operations teams. For this reason, the operation of the Kubernetes cluster is usually outsourced to a separate team consisting mainly of operations specialists—albeit supplemented with people who know the application and its (container) construction well.
>
> If your team or company is too small to manage your own data center with a Kubernetes cluster yourself, cloud service providers offer a good, albeit rather expensive, alternative. Do not underestimate the administrative effort that Kubernetes entails. A managed service in the cloud can reduce a lot of work in operations, but even then you and your team need a basic understanding of the components that work there.

Components

There are two categories of components that make up Kubernetes: the infrastructure of the cluster itself and the components responsible for operating the applications.

Control Plane and Node Components

In abstract terms, the Kubernetes infrastructure consists of two parts: the *control plane* and the *worker nodes* that are managed by it. (A controller node can also be a worker on which the payload is also executed. This is common for smaller clusters but is a compromise.)

The components of the control plane run on *controller nodes* and are necessary for the Kubernetes cluster itself. Services such as etcd, kube-apiserver, kube-schedule, and kube-controller-manager and a few more components run there.

Every node that provides a regular workload requires at least the *Kubelet* and *Kube proxy*. The Kubelet is the agent that registers the node on the kube-apiserver and introduces it to the cluster; it is basically responsible for ensuring that the containers in a pod actually run. The Kube proxy abstracts the network access and ensures that the nodes can communicate with the controllers and with each other. In addition, a *container runtime* is required on each node, as the containers have to be executed and started somehow.

Containers, Pods, Services, Ingress, and Everything Else

The Kubernetes world brings with it a multitude of terms and designations that are important for understanding the tool, but unfortunately, they are not self-explanatory, are often ambiguous, and are constantly changing. We'll go through some of these terms by looking at the path that traffic takes from the outside world deep into the cluster (see Figure 8.11). This should give you an overview of the typical Kubernetes objects and their uses. Note that I am simplifying some of these points quite a bit.

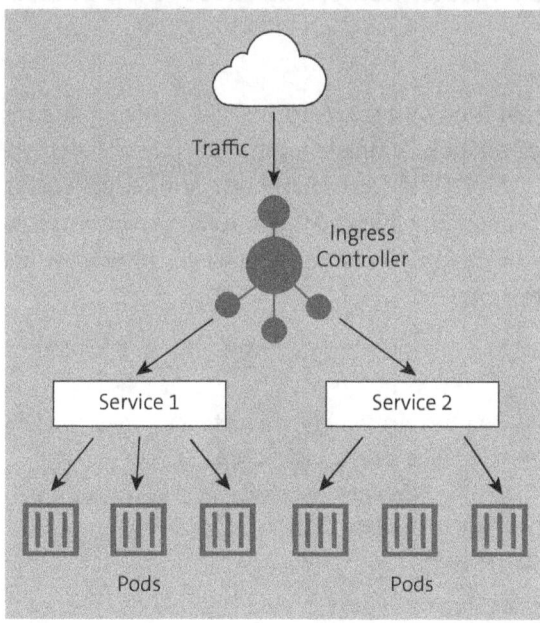

Figure 8.11 Flow of Traffic in a Kubernetes Cluster

When a user calls a URL, the call first ends up at the load balancer or a reverse proxy. Traefik (*https://traefik.io/traefik/*) is a widely used reverse proxy and load balancer in the Kubernetes environment.

However, an ingress *object* must be created so that Traefik knows which route it should listen on and where the traffic should be routed to. This ingress object defines which URL and which path the traffic should be forwarded to which port of which service.

The *service* can also be accessed by other applications within the cluster. In the definition, almost only the service is then linked to the *deployment*.

The term *deployment* in the context of Kubernetes can be a little confusing, as the term also exists outside of Kubernetes for, well, deployments. The specification of a deployment states which containers should run, which and how much storage is required for the application, and how many instances should run. If special configurations such as access rights are required, *ConfigMaps* and *Secrets* must be included there. ConfigMaps are configuration files that are integrated into a container. Passwords and other keys are managed in Secrets.

In addition to deployments, there are also *ReplicaSets*, *StatefulSets*, and *DaemonSets*. For example, a ReplicaSet is created automatically via a deployment to manage the number of replicas. *CronJobs* can also be defined for recurring tasks.

A ReplicaSet creates a *pod*. A pod contains one or more containers that are required to run the application.

And finally, there is also the question of storage. *Persistent volumes* and *persistent volume claims* can be defined for persistent storage, which are mounted by the containers.

Deployment of Applications

A major advantage of Kubernetes is that it offers a standardized API for deployments. While systems without Kubernetes often look completely different for each deployment, Kubernetes ensures some uniformity. This means that everyone on a team using Kubernetes can understand the deployment pipeline and the actions that are required to create it. Having a good overview of the processes can help when developers are familiarizing themselves with new projects.

> **Attention: Access Rights for Clusters**
>
> As the cross-functional teams deploy the application themselves, they must have access to the clusters. However, you should make a strict distinction between the individual deployment tasks, as there are usually different clusters for staging, reviewing, and testing, and production. In this way, you can create visibility, as all team members would have an insight into the internal areas.
>
> Access to the cluster for production systems should be restricted. This can be done via RBACs; OpenShift has capabilities for this type of restriction. In addition, individual areas of the cluster can be demarcated using *namespaces*. Namespaces can be used to logically separate the various objects from one another, which can also be broken down to the authorization structure and thus isolate the teams and their applications from one another.

> Changes to the production cluster must be closely monitored and restricted, but also pay attention to the other clusters, especially if you are working in the cloud: Incorrect configurations can cause a whole cluster to go down, and automatic scaling then wastes a lot of resources. This can quickly become very expensive. Nobody should be allowed to make changes to a cluster without a code review and appropriate change management.

(Automatic) Scaling

In classic systems, scaling resources is a major problem. Custom scripting is often used in classic systems to create virtual machines on the fly and discard them when not in use. Of course, this works only if monitoring systems provide the relevant information directly, which, in my experience, is not usually the case. Instead, those who need to take such an approach to scaling resources in classic systems often have to fly blind.

Kubernetes can handle the work of scaling resources for you through its *HorizontalPod-Autoscaler*, which is used to scale a deployment or a stateful set up or down depending on how busy the system is (see Figure 8.12).

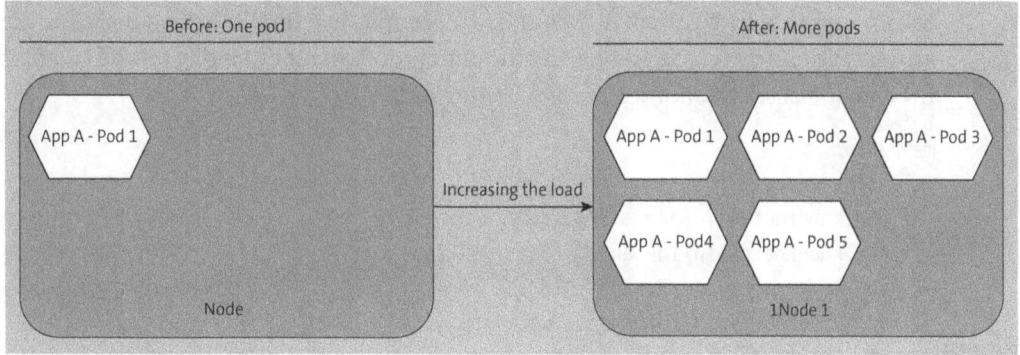

Figure 8.12 Increase of Pods with Higher Loads

Horizontal scaling means that multiple copies of the application are launched. (In our department store example, this means more cash registers are opened.) For this to work properly, your applications must also be designed and written accordingly (see Chapter 9, Section 9.3.4, on cloud-native software). Simply lifting and shifting a monolithic application only leads to an additional layer that makes operation and deployment more complicated.

The practical features that Kubernetes offers provide an advantage only if you are also willing to adapt the architecture accordingly. And in the end, this also includes implementing real CI/CD and DevOps principles.

Updates and Rollbacks

One of the great strengths of Kubernetes is its features for the management of updates and rollbacks. As it manages traffic via load balancers, it is (relatively) easy to configure the way new versions of your application are rolled out. You can find an overview of the different strategies in Section 8.4.

If errors occur after a deployment, you can execute a rollback to the previous version. This is usually quicker and less painful than running the pipeline again. Kubernetes saves the history of deployments in the etcd database, at least temporarily, which you can use to undo errors.

Closely related to this is the *self-healing* abilities of clusters. A configuration can become very complex, as almost everything can be configured with great precision. The basic idea of self-healing is that containers are automatically restarted if applications crash or log too many errors. This ensures that at least the application continues to run, even if errors are not corrected immediately.

Labels and Annotations

Kubernetes uses *labels* and *annotations* in the manifests to describe resources. Annotations in particular take some getting used to at first and can be somewhat confusing, as they are used to perform automation. These relationships are often opaque and difficult to understand.

Labels can simply be used like standard descriptions. Simple key-value pairs are attached to objects to document what the objects do, what they belong to, and whether they are managed by another tool (see Listing 8.2).

```
metadata:
  labels:
    app.kubernetes.io/name: redis
    app.kubernetes.io/instance: redis-dev-testing
    app.kubernetes.io/version: "7.2.1"
    app.kubernetes.io/component: database
    app.kubernetes.io/part-of: store
    app.kubernetes.io/managed-by: helm
```

Listing 8.2 Labels in Kubernetes

You can also enter the name of your team and a contact person to a label to make it clear who cluster admins should contact if they have any questions.

Annotations are much more exciting, as they are used to "hook" into other tools. With such hooks, you can, for example, ensure that the Prometheus monitoring system knows which systems it has to monitor and how. You need to set corresponding annotations on the services that Prometheus listens to so that it knows at which endpoints it can collect the data. Annotations can also be used for the management of ingress routes.

As with labels, annotations can help teams have full control over their application and connect to other tools in the Kubernetes ecosystem.

The Ecosystem

In recent years, a whole ecosystem of additional tools has formed around Kubernetes, almost all of which are cloud native. These tools are not part of the core of Kubernetes (which is already complex enough) and often take on central tasks in cluster management.

Package Management with Helm

Helm is a package manager for Kubernetes that enables you to deploy projects easily. Many applications from the cloud-native environment provide a *Helm chart*, a collection of files that packages a cluster's resources an application, which can be managed by the projects themselves. Helm can also be used to deploy your own application by collecting the manifest files, Kubernetes deployment, ConfigMaps, services, ingress controller, and so on.

Helm is actually a mechanism for templating application deployments. With Helm, YAML manifests can be created as reusable templates so that the same application can be deployed multiple times with different configurations. The configuration for an instance of an application is provided in a *values.yaml* file. This is particularly practical if you want to deploy an application to several environments in addition to the more static production and staging environments, such as temporary review environments.

Other Relevant Tools

The more you use Kubernetes, the more tools that are indispensable in modern infrastructures you will find. These include monitoring and observability tools such as Prometheus and service meshes such as Istio and Linkerd (see Chapter 10). Argo CD and Flux, which we will look at in the following sub-section on GitOps, help with complex deployments.

HashiCorp Nomad

HashiCorp Nomad is an alternative to Kubernetes. Nomad aims to be easier to use and manage; if you are familiar with Kubernetes, you will understand where this desire comes from. One catch with Kubernetes is that you can use it to deploy only containerized applications. This is different with Nomad: it supports both containerized and non-containerized applications.

Otherwise, the core aspects of Kubernetes and Nomad are similar: They both feature efficient resource utilization, self-healing functions, and the possibility of zero-downtime upgrades.

Nomad is quite lean compared to Kubernetes. It consists only of a single nomad binary that needs to be installed on the servers and serves as an agent.

Whether Nomad can be an alternative for you as an orchestration tool depends on whether you already have Kubernetes know-how. If so, you are better off opting for Kubernetes, as it is the much larger, more widespread and established system. On the job market, it is easier to find people with Kubernetes knowledge than with Nomad knowledge. If you are continuing your education, you are more likely to score points with Kubernetes expertise than with Nomad.

Apart from that, Nomad is well suited to small to medium-sized environments and copes better than Kubernetes with traditional software projects or a mixture of new and old projects. If there is nothing to prevent you from relying on a niche solution, Nomad can be a very sensible alternative.

GitOps

The term *GitOps* is a combination of the terms "Git" and "operations." It refers to the operation of an infrastructure with the version management of Git.

Depending on who you ask, you will hear different definitions of GitOps. The term itself is relatively new; it comes from the Kubernetes world and describes the way in which applications are deployed and the associated code is defined and rolled out. The term was coined (and marketed) by the (now defunct) company Weaveworks in 2017. There are now various projects and products that fall into this relatively new GitOps category.

At that time, Weaveworks defined GitOps using these four principles:

- The target state should be defined declaratively.
- The files should be versioned in Git.
- Approved changes should be applied automatically.
- Agents should ensure that the target status is maintained and raise the alarm if necessary.

The idea is that all deployments are defined in YAML files, which are then used by Kubernetes in the cluster. The agents, which also run inside the cluster, are responsible for this.

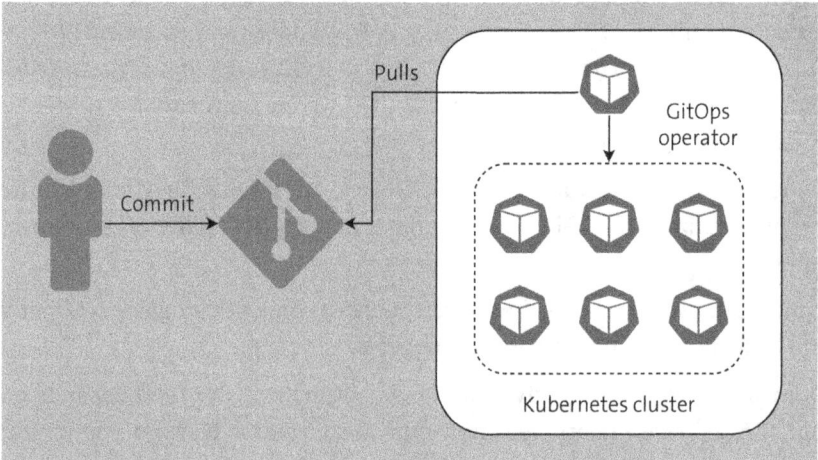

Figure 8.13 GitOps Pattern

Therefore, an agent that pulls the declarative code from a repository and applies it to the system is required. It is important to note that this is a pull principle and not a push principle (see Figure 8.13). (It is the same principle as configuration management with Puppet—see Chapter 9, Section 9.5; so basically, the idea is not new.)

> **Note: Pull versus Push**
>
> The question of how changes are transmitted and processed has a major impact on the setup, especially with regard to network shares. A distinction is made between *push-based* and *pull-based* deployments:
>
> - With push-based deployments, the artifact is copied to the target systems from outside. This happens automatically in a pipeline.
> - With pull-based deployments, the target system itself retrieves its own configuration and required artifacts and carries out the deployment.

A few years later after Weaveworks introduced GitOps, a working group called OpenGitOps was formed and revised the GitOps principles (*https://opengitops.dev/*). OpenGitOps revised the four principles as the following:

1. **Declaratively defined**
   A system managed by GitOps must have its desired state expressed declaratively.

2. **Versioned and immutable**
   The desired state is stored in a way that enforces immutability and versioning and retains a complete version history.

3. **Pulled automatically**
   Software agents automatically pull the desired state declarations from the source.

4. **Continuously reconciled**
   Software agents continuously observe actual system state and attempt to apply the desired state.

Other interpretations of GitOps beyond the definitions given by Weaveworks or OpenGitOps exist. GitLab, for example, understands GitOps as a combination of IaC, merge requests and CI/CD.

Whether the mechanism is pull- or push-based is irrelevant, as both variants ultimately roll out the software. For GitLab, both Terraform and Ansible are GitOps tools, although there is no automatic pull and no automatic synchronization.

Whatever the specific technical implementation looks like, it is important to work with code reviews and to track the history in a Git repository. There is one important point that I have not addressed so far: A cleanly defined GitOps repository for multiple applications deploys everything in one go, whereas with a classic *push* deployment via a pipeline, several pipelines have to be restarted for each package in order to make the deployments. In the event of a major failure, it is quicker and easier to restore the original state.

However, GitOps also has some disadvantages, as it requires a separation between development and operations. So be careful to use GitOps only to solve certain problems avoid building up silos again.

Flux and Argo CD, mentioned earlier in this section, are two GitOps tools that are particularly relevant in the Kubernetes world. Both are agents that run within a Kubernetes cluster and pull configurations to be rolled out from a repository. We'll discuss each in the following sections, but note that you can achieve your goal with either tool.

Flux

Flux (*https://fluxcd.io/*) was originally developed by Weaveworks but is now part of the Cloud Native Computing Foundation.

To use Flux, you must set up a Git repository according to a specified schema. Flux supports pure Kubernetes resources in YAML files and Helm charts. If you use Helm charts, your development team can complete control the rollouts carried out using Flux.

A separate repository is used to define which version is to be rolled out. There are various options for how deployments can be carried out. One possibility is that each version requires a separate commit in which the version number is incremented. However, this is additional work that may not even be needed.

Depending on the environment and requirements, it may also make sense for minor versions to be updated automatically, while major versions can be deployed only after approval. In this case, a new commit would have to be made only for a new major version.

Flux essentially consists of two parts: the agent itself, which runs in the cluster, and a command line tool to manage the agent. However, there is no web interface in the free version, so you still have to use command line tools such as kubectl.

Argo CD

Argo CD (*https://argoproj.github.io/*) is also a project of the Cloud Native Computing Foundation.

In contrast to Flux, Argo CD has a web interface in which the cluster structure can be displayed visually, and it has a somewhat wider range of functions. Also, whereas the Flux agent has to be installed on every cluster that Flux is to manage, Argo CD has a central server that manages deployments to one or more clusters.

While Flux is very much aimed at the operations workflow (partly because it lacks a web interface), Argo CD clearly focuses on the developer workflow. It also has a number of user management functions that are missing in Flux.

8.5.2 Orchestrating Deployments at nicely-dressed.com

The team at *nicely-dressed.com* wanted to make deployments much more flexible and to use modern features such as rollbacks and feature flags. The team therefore decided to implement Kubernetes. However, this meant that a great deal of training was required, as the use of Kubernetes changed the team's way of working significantly.

The team had to change the infrastructure so that everything in it could be scaled up and down automatically; we will encounter this problem again in Chapter 9. First, the team configured horizontal scaling at the Kubernetes level so that more pods were started when the load increased. If there was too much load on the cluster, a new worker node was automatically provisioned and added to the cluster so that the load could be distributed even more widely. This allowed the team to be able to gradually implement techniques such as canary deployments and A/B testing.

The changes to the infrastructure didn't happen overnight; it took some time until it was technically possible to implement Kubernetes. The team also needed to change their way of working.

> **Note: Reflection**
>
> As you can see, the technical implementation of continuous delivery is a complex task.
>
> In particular, while the use of Kubernetes can be exciting, it can be relatively complex. But that depends on whether the problem Kubernetes is being used to solve is also complex. For example, if the load distribution across services is different at all times and a uniform deployment has to be run, then Kubernetes is not such a complex solution. Therefore, you should bear this ratio in mind.
>
> However, if you don't have complex problems to solve, then Kubernetes is probably not for you. My advice is to take a look at your problem and ask yourself what is really relevant. Not every company needs A/B tests and canary or blue-green deployments. Not every software needs to scale strongly. Only by asking the right questions can you find the most appropriate solution.

8.6 Summary

Continuous delivery with DevOps principles is a science in and of itself. Implementing proper continuous delivery is anything but trivial, as many technical requirements have to be met.

This chapter should have given you a good insight into the challenges and opportunities you can expect when implementing continuous delivery from a technical perspective.

There is usually no way around Kubernetes, and this is unlikely to change in the coming years. But it may not be the right solution for every team, so it is important to weigh the options: Do you really need the functions of Kubernetes, and can you operate and support it reliably?

One thing is important in technology: Even the best tools and techniques won't help if employees have little or no experience. Without the necessary experience, the probability of failure is high. And you should always bear in mind that your team may change in the future: Some experienced people will leave, and some newcomers will arrive with different levels of experience. Therefore, it is important to ensure your team has the training to understand and effectively use these tools and techniques.

Chapter 9
Operating the Service

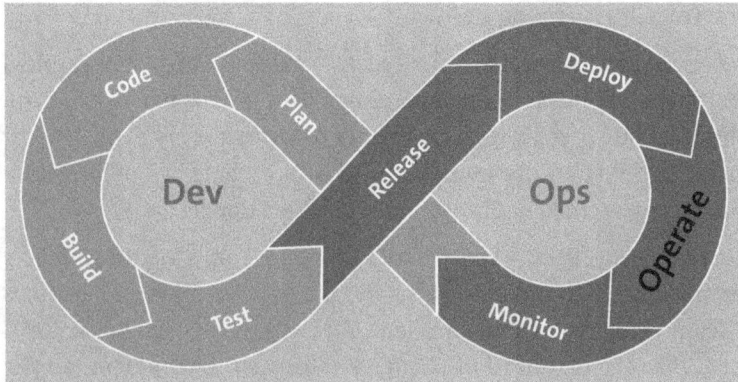

Figure 9.1 The Operation Phase of the DevOps Software Development Lifecycle

Chapter 8 covered the process of completing releases and deploying applications. You have probably already realized that the boundaries between release and deployment and operating the application are fluid.

In this chapter, we discuss the operation of the application. Compared to traditional standalone operations teams, the responsibilities of operations specialists working alongside developers on DevOps-focused teams are different. Let's take a look at how the role of an operations admin changes when a team transitions to a DevOps environment, and what practical impact DevOps principles have on day-to-day operations, for both developers and admins.

9.1 Typical Problems with Operating Services

So far, we have covered the fact that when a team transitions to a DevOps environment, the operations admins are integrated into new cross-functional teams alongside the developers. This leads to an increased merging of development and operations. In this chapter, I focus on the tasks from the perspective of the operations team, which are closely linked to those of the infrastructure team that provides the required infrastructure.

> **Note: Day 0, Day 1, and Day 2**
>
> In the DevOps world, Day 1 is the deployment—that is, the time from which the service is operated. Everything that needs to be done after that falls under Day 2 tasks. What comes before deployment—that is, the requirements analysis and the actual development—are part of Day 0.

9.1.1 Lengthy Infrastructure Planning

As the *nicely-dressed.com* team always planned for six months in advance, this resulted in long-term and cumbersome infrastructure planning. Due to new marketing campaigns and the general increase in sales, it was necessary to estimate very early on what the server infrastructure would look like in the future. Requirements had to be formulated in good time, and servers could not be provided on short notice.

Estimating how much load was to be expected and how the store's visitor numbers would develop over the next six to twelve months was not easy, as the crystal ball was often broken. Each team made various assumptions about these metrics, but none of them used real data. The exact utilization changed with each deployment, so the annual planning process resulted in guesswork.

There were several reasons for this. The monitoring of the systems was just as fragmented as the team structure, so the finance team, the business analysts, and the developers could not obtain the proper data to understand the requirements that would arise from the planned growth.

The infrastructure team at *nicely-dressed.com* was responsible for managing the hardware. The hardware was operated in the company's own data center. In addition to procuring and installing the servers, the infrastructure team was also responsible for replacing defective hardware (i.e., servers), parts of the network infrastructure, and redundant power supply devices.

The infrastructure team worked closely with the operations team, which took over the hardware as soon as it was fully installed and installed the base operating system on it.

As part of the annual planning phase for each new version, a plan was always drawn up for the operations team and the infrastructure team. Every year, support for some of the hardware expired and had to be replaced. As ever higher visitor numbers were expected and new features required more computing power, new hardware always had to be procured.

9.1.2 Hardware Exchange with Obstacles

The infrastructure team had a sufficient budget for the procurement of new hardware, which the team ordered and installed over the course of the year. However, before it could install the new hardware, the team had to replace the old, obsolete hardware first. This required close collaboration with the operations team to ensure the online store could continue to operate during this process.

This collaboration between the two teams frequently led to conflicts, although at first glance it seemed as if both teams were actually pursuing fairly similar goals; both were supposed to ensure that the online store was operated properly. For the infrastructure team, this meant looking after the hardware, and for the operations team, it meant looking after the software before the application was deployed to the servers.

It would have been easy to collaborate effectively to achieve their goals if the two teams hadn't had to deal with different support periods:

- The infrastructure team had to work with the hardware support periods; in some cases, there was no support at all, which meant that replacements were no longer available in the event of hardware failures.
- The operations team had to keep an eye on the operating system's support periods. Over time, more and more security vulnerabilities became known and could not be corrected because operating systems were being used that had almost reached the age of maturity.

It was also common for operating system updates to be installed manually, as there was no automatic patch management; updates were installed only once an admin had logged onto the system—for example, to install a deployment or investigate a bug. It should hopefully be clear that this is not a good idea for security reasons.

The different support periods led to chicken-and-egg problems: The operations team could update the basic operating system only when new hardware was available to ensure the operation of the online store. In turn, the infrastructure team could install new servers only if old servers had previously been decommissioned.

And that wasn't the only dependency. There was also a related conflict between the development team and the operations team: The middleware used could be updated only if the base operating system was also updated. However, this also required the application to be adapted, which then only ran on newer machines with a newer operating system and the newer middleware.

In case you think this scenario is a little too far-fetched, I observed this exact situation in a large company. High walls and questionable architectural decisions led to deadlocks on important updates, as complex (or less complex) dependencies were not resolved. The whole situation was made worse by the fact that no one felt responsible for addressing this problem effectively.

9.1.3 Unfavorable Server Utilization

Those responsible for operations at *nicely-dressed.com* also always had to keep an eye on the development teams' deadlines to ensure that work on the servers was completed before the next major deployment. This always caused a lot of stress, as the few large deployments were planned completely inflexibly and, therefore, depended on the servers' being ready on time. This meant a lot of manual work, as there were very few automated mechanisms to install the servers or even to ensure daily operation.

Another problem for the operations team was the uneven utilization of the servers: Some systems were oversized for their tasks, while others were far too small for the load, which customers also felt when the store was slow to respond. The servers were all dedicated; you could tell from the host name of each server what it was set up for.

The fact that all the hardware was already planned at the beginning of the year meant that the systems were ultimately always running behind the changes. This regularly led to tricky situations in which the entire system or important parts of it failed. For the customers of *nicely-dressed.com*, this meant that they were unable to place their orders, and for the company, it meant that it was unable to generate any revenue.

Some of the increased load was due to changes resulting from the further development of the store. This was very tedious for the operations team, as they were always surprised by changes to the software. In the days and weeks following a major release, this always meant more work and many late-night calls for the operations team because something wasn't working as it should.

> **Attention: Conflicts between Admins and Developers**
> All these problems always led to frustration and finger-pointing. The main point of criticism from the operations team was always that it had received too little or incorrect information about the operation of the application. As each development team used different programming languages and tools, the operations team had to familiarize themselves with the peculiarities of the software and the runtime environment each time a problem arose.

9.1.4 Common Outages during the Night

The worst thing that can happen to an online store is unscheduled downtime. If the store doesn't work, no sales are generated and customers lose trust in the brand and the company.

The operations team at *nicely-dressed.com* was therefore responsible for ensuring that the online store was always up and running. This usually involved an on-call service outside normal office hours. It was not uncommon for the admins to be woken up during the night to address an issue that the monitoring system detected. Let's take an example scenario that is representative of a frequent problem: A service on a server was no longer running. As it was a highly available setup, this particular failure was not critical at the time, as the other services were still running. Nevertheless, an admin had to investigate the problem and correct it during the night. It was particularly important that the error did not also occur on the other servers and cause the online store to fail.

The reason for this issue was often quite trivial: A closer look at the failed service revealed that the filesystem of a server had filled up. After making a little space on the server, the operations specialist was able to restart the service. And fortunately, only this one server was affected and the admin was able to go back to sleep after solving the problem.

He was woken up again the next night: The service was down again. This time, however, the alarm came from a different server. Once again, the disk was full, and once again, the admin was able to restart the service after making a little space on the disk.

The real mistake, of course, was not investigating why the disk was filling up. The admin did only the bare minimum to stop the warnings from happening; the next morning, the actual problem was no longer visible, so it was difficult to find the real cause.

When the error occurred again the next night, the admin finally took the time to get to the root of the problem. The admin discovered that the application had been writing far too much data to a log file in certain cases since the last deployment, which was filling up the hard disk space. The admin was annoyed, as he was woken up for the third night in a row with this problem.

The next morning, he opened a ticket with the development team with the observations he had made. However, because he kept temporarily solving the problem by quickly making space on the disk, there was only his eyewitness report and no log files for the development team to look at. So the development team could not reproduce and fully understand the problem, and it took no action to correct it. Although the development team was informed of the issue, there was nothing it could do about it without further information.

The admin was woken up again the next night, and the same story started all over again: An error occurred, but the developers did not receive prompt, complete feedback to find the cause. Ultimately, such situations only ever led to frustration between the operations team and the development team, as each team always blamed the other.

Of course, both teams played their part in these situations. Any problems that arise should always be tackled together.

> **Note: Reflection**
>
> If you look around your company, you may find that some of the problems encountered on the *nicely-dressed.com* team also apply to your team. Depending on your role, it is likely that some of the problems exist but are not visible to you.
>
> This is particularly the case if you are from the development team and have no insight into the operation of the application. This is often partly due to the fact that the application is running directly at the customer's site and the hurdle is therefore high; however, the problem is also often partly due to silos.

> If you have little or no insight into the operation of the application, simply speak to someone involved in operations and ask them to explain how the application works. It is important to ask questions and to listen.
>
> This also applies in the other direction: If you are responsible for operations, a brief visit to the development team can help to uncover some problems at an early stage, even if your company is composed of separate development and operations teams.

9.2 Breaking Up the Highly Coupled Infrastructure Architecture

In Chapter 8, we discussed how the deployment process can be made faster and more flexible. With these insights, you can start to break down rigid infrastructure planning and create systems that can be scaled and utilized more flexibly.

As is always the case when it comes to infrastructure, fundamental problems cannot be solved overnight. Again, it is important that you look at the whole system and make decisions based on whether they fit into a larger strategy and improve the overall result. It is counterproductive if you rent wonderfully flexible cloud infrastructures but then operate them with the same processes and applications you always used. Simply copying a cumbersome infrastructure to the cloud—an approach known as *lift and shift*—only leads to higher costs.

Automation is all well and good, but if you simply automate bad processes and retain a bad culture, you have gained nothing. Take a step back, look at the actual root of the problems, and find a holistic solution.

At *nicely-dressed.com*, the very strong coupling of the application, operating system, and servers was the main problem. Different support periods, inconsistent configurations, and generally cumbersome planning made rapid upgrades to the infrastructure very laborious. The teams have already improved this situation by significantly reducing lead time by implementing faster and shorter deployments, enabling a more flexible response to changes. This made it easier for the operations team to plan downtime for hardware upgrades. Orchestrating the deployments with Kubernetes (see Chapter 8, Section 8.5.1) simplified the scaling of the application, and the adjustments to the software architecture toward microservices led to fewer load problems.

9.2.1 Cattle, not Pets

One of the problems at *nicely-dressed.com* was that the infrastructure was very rigid and not very agile. The servers had names based on the software deployed on them. The servers were provided with software updates over their long support period, more or less.

In a DevOps environment, developers follow the principle of "cattle, not pets". The basic idea is that servers should not be treated like pets, but like cattle. A server treated like a pet will likely have its own name, and developers will care for it to the best of their ability—just like we regularly feed and pet our pets.

From a DevOps perspective, the infrastructure is instead treated like livestock: In a herd of cattle, none of the animals have a name, and if one of the animals is sick, it is shot instead of nursed back to health. Infrastructure should be treated in the same way: Everything is transitory. If a server is broken, is the development team simply rebuilds and replaces it instead of repairing it, at great cost.

> **Note: "Cattle, not Pets" Vegan Edition**
> If you prefer an animal-friendly alternative to this principle, you can imagine the same scenario with old oak trees and a cornfield: Oak trees live to be several hundred or even a thousand years old and should be protected. Cornfields, on the other hand, are replanted every year and harvested at the end of the season. The same applies to infrastructure: Instead of protecting your servers the same way you'd protect an old oak tree, treat them like a cornfield, which is constantly being replanted.

This may sound a bit harsh—and it is for the animals—but the principle has many advantages: Both the infrastructure and the applications should be reusable and easy to reconstruct, not unique and irreplaceable. Full automation in the code makes it possible to manage the stack efficiently.

By using this principle, if a server does not work, it can easily be rolled out again, which significantly improves the infrastructure's exchangeability and maintainability. If faulty configurations have been delivered, perhaps even on hundreds of servers, the team can roll out a correction much more quickly than repairing each individual server.

> **Note: Throw Away the Server**
> When people talk about throwing away servers, they do not mean that physical servers should be thrown away. That would not make much sense ecologically or economically, nor could it be automated in any meaningful way. The term *throw away* should be understood abstractly as throwing away virtual machines or containers. But flattening and reinstalling physical machines is also done in certain use cases.

All the changes the *nicely-dressed.com* team made to break up the tightly coupled architecture made the work of the operations team much easier, even though they were not directly related to the operation of the application. It was only by looking at the big picture and working more closely together could the team solve these problems. Because even if the company had been given a bunch of new servers with new operating system licenses, team could solve the actual problems only by working together.

9 Operating the Service

Let's take a look at the details. The next question is where to host the software. You can choose between physical servers in your own data center—known as *bare metal*—a virtualized infrastructure, and modern container systems.

> **Note: The Choice of Infrastructure Depends on Many Factors**
>
> I am limiting my discussion of the choice of infrastructure to the technical implementation details. In practice, however, the decision does not depend only on what is technically feasible. For example, if you work with data that is particularly sensitive, then a separate server environment may be the only possible solution. Even if you need to use special hardware, cloud offerings will not be an alternative.
>
> It can also make sense to find a suitable solution depending on the different skills on the team. After all, regardless of the specific solution, it is important that the team understands the entire system in order to be able to act quickly in the event of problems.

9.2.2 Abstracting the Infrastructure

Figure 9.2 provides an overview of the infrastructure options. The simplest option is a bare metal server with an operating system installed directly on it. If the server uses virtual machines, the hardware is abstracted so that several virtualized systems can be used. In the example on the far right of the figure, the virtual machines have also been equipped with several containers. In this case, the link to the hardware is virtually eliminated.

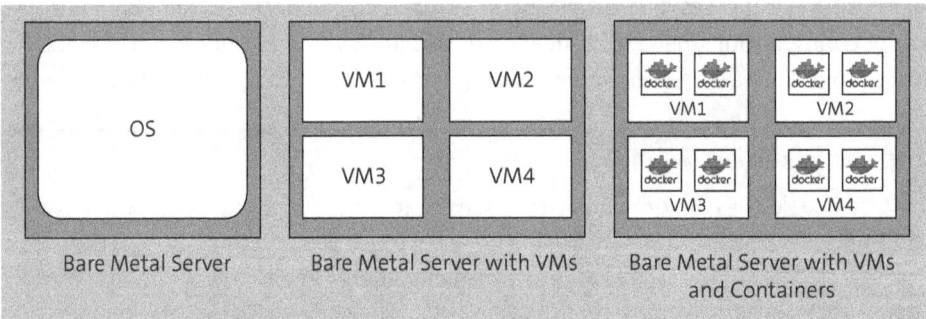

Figure 9.2 Types of Infrastructure

Provisioning and Support of the Hosts

Bare metal servers are the starting point for all infrastructures. Companies use them in their own data centers, but many cloud providers also offer non-virtualized servers for rent. Your servers are then not located on your server rack, but in a data center managed by another company.

Let's first take a look at the on-premises data centers. If you're working with bare metal servers, you have to do a lot yourself, from ordering and installing the servers to all tasks related to the network and storage. You also have to think about using an uninterruptible power supply and good air conditioning so that the servers don't get too hot.

You should automate at least the installation of the basic operating system. To do this, a number of preconditions that take place on the network side must be met. You can then use a tool such as Foreman (*https://www.theforeman.org/*) to provision the bare metal servers with the operating system.

If you want to save yourself the hassle of dealing with hardware and your own data center while retaining the greatest possible control, the use of bare metal servers in the cloud or in a *colocation*—a rented third-party data center—is an option. The main advantage to this option is that a tool stack is provided for provisioning, which can be provisioned with the API and the tools used. This is significantly less maintenance-intensive and more flexible than typical on-premises data centers.

Why does bare metal still make sense as a deployment target for applications in the cloud age? Although virtual machines or containers can be used much more flexibly, real hardware is required for some use cases. This is particularly the case if a separate virtualization stack or special hardware that cannot be rented in the cloud is required. You also don't have to worry about *noisy neighbors*—that is, virtual machines running on the same host that could disrupt operations.

Bare metal servers give you maximum flexibility in their deployment and full control over their performance. This comes at the cost of increased server administration costs and less portability, as everything is relatively hardwired.

Virtual Machines

Virtual machines are the first step toward breaking the close link between deployments and hardware. Rather running directly on bare metal servers, the application is installed on a virtualized computer. These machines are managed by a *hypervisor*, which represents the abstraction layer. Figure 9.3 shows the basic principle.

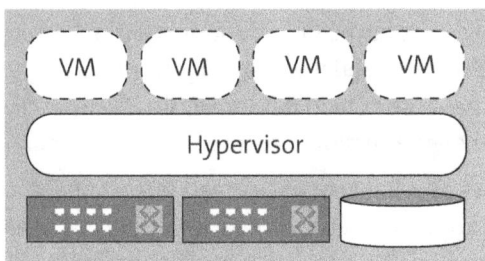

Figure 9.3 Virtual Machines Managed by a Hypervisor

The use of virtual machines makes workloads much more portable. If an underlying server needs to be replaced, the virtual machines running on it only need to be shut down, moved, and started up again somewhere else. The more flexibly the application supports this, the better and simpler the management.

Virtualization can at least solve the problem of servers that don't have much to do. Applications in virtual machines can be operated side by side on one host, which saves costs. This also has a direct impact on the provisioning of the infrastructure and the deployment of applications, as it enables automation that rigid bare metal servers do not allow.

One of the main disadvantages of virtual machines is that you still have to take care of the operating system. Booting up virtual machines and installing and managing the required packages takes time. This can be automated, but the minutes add up. Imagine you want to offer only a small server service, but an entire operating system still has to be kept in a virtual machine and your service has to be installed in the traditional way. This is where containers come in, which we discussed in Chapter 8, Section 8.3.3 and whose operation we will look at in Section 9.2.3 of this chapter in more detail. First, however, there are still a few infrastructure topics to discuss, such as which operating system should be used to run the applications.

The Operating System

The question of which operating system should be used for servers hardly ever arises. Linux has a dominant market position here, while Windows plays practically no role apart from small niches, such as .NET. Knowledge of Linux is therefore very important in the DevOps area.

Linux is offered in many different distributions. For classic servers, the support period of the distribution offering plays the biggest role in the decision. In the corporate environment, Ubuntu, SUSE, and Red Hat and its derivatives are widely used, as they offer reliable support over predictable periods of time. If you can live with restrictions in support, other distributions such as Debian are also suitable, as the technical differences between them and the previously mentioned distributions are small.

For the host system, you can't go wrong with any of the distributions mentioned. The main consideration is that you install the security updates promptly throughout the chosen distribution's lifetime (ideally on the same or next day, or within a week at the latest) and that you have an overview of when the versions are no longer supported.

However, virtual machines and containers also have an operating system. And you already know that containers in particular are very volatile and are never actually used for long. With every run of the CI/CD pipeline, new containers are built and deployed, which are then replaced after a short time. Does it still matter whether the operating system is supported for seven or 10 years if each container lives only for a few days or weeks? The time during which an operating system receives updates is becoming less important in the DevOps world. In fact, many people in DevOps are even moving in the opposite direction, stating that the operating system for containers should be *immutable*—that is, that it cannot be modified. If there are problems with the operating system, they are not repaired; instead, the operating system is thrown away and rebuilt—the "cattle, pets" principle.

Infrastructure for Application Deployments

Once the hardware is ready and the operating system is installed, the application can be deployed together with the required runtime environment.

The challenge of using bare metal servers with regard to DevOps is that their rigid and rather inflexible structure means that it is very difficult for an application running on them to scale. One of the main disadvantages of bare metal is also that the entire hardware is used, even if much less performance is required.

Many applications do not run at full performance, so they do not utilize the available resources of the servers, especially considering that bare metal servers are usually quite large. These servers naturally cost a lot of money, but most users get very little out of them. The setup is not flexible, as applications are assigned directly and permanently to the servers. This means that working according to the "cattle, not pets" principle is not common, and arguably not possible.

Because applications are hardwired to the servers they run on, the various layers of hardware, base operating system, runtime environment, and application are very tightly coupled, which leads to the problems described in Section 9.1. It is not possible to simply replace a server, as the application must first be relocated at great expense.

And since DevOps specialists do not want to have fixed wiring at the technical level, in order to remain flexible, hardly any bare metal servers are operated these days.

This is softened a little in a public cloud, because there are virtually any number of computers that can be used. But if you use a public cloud in almost the same way as your own data center, then you haven't really gained anything—you are repeating the same structural problems, just in a different location.

9.2.3 Containers for Faster Deployments

Software containers have taken the IT world by storm in recent years, as they make it (relatively) easy to package an application with all its dependencies in a standardized format. The package can be rolled out easily and then runs in isolation from the host system and other containers. Such decoupling from the host system is impossible to achieve with bare metal servers; it would be easier to achieve with virtual machines, but they lack the simple options for building such images. (When working with virtual machines in a cloud environment, you should also set up the infrastructure in such a way that the virtual machines can simply be rebuilt, almost like oversized containers.)

You have already learned about containers in Chapter 8, Section 8.3.3, when we discussed the process of building and packaging the application. However, the build process is just one challenge that can be simplified by using container technologies. Nowadays, there are also ways to operate containers that are difficult to implement with direct server hosting or virtual machines. This involves what is known as *orchestration*, or the interaction of many different container services. The most important tool for this is Kubernetes (see Chapter 8, Section 8.5.1), but you can also plan deployments with the home remedies Docker Compose and Docker Swarm if you do not need the features that Kubernetes offers.

The most important advantage of container systems in operation is scaling. Depending on the requirements of the application, you can start more or fewer containers to ensure that load peaks are absorbed and high availability is guaranteed. This gives you better control over the resource utilization of the individual applications, and you can also limit CPU use and memory resources by simple means so that a container does not interfere with other running services on the same host. This is not so easy with traditional hosting; however, the hardware must also play along with container systems and allow scaling—even containers cannot override the limits of physics. This is discussed in more detail in Section 9.3, which discusses deployment in the cloud.

Another core component of containers are health checks, which can be defined when both building and deploying the containers. With health checks, a container can independently check whether the application is running correctly. In the case of a web service, for example, a simple health check can check whether the HTTP status code 200 is returned, throwing an error if not. In the case of an error, the result is used to restart the container to fix the issue. And as long as the health check is not functional, only the healthy containers are served with traffic. Ideally, the end user will not be aware of individual faulty containers. This is helpful, for example, if one container host has problems but the others continue to function normally.

Health checks are one way of dealing with errors. Depending on where the containers are executed, there are a number of other options, such as triggering an automatic rollback to an older container image if health checks fail.

However, these functions of orchestration platforms such as Kubernetes can be used only if the foundation is sufficiently flexible to juggle containers really well and quickly. This flexibility can be found in the cloud, where the application is completely decoupled from the hardware.

> **Note: Usage of Container Orchestrators**
> In the 2022 State of DevOps Report, one thing was clear: Container orchestrators were the most popular deployment target. This includes tools such as Kubernetes and Docker. In the group surveyed, 54% stated that container orchestrators were their primary deployment target 42% indicated virtual machines.

9.3 Cloud Computing

The infamous cloud is often a solution chosen to drive infrastructure automation. What the cloud is, what it is good for, and what it is not good for are subjects of controversial debate.

This section looks at the role of the cloud in DevOps and the improvements that can be made to strengthen team collaboration that were not previously possible without the cloud.

> **Attention: The Costs of the Cloud**
>
> Cost is not a major topic in this book, as it is difficult to estimate and depends on many factors regarding on how the cloud is used. But it is important to note that if you use the cloud only as a copy of your own data center, you are giving away the benefits of such a solution and incurring fairly high costs. Exceptions prove the rule.

9.3.1 What Is the Cloud?

But what is the cloud anyway? The term *cloud computing* is used in an inflationary way, and whether it is a good idea to move your own infrastructure to external servers is a controversial issue. The answer to the question of whether the cloud is *worthwhile* is therefore quite abstract: It depends.

To answer this question for your specific situation, you need to understand what is meant by the cloud in the first place. The definition given by the National Institute of Standards and Technology (NIST) in 2011, which you can find here, is still useful: *https:// nvlpubs.nist.gov/nistpubs/Legacy/SP/nistspecialpublication800-145.pdf*.

The definition basically concerns three aspects:

- Essential characteristics
- Service models
- Deployment models

The definition lists five essential characteristics:

- On-demand self-service: The user can unilaterally request both computing and network services without having to do anything manually on the provider side.
- Broad network access: Access should be available via the regular network so that it can be used by normal end devices.
- Resource pooling. Several users can use the same resources, even if they are running on the same hardware. The resources are isolated from each other. Depending on user requirements, these resources can also be discarded or moved accordingly. It is not clear to the user exactly where their software is running, apart from rough key figures such as the country, city, and data center.

- Rapid elasticity: Resources can be scaled up and down quickly and, in some cases, automatically, depending on demand. In many cases, this feels unlimited and should be possible at any time.
- Measured service: The use and consumption of resources are measured, and reports are generated from these measurements. They can be used for billing purposes.

9.3.2 Cloud Models

Where does the cloud actually run? Cloud computing is often equated with *public cloud* services, offerings from large service providers such as Amazon, Microsoft, and Google, where anyone can become a customer. Billing is based on performance or time used. However, the public cloud is only one of several models.

Some organizations go the *multicloud* route to avoid *vendor lock-in* and prevent outages of a single cloud. Although this is a good idea, in practice it is very complex and expensive. The data is stored in different data centers of different cloud providers, and the different APIs require additional development work. Also, moving data between data centers is time-consuming and comes with high latency and costs.

In addition to the public cloud, there is also the *private cloud*. A private cloud runs in an organization's own data center and fulfills the aforementioned characteristics. The users of a private cloud are individual teams or departments within the organization running it. From a technical perspective, it doesn't really matter whether you work in a public or private cloud. The only thing that matters is that you can use the automation of the cloud, regardless of whether it is provided internally or purchased externally. When it comes to data protection and costs, however, the two offerings differ significantly.

In a *community cloud*, there are several operators who jointly pursue a specific purpose and provide the cloud for this purpose; this model is rare.

Last but not least, there is the *hybrid cloud* model, in which a private cloud is combined with a public cloud. As a rule, the computing power of the private cloud may be sufficient, but if more power is needed, the resources of the public cloud are used.

> **Note: Cloud Usage**
>
> According to the 2022 State of DevOps Report, cloud usage is growing steadily, while the "no cloud" option continues to decline. "Cloud usage" refers to the use of not only the public cloud but also a hybrid cloud, multi-cloud, and private cloud models.
>
> In the future, there will be only a few islands in IT that do not use resources in a third-party data center in some form. You should therefore be familiar with the basic concepts, even if you don't want to send all your data and applications to Microsoft or Amazon at the same time.

9.3.3 Service Models

In cloud computing, a distinction is made between three types of service models (see Figure 9.4):

- Infrastructure as a service (IaaS)
- Platform as a service (PaaS)
- Software as a service (SaaS)

As we are looking at cloud computing from an infrastructure perspective, the focus here is on IaaS and PaaS.

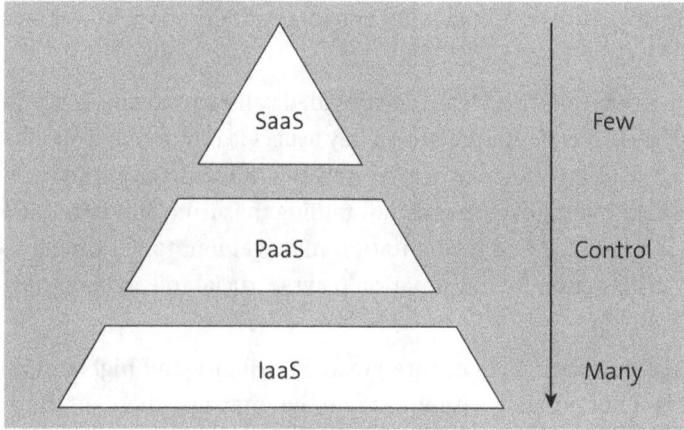

Figure 9.4 The Three Types of Service Models

Infrastructure as a Service

IaaS means that the IT infrastructure, such as the servers, network, and storage, is provided dynamically. This makes it possible for the cross-functional teams to manage almost the entire stack—from the infrastructure to the application—themselves. This applies not only to the production environment, but also to the fast and uncomplicated provision of staging and review environments, which have already been discussed in Chapter 8, Section 8.2.2. Because cross-functional teams can administer the infrastructure themselves, IaaS is an exciting option from a DevOps perspective.

The core idea behind IaaS is that teams should be able to work as autonomously as possible. The best way to achieve this is to rely on complete automation by defining everything as code.

Therefore, teams using IaaS should manage the infrastructure entirely via code as far as possible, including the use of reviews if changes to the infrastructure are necessary. The development workflow described in Chapter 5 is used to manage the infrastructure (see Section 9.5 of this chapter for more detail). The strong coupling of the individual layers is removed; with IaaS, the infrastructure, the operating system, the middleware, and the application are only loosely coupled.

To get this to work, you need to keep the first of the Three Ways in mind: The team must understand the entire system.

Implementing Infrastructure as a Service

Three of the biggest American companies dominate the IaaS sector: AWS, Microsoft Azure, and the Google Cloud Platform (GCP).

From a technical point of view, the provider you choose is largely irrelevant, as the basic concepts behind the offerings are the same. At a higher level, the various cloud providers are fairly interchangeable.

Generally, all of these providers offer the option of starting virtual machines in various configurations at the infrastructure level, including network configurations and shares between systems.

It is more important to consider how the features provided can be used efficiently in order to implement DevOps principles in practice. A key aspect is how much responsibility the teams want to, should, and can take on themselves. With IaaS, the entire hardware operation is outsourced, but managing and maintaining the virtual infrastructure still requires work. This is where IaC and configuration management tools come into play, which we will look at in Section 9.5. Automation plays a crucial role in being able to utilize the benefits effectively.

If cloud offerings are used correctly, they ensure greater flexibility and high service availability. However, this is not possible without a team that manages these offerings and has the relevant knowledge to do so. Use of the cloud does not make operational specialists obsolete; ideally, it relieves them of administrative tasks.

For example, by using a cloud offering, there is no need to manage the hardware. Thus, other expertise will be required; it is more important to be able to write automation scripts that interact with the cloud APIs than it is to be particularly handy with a screwdriver.

As far as scaling is concerned, almost all cloud providers advertise their services as being able to scale up and down dynamically. But do you actually need (automatic) scaling of the underlying infrastructure for your application? This often makes sense, but it is not uncommon to see an overly complex and expensive infrastructure with minimal utilization.

Platform as a Service

The alternative—or supplement—to IaaS is *PaaS*. With PaaS, the team responsible for an application does not have to provision and manage the infrastructure itself, meaning it does not have to manage the hardware, operating system, or middleware. Whether the hardware and stack are managed by an internal team or an external provider is ultimately irrelevant. In a private cloud, this task is performed by the platform team; in a public cloud, the responsibility lies with the service provider.

As these tasks are no longer required, the team can focus entirely on the development and provision of the application. The link between application and platform is still there, but the dependency on the platform and infrastructure is abstracted away. Therefore, with PaaS, it doesn't matter where an applications actually run, as long as the team can create and use their resources automatically.

Exactly which managed services are offered and what the services are called differs slightly across providers, of course, but each offering comes with the basics. For example, AWS Elastic Beanstalk offers many different programming languages, such as Java, .NET, and Python. It also comes with practical functions such as automatic scaling, which means that no manual work is required when the load increases, and high availability is guaranteed at the same time. It also supports other deployment methods such as blue-green deployments.

Persistence of Data

Applications often need to persist data. There are different types of storage systems, each serving different purposes. The most important are block storage and object storage:

- *Block storage* divides the memory into blocks of bytes that are integrated into the operating system, similar to a physical hard disk.
- *Object storage* enables applications to store data via an API without having to reserve the entire file system size. This is more cost-effective but a bit slower.

One object storage system is AWS Simple Storage Service (S3), through which AWS manages the scalable storage of data. The API of AWS S3 is also implemented by other storage solutions, which means that applications that can communicate with an AWS S3 API can also manage data outside of AWS S3. An open-source alternative with an AWS S3–compatible API is *MinIO*.

Another important aspect of storage is databases, which are offered by various cloud providers as managed services for different dialects such as MySQL and PostgreSQL; there are also NoSQL databases, such as MongoDB. Setting up highly available database clusters requires a lot of work, as does configuring backups (and restores), even if you purchase a database as a managed service. High availability requires deep knowledge, especially if you implement the database yourself. Managed services cannot replace dedicated database administrators or database reliability engineers, but they can certainly help them with their work.

Operating Containers

Different cloud providers also offer different solutions for operating containers, so you don't have to worry about the underlying platform and infrastructure. For example, AWS offered eleven different products related to containers in the fall of 2023. It is important to carefully consider which solutions meet your requirements and which may not be necessary.

Manually running your own Kubernetes cluster is demanding, which is why a Kubernetes cluster managed in the cloud can be a viable alternative to reduce the team's workload and focus on the essentials. Cloud providers offer automation to deploy and scale Kubernetes clusters so you don't have to do these tasks manually. However, this makes you quite dependent on your cloud service provider; switching to a local cluster is possible only with great effort. In addition, renting such a cluster is very expensive. You can only find the best way to operate containers for your situation by evaluating your own requirements. You can find some tips on this in Chapter 8, Section 8.5.1.

Challenges with Platform as a Service
PaaS comes with a number of challenges that should be considered. The most important is vendor lock-in; many PaaS services are heavily customized to the respective cloud environment. Once you have made yourself comfortable there, the effort required to leave this environment can be immense. Price increases and changes to service conditions will hit you hard, and if your service provider discontinues the service completely, you will have a problem.

Vendor lock-in can be mitigated by using only PaaS services that are based on open-source projects. A *database as a service* (DBaaS) based on PostgreSQL, for example, is less susceptible to hard vendor lock-in than proprietary solutions.

However, there is always a certain amount of lock-in. There are also always pains associated with a possible migration, whether with proprietary solutions from a specific cloud or with a self-managed solution in your own infrastructure.

In the end, you have to weigh many factors when deciding on a PaaS provider. One key factor should be the use of an open standard.

Lift and Shift and the Question of Software Architecture
One topic that I have not yet covered in detail is software architecture. Many assume that DevOps does not work with monolithic projects or that it works only with a microservices approach.

If you have read the book up to this point, you should realize that software architecture is not so relevant to the overall success of a DevOps-focused team. On a technical level, of course, architecture is important, and it also has an impact on the team structure. But this is just one factor of many that does not affect the general concepts behind DevOps.

However, when discussing a possible migration to the cloud, there are a few points that should not be neglected. You hear a lot of talk about the "lift and shift" approach when migrating an application to the cloud. The approach is simple: The application as it currently functions and is operated is simply "lifted" into the cloud.

With this approach, you don't benefit from the advantages of the cloud. You do avoid all hardware-related work, but the other advantages fall by the wayside. One of the practical features of migrating an application to the cloud is that the systems can be scaled up and down more easily if required. However, this is possible only if the architecture of the application allows it.

If you simply choose a "lift and shift" approach, you will initially only have significantly higher costs without any real added value. In the short term, this is a good approach, but in the medium and long term, the architecture must be converted to a cloud-native architecture so that the benefits of the cloud can be used more effectively.

Software as a Service

SaaS goes one step further than PaaS. With SaaS, the software is offered directly to the customer. For example, instead of operating an online store yourself, you rent a third-party store. This approach is less flexible than in-house development, but it can be a suitable solution if you do not want to worry about the infrastructure and its provision.

For an online store like *nicely-dressed.com*, the big advantage of this model would be that resources are used more sparingly through clever automation, as the load on an online store fluctuates greatly. During seasons with an increase in offers, such as Christmas, and during marketing campaigns, the store always has more traffic than during normal times of the year. An SaaS model can handle fluctuations in load well, as the technical implementation is carried out by the provider of the SaaS service. The disadvantage is, of course, that you have to relinquish control and, therefore, lose flexibility.

> **Attention: Do Not Scale Too Much**
> A common mistake in using SaaS is expending too much engineering effort to enable scaling to what feels like infinity. In most cases, this level of scaling is not needed because there is not that much load on the system. Do not over-engineer until you are pretty sure that you need it.

However, it is rare for a company's entire core business to be outsourced in this way. From a DevOps perspective, it is better to consider purchasing certain SaaS services than to use SaaS for all aspects of development. For example, consider whether certain SaaS services like source code management, CI/CD, and monitoring and observability solutions would suit your team, or whether those services would be better operated in-house.

Determining whether to use SaaS for certain services and which ones is not easy. Provided that data protection and privacy reasons do not play a role, it can make sense to leave hosting to the experts and purchase these services. It is crucial that the costs and risks are carefully evaluated. If the DevOps toolchain is unusable, the entire company comes to a standstill, as no changes can be implemented and deployed.

My advice is to start by analyzing the status quo of your team's day-to-day work and make the decision based on how much pain your team is experiencing with the current setup. The bigger the problems your team has in keeping the CI/CD infrastructure running and up to date, the more sense it may make to go for an SaaS solution.

Figure 9.5 shows an overview of the different service models we discussed in this section and the distribution of tasks for each.

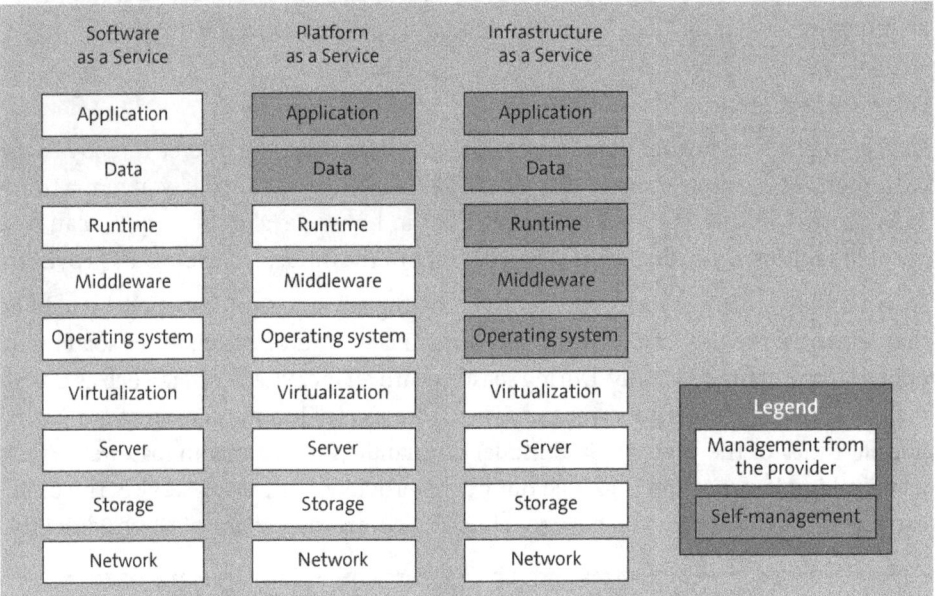

Figure 9.5 Distribution of Tasks with IaaS, PaaS, and SaaS

9.3.4 Cloud Native

I have already explained that lifting an application with a monolithic software architecture into a cloud via lift and shift only causes higher costs and does not provide the benefits of operating an application in a cloud. Software that is explicitly designed for the cloud is called *cloud native*. It is easier to implement the technical aspects of DevOps principles when working with cloud-native software. These services can then be developed more quickly and can be built, tested, rolled out, and rolled back using CI/CD pipelines.

The design elements of cloud-native applications are scalability, elasticity, and flexibility—precisely the advantages that a cloud offers. In very abstract terms, this means that cloud-native applications should not be planned as large, monolithic units, but should consist of many small parts, so-called *microservices*. A DevOps team should be able to develop and operate microservices independently of each other; their interaction is what makes the application.

In a web store, for example, the billing backend should run independently of other parts of the software. If the payment process takes too long because too many customers are placing orders, the team should be able to scale this process flexibly. If there is a really big rush, the team can multiply the instances of the web store and use a load balancer to ensure that the customers are distributed evenly. However, this is possible only if the team already considered how sessions are managed across different instances when designing the application.

Containers are often used because they are technically easy to start and stop and can be scaled up and down quickly on a large number of machines. It makes sense to use containers only if the application can be easily monitored. While applications and servers are predefined in traditional monitoring, instances of the application are constantly being created and disappearing in the cloud-native world. The monitoring system must constantly search for new targets (see Chapter 10 for more detail on monitoring and observability).

Converting an existing application to a cloud-native architecture is very time-consuming and also affects the application's operation. It is usually easier to start directly with a new development that was designed from the outset for flexible use in the cloud.

> **Note: The 12 Factor Application**
>
> The website *https://12factor.net/* by Adam Wiggins provides a very good overview of the principles and methods of developing a cloud-native application.

Before you rush to throw your entire application overboard, know that a microservice architecture of this kind will really come into its own only if the benefits of the cloud are consistently utilized and there is a need for scaling and flexibility. In addition, the entire development and deployment workflow must fit in. The cloud-native world is very fast-moving, and there are often breaking changes that have a direct impact on other applications. So even more care needs to be taken to ensure that there are no breaks in the API between the various services so that they can continue to be managed and developed cleanly. I would argue that you can implement a good DevOps culture without microservices and cloud-native applications, but development of cloud-native applications without DevOps will not work.

> **Note: Cloud Native Computing Foundation**
>
> If you want to learn more about the area of cloud-native computing, look into the Cloud Native Computing Foundation (*https://www.cncf.io/*), which is committed to the interests of cloud-native projects. It is part of the Linux Foundation and is the umbrella organization for several conferences such as KubeCon and CloudNativeCon.
>
> The Cloud Native Computing Foundation assesses the projects under its umbrella with the maturity levels "sandbox," "incubating," and "graduated." The sandbox category contains a large number of fairly new projects. After a certain development period, these projects end up in the incubating category. The most important and mature projects end up in the graduated category, the best-known of which are Kubernetes, Prometheus, and Helm.

> If you look at the Cloud Native Landscape at *https://landscape.cncf.io/*, you will feel almost overwhelmed by the number of projects that are relevant to the cloud-native world. All of these projects work in and with a cloud. Of course, you don't need to know them all, and not all of them were cloud-native developments from the outset. However, you can get an overview of what is happening in the field of cloud infrastructures there.

9.3.5 The Cloud at nicely-dressed.com

Management at *nicely-dressed.com* had a heated debate about what the company's cloud strategy should look like. Many focused mainly on whether the company should go with one of the large American cloud providers or a smaller niche provider. The question of cost was also a recurring issue.

A large part of the online store's business takes place during the Christmas period. During this time, the online store runs a number of offers, so it makes far more money than it does during the rest of the year. The top seller every year is that year's new ugly Christmas sweater.

This peak time has had consequences for the infrastructure: For most of the year, the load on the servers in the company's own data center was low, but every year as Christmas came closer, the load increased. At peak times, the servers' capacities were regularly overloaded, which resulted in slower ordering processes and caused customers to leave the site out of frustration.

By moving the online store's infrastructure to the cloud, the company should solve these problems. Ideally, only as much infrastructure as needed should be available at any given time. The infrastructure needed is significantly less overnight than in the evening, and less in the summer slump than in the pre-Christmas rush. Automated scaling should therefore be helpful in terms of both saving costs and absorbing peak loads. The cloud provider chosen by the company is actually irrelevant, as it depends only on the tools and architecture used. As part of the company's modernization efforts, a lot was switched to container technologies, and slowly but surely a fundamental conversion of the store into a cloud-native application was also considered.

> **Note: Reflection**
>
> Provisioning and managing a cloud infrastructure is a science in itself; you need to very carefully evaluate what your requirements are and what leaps you trust your application and work processes to make. Heaving a cumbersome application into the cloud just once will result only in unnecessary costs and will not change much.
>
> Yes, cloud providers cost money, and a lot of it. However, operating your own hardware also costs money. On the other hand, costs can be optimized in the cloud if you take

> advantage of the benefits that come with using the cloud. It's all about automation and scaling according to demand.
>
> Especially if you are interested in the PaaS and SaaS models, the cost aspect becomes more complicated. Is it cheaper to build up internal expertise in cloud computing—that is, to hire and train a team? Or do you pay a cloud provider a lot of money to get everything ready-made off the shelf? Don't forget that the cloud services also need to be managed, albeit with significantly fewer staff.
>
> Basically, there is always a certain dependency: You are either dependent on your internal colleagues or on the cloud provider.

9.4 Stronger Collaboration between Development and Operations

Back to the topic of DevOps culture! As with the other aspects of DevOps, the operation of services is also about improving collaboration between the different teams. Though there is overlap between the areas of responsibility of the teams that compose a cross-functional DevOps team, different people continue to play different roles and manage their respective tasks (see Figure 9.6).

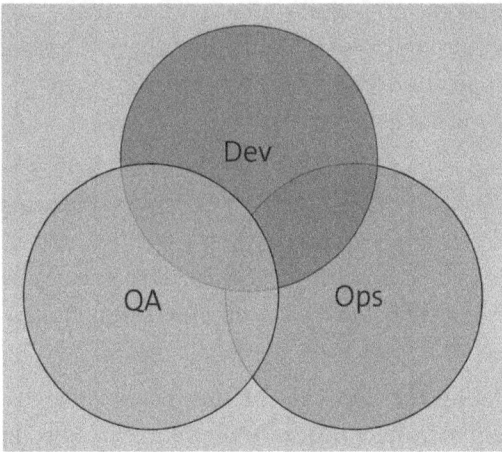

Figure 9.6 Overlap between Development, Operations, and QA

At this point, the following questions arise: What happens during when an application goes down? And how does the team deal with errors?

9.4.1 Everyone Should Be Ready

When a new cross-functional team is formed at a company, you often hear from the operations specialists that, because developers are now expected to carry out deployments, they should be have to be on call too.

And yes, that is correct! On a cross-functional team that works according to DevOps principles, it is essential that developers also take on the responsibility of being on call. It's not uncommon that you hear, "You built it, you run it, you own it." It is entirely in the interests of the team that those who develop the project also run it and are therefore responsible for it.

At *nicely-dressed.com*, a person from the operations team was always woken up when the file system was full, which happened regularly. But once the operations and development teams combined into the new cross-functional DevOps team, the developers were also expected to take on this role.

In a DevOps context, other responsibilities are also distributed throughout the team. In particular, all team members should be able to act quickly in the event of security incidents.

> **Note: Contractual Basis**
> The fact that people take on on-call duty may also have an impact on their employment contract. This issue is important and should not be forgotten. This also applies to working time regulations and adjusted pay.

However, just because there are several people on standby, that does not mean that they all have to be notified when an issue occurs. It is still in the interest of the team that someone who is responsible for the operation is alerted first before they then pass on the job, depending on the given need and the given errors.

Let's go back to our previous example and look at a specific instance of a full file system. An operations specialist on call investigates an error that was reported because the file system was full. They discover that the application is writing a log file that leads to the error. Because the team's operations side is now much closer to the software due to the merging of tasks on the new cross-functional team, the employee also looks at the contents of the log file.

So instead of jumping straight to deleting the contents of the file system, this employee can better assess the fundamental problem because they were involved in the development work. However, just because is the employee has greater familiarity with the source code and the project, that does not necessarily mean that they can solve the problem directly. As first step, it is important to recognize whether it is a serious problem.

If, for example, the log file shows that errors are thrown when customers try to add goods to their shopping cart and are therefore unable to complete their purchases, this error leads to lost sales and dissatisfied customers. In this case, it would be fatal not to alert the developer on standby. This problem can only be solved together and, ideally, a release can be carried out directly to solve the problem.

This would also be possible without DevOps principles, but only if there is open and transparent communication within and between the teams. Without such communication, the teams cannot have meaningful discussions on how to solve such problems. And that is precisely why so many operations teams have to deal with being woken up by problems that should be corrected at the root instead of just treating the symptoms.

On-call service for all has the significant advantage that the team can work on different possible solutions to problems together.

This level of collaboration is usually not as common when development and operations teams are separate and in silos, as these teams usually don't know or work with each other and their focus is on their own team goals instead of the goal of the organization. After all, you are more likely to help a friend than an unknown person.

To summarize, it is important that people with different backgrounds and knowledge are always expected to be on call. DevOps-focused teams should also clearly define who should be informed and when in the event of an incident, in order to deal with such incidents efficiently. This also includes a healthy rotation of the on-call service.

9.4.2 Blameless Post-Mortems

Even when you follow DevOps principles, there will always be incidents that cause downtime. In fact, the question is not whether there will be downtime, but when it will occur and how your team will deal with it. (Incidentally, the question of "how" is completely independent of whether DevOps principles are followed.)

When downtime occurs, the only important questions are why the problem occurred and how it can be avoided in the future. In keeping with DevOps principles, it is important to deal with mistakes openly, regardless of who is ultimately responsible for the issue. Ideally, the team culture is such that everyone knows from the outset that it is perfectly fine for mistakes to happen. An open error culture is absolutely essential for effective team collaboration. However, it is not uncommon to see that when mistakes do occur, they are dealt with in a makeshift manner and then swept under the rug. When dealing with errors in this way, the team can neither find and fix the root of the problem nor learn anything new from the error.

Even though a healthy open error culture is beneficial, bear these two principles in mind:

- Mistakes that have already been made should not be made again.
- It should be possible to trace in detail why an error occurred in the first place so that you can learn from it.

In the DevOps world, this process is called *blameless post-mortem*. It involves documenting what exactly happened after downtime or another serious error, why the error occurred, and why these problems were not noticed beforehand. It is very important that no one is blamed in this process, as it is only human to make errors.

Blameless post-mortems can be written for any type of incident, including not only errors resulting in downtime but also security incidents. A blameless post-mortem is ultimately a formulated document about an incident that recapitulates what happened. Writing a post-mortem is not a punishment but a retrospective to gain new insights.

But what exactly should a blameless post-mortem say? In the first step, it is important to describe the impact that the incident had and the steps that were necessary to correct the error. It is essential that the root cause of the error is found and documented. If a root cause cannot be found in a blameless post-mortem, then it is not a good post-mortem. The necessary follow-up actions should also be defined so that such an incident does not occur again and can be prevented at an early stage.

When exactly should a post-mortem be written? You can define different trigger points, such as the following, that your team can use to determine whether a postmortem should be written:

- Downtime that is also felt by end users
- A security incident in which data was tapped
- Data loss
- A failure of core components of the internal infrastructure

There may be many cases in which a downgrade is made because errors have occurred. It is not necessary to write a post-mortem for every case, because nothing is more unnecessary than doing extra work when there is no apparent benefit. It would be more problematic if a rollback does not work. In this case, a post-mortem may be necessary.

It is important to define proper criteria for your own organization as to whether a post-mortem should be written for certain incidents. These criteria should never be set in stone but, like the source code in general, should always be in constant flux. For example, you can revise these criteria after you realize you've written an unhelpful post-mortem

Many companies and organizations also make their post-mortems public, even though some post-mortems could be categorized as "embarrassing." But that is precisely the spirit of blameless post-mortems: even supposedly embarrassing mistakes are mistakes that shouldn't have happened, but did.

> **Note: A Post-Mortem at GitLab**
>
> A prime example of a public post-mortem was published by GitLab in February 2017:
>
> https://about.gitlab.com/blog/2017/02/10/postmortem-of-database-outage-of-january-31/.
>
> It's about an incident that occurred on January 31, 2017. A database administrator wanted to correct an issue with the synchronization on the secondary server. To do

this, he deleted the directory on the file system with the intention of starting the synchronization from the beginning. Unfortunately, he did this on the primary server instead of the secondary server. The result was that the GitLab website was immediately offline and backups had to be imported.

That incident alone would make a good example of a post-mortem. However, the story does not end there. The backups could not be imported. Although there were different types of backups, many of them did not work because they had never really been tested. In the end, the website was offline for over 18 hours, and there was a data loss of a few hours.

This incident happened a few years before I became a GitLab employee myself. I watched the whole incident unfold from the outside that day. Many of my colleagues from various other companies and organizations were also following the outage. After reading the post-mortem from this incident, many of my colleagues were inspired to check their own backups so that they wouldn't end up in such a mess, too. Sharing post-mortems publicly can therefore be beneficial for both internal and external readers. A blameless post-mortem can be a way to share knowledge and avoid future mistakes at the same time.

9.4.3 Communication Solutions and ChatOps

In this book, I have written about the problems of teams in silos and the importance of effective communication. So far, however, I haven't gone into too much detail about solutions for internal communication within the organization.

It is important to emphasize that communication should not be limited to the use of ticket systems, code reviews, and similar tools. Tools such as Slack, Microsoft Teams, Mattermost, and Rocket.chat should not be seen only as chat platforms. They can be used in distributed organizations to allow teams to other teams questions and to foster collaboration across team and role boundaries. Unfortunately, this aspect of such tools is often neglected in many companies.

Many companies often only consider making a chat tool and perhaps even a video conferencing platform available. Although availability of such tools is undoubtedly important, a healthy communication culture goes beyond simple chatting and video conferencing. Also, there are many potential sources of error in using these tools. I often see companies that create only closed channels and either avoid or even forbid public channels. This reinforces the walls between teams, as it prevents people from simply switching to a channel to ask a question.

Another common mistake is the use of a chat tool as a source of documentation for work processes. Although these tools have a search function, it is often not efficient enough to find and process important information. A chat tool is not a place where decisions and work processes should be documented!

It is even more important that space is created for the exchange of technical and non-technical discussions. There are often people in companies with diverse knowledge of programming languages, frameworks, and tools that they acquired through previous positions or hobbies; although this knowledge may not be currently used in the company, it could be helpful in making future decisions or solving future problems.

If a company promotes an open culture and offers space for the exchange of information outside of its direct business activities, this can also be important from a DevOps perspective.

> **Note: Creative Uses of Chat Tools**
>
> Space should also be created for non-work-related topics or themes. My favorite example is a Slack channel called #all-caps. There is only one rule in this channel: You're only allowed to use capital letters. A channel like this gives people the space to write about trivialities without restraint; they can complain about little things that are going wrong or write about small mistakes they have made and lessons they have learned from them.
>
> By using capital letters, the communication comes across as shouting, but everyone knows that it is not meant to be taken seriously. By chance, processes with stumbling blocks that someone is annoyed about and that should be adjusted often come to light. This is my absolute favorite channel for this reason.

However, communication via chat is not the only thing that a useful communication solution can be used for. Chatbots that can be used to implement ChatOps can also be practical.

ChatOps

The idea behind ChatOps is to use chatbots to simplify the administration of infrastructure and applications. For example, you can use chatbots to trigger deployments via the chat tool of your choice, and also to generally improve visibility of information. As it is very easy to access chatbots, they can be used to enable visibility into systems and processes. Not everyone will be willing to constantly check the CI/CD tool to see if a deployment is in progress, but they may be more willing to quickly get that information from a chatbot. ChatOps can be used as a method of sharing information within a team, simplifying the communication structure.

Such an approach can also be useful for nontechnical employees. These employees may need to obtain information about a deployment even though they may have no experience with the technical implementation of the CI/CD pipeline. With chatbots, they can stay informed about deployments without having to look into a tool with which they have little familiarity.

Typically, alerts from the monitoring system are integrated into specific chat channels that provide some information as well as links to further details. A somewhat more advanced use of chatbots is to trigger and adjust canary and blue-green deployments so that the team always has an overview of the various deployments in the chat tool. This also applies to rollbacks. The chatbot is only the trigger here; permanent logging and configuration take place in the deployment tool to ensure traceability.

> **Note: Reflection**
>
> Tearing down silos and automating deployments and infrastructure are core components in the operation of applications and the infrastructure on which they run.
>
> If you are considering using a cloud, you should also make sure that you effectively use the scaling options. Migrating an application to a cloud without customization is necessary as a first step, but it does not make sense to do so in the medium and long term.
>
> The industry is increasingly moving toward cloud computing. This applies not only to in-house applications but also to third-party services that can be offered and used as SaaS, for example.
>
> However, the cloud is only one aspect of operating an application. Mistakes happen during operation, and you should not sweep them under the rug, but distribute information about them throughout the company via blameless post-mortems, for example. If you are afraid of facing consequences for making mistakes known, you should perhaps start on a small scale: First distribute this information throughout your team, then throughout the larger department, and so on, gradually increasing the number of recipients.
>
> Start by reporting small errors and stumbling blocks on a trial basis so that you can point out larger problems later. This will give you a better feeling for how the company reacts to such error reports.

9.5 Configuration Management: Everything as Code

Throughout this chapter, I have repeatedly stated that infrastructure should be automated as completely as possible. The definition of the target environment should always be available "as code." Ideally, all workflows and environments should be defined as code. This applies to both the infrastructure, with a tool like Terraform, and the configuration management, with a tool Ansible or Puppet, as well as deployments to Kubernetes with a GitOps pattern.

There are two important reasons for defining all workflows and environments as code:

- Doing so allows the target status to be clearly defined. Everyone can see the current status and contribute to a change via pull or merge requests.
- Configurations can be analyzed and reproduced as required.

A common problem with manual or only semiautomated configuration management is that, although many services have been rolled out on many different servers, manual intervention is always required to correct any problems that occur.

Although manual intervention is often a quick and easy solution to the problem in the short term, the changes are almost never applied to other instances of the service on other servers. This leads to inconsistencies—the *configuration drift*. This means that systems that are actually the same behave slightly differently, which makes troubleshooting very complicated.

Instead, configurations should be versioned and rolled out uniformly. If a problem is solved, then it should be fixed not only on the specific server but also centrally via configuration management, which then takes on the task of distributing the target status to all monitored systems.

Otherwise, inconsistencies inevitably arise, which are not only annoying but also inefficient: The same work is done over and over again. An error occurs, it is investigated, the problem is fixed, and then the same error occurs again. That's no fun.

Depending on which configuration management solution you use, the code is either more declarative or more imperative. *Declarative* code describes a target state, whereas *imperative* (scripted) code describes the work steps that are to be carried out. Both approaches have advantages and disadvantages; I prefer a declarative approach because it makes *idempotence* easier to achieve. This means that the code always leads to the same target state when executed multiple times. For example, if the code defines that software should be installed on the target system, the declarative approach means that nothing happens if the software is already present—the target state has already been reached. With an imperative approach, the installation would still be executed again, because that was the command.

The definition of the target environment as code also ensures collaboration between teams: If one team requires a configuration change, it can make a change proposal directly in the repository with the code.

It might make sense to roll out such changes continuously. If that is not the case, people will tend to tinker with the systems manually, which should be avoided.

9.5.1 Infrastructure as Code with Terraform

Terraform can be used to define and create IaC. There are also various other tools, such as Pulumi (*https://www.pulumi.com/*) and Crossplane (*https://www.crossplane.io/*). Ansible, Puppet, and Saltstack, which are mainly known for configuration management, can also be used for IaC, albeit with some limitations.

HashiCorp released Terraform in July 2014. To define infrastructures, the proprietary HashiCorp Configuration Language (HCL) is used. This language was developed in-house and is both machine-readable and human-readable. It is easier to read than JSON and more powerful and less error prone than YAML.

> **Note: OpenTofu**
>
> In the autumn of 2023, the company HashiCorp switched the licenses of its projects—including Terraform—from the open-source license Mozilla Public License (v2.0) to the Business Source License (BSL), which is only "source-available." This move was not well received in the open-source community, so forks of Terraform quickly formed. Work is now underway on a Terraform fork called OpenTofu (*https://opentofu.org/*) under the umbrella of the Linux Foundation. A lot is likely to happen here in the future.

The required resources are defined declaratively in an HCL file. To define the resources, Terraform requires *providers* that translate the HCL code into the instructions of the underlying providers, so to speak. There are numerous providers for all major and many smaller cloud providers.

The HCL file defines the virtual machine on a cloud provider, the use of the cloud provider's database as a managed service, the storage, the DNS and, the virtual networks. Ideally, the entire infrastructure is described in the HCL file so that everything can be provisioned and managed with it.

Terraform projects are then treated like normal software development projects, even if it is mainly the admins who use Terraform. They work in Git repositories and use Git workflows to contribute and review changes. This process brings all the advantages of source code management into play, as the repository not only allows you to see what the infrastructure looks like, but also provides you with a history of how the infrastructure has developed. Every change is documented.

The prerequisite for this is that only Terraform can be used to implement changes to the infrastructure. A pipeline that executes Terraform must therefore also be written in such projects in order to roll out the corresponding changes.

In addition to the pure code, Terraform also requires a backend in which the *Terraform state* is stored (see Figure 9.7). The current configuration is persisted in this file. This allows Terraform to track any changes that have been made compared to the previous version. By default, this file is saved locally and should be accessible to all members of the team to enable collaboration.

In addition to Terraform, other tools are available to implement IaC. Each tool offers a unique approach. Pulumi (*https://www.pulumi.com/*) supports a large number of programming languages and constructs such as loops, so it can be used very flexibly. However, these features also mean that there are many different ways to reach a goal; it usually takes a little while to find your way around in unfamiliar environments and understand what is happening.

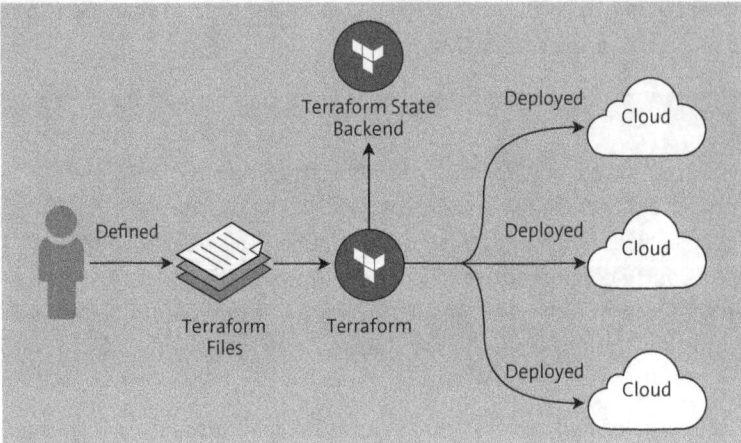

Figure 9.7 How Terraform Works

Getting started with Crossplane (*https://www.crossplane.io/*) is even more difficult. Crossplane relies on Kubernetes and uses its features to pull up the target environments. This is quite complex, but it has the advantage that changed source files are executed directly without having to deal with state backends or pipelines.

> **Attention: Delete Resources**
>
> One problem is how to deal with deleted resources. If you delete files that map the infrastructure, should the infrastructure behind them also be cleared away?
>
> What exactly happens depends on the tool that is being used. For good reasons, some tools rely on the fact that the absence of information does not mean that the infrastructure is automatically removed: Deletions must also be formulated as an order.
>
> Instead, if you work with setups in which missing information is interpreted as a deletion request, you must work very carefully, as errors can have serious consequences. All it takes is a changed name used to identify data access points for data to be deleted in no time at all.
>
> And data loss can occur again and again. That's why a smart backup and restore process is essential. These are typical operations tasks that often fall by the wayside on a DevOps team. If backups are created, they should also be tested regularly. Getting the infrastructure and the configuration of the applications running on it working is all well and good, but if the required data is then missing, you have a serious problem.

Once the infrastructure is in place, the systems that have been provisioned need to be configured and equipped with software. Whether this involves bare metal or virtual machines is irrelevant, as both require additional software with a configuration on their installed operating system.

In the past, such work was done manually, or scripts were written that installed the applications on the systems, rolled out configuration files, and started services in the correct order.

Although this approach is better than using no automation at all, the whole process is time-consuming to maintain and prone to errors. A configuration management tool that abstracts the work with the configurations makes the process more efficient and less error prone. Although such tools bring with them a certain degree of complexity, they can be used efficiently after overcoming a learning curve; then, hundreds of mouse clicks can be replaced by one job. This also makes it much easier to rebuild the infrastructure, as its target status is managed and reviewed.

> **Note: Configuration Management for Infrastructure and Applications**
>
> In principle, you can use tools such as Ansible or Puppet to manage your infrastructure and configure the actual application. With the increasing use of containers and orchestration and scaling tools such as Kubernetes, at least the application configuration is becoming less important, as this is no longer done with classic configuration management. However, the situation is different for the configuration of the base operating system; tools such as Puppet and Ansible are still required for this task.

9.5.2 Ansible versus Puppet

Ansible and Puppet are the most widely used configuration management tools. Of course, there are many others, including Chef and Saltstack. However, because Ansible and Puppet are the most commonly used, I will concentrate on these two in the following sections.

Ansible

Ansible is probably the most widely used tool for configuration management these days; most developers of IaC projects now insist on using it. There are good reasons for this, but Ansible is not always very easy to use. But first things first!

Ansible is available under an open-source license, is written in Python, and has existed since 2012. In 2015, the company that developed Ansible was acquired by Red Hat, which has since integrated the tool into its portfolio.

If you take a look at the concept of Ansible, it should quickly become clear why it has become so successful and widespread. Ansible is based on a very simple architecture (see Figure 9.8), where everything revolves around *playbooks*, which can be easily managed and versioned, and the servers are configured via the Ansible control node.

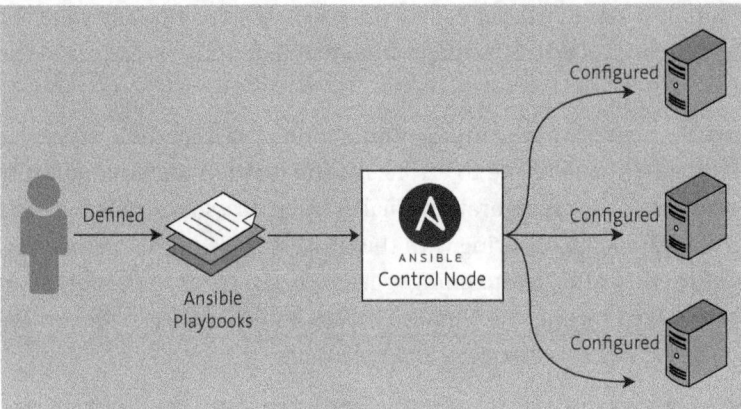

Figure 9.8 Ansible Architecture

The playbooks describe in YAML code the instructions that are to be executed on the systems. YAML is generally very easy to read, but it can become more difficult if too many indentations are needed. When such playbooks are executed, the instructions are processed from top to bottom.

```
- name: Install apache httpd
  ansible.builtin.apt:
    name: apache2
    state: present
    update_cache: yes

- name: Start service httpd, if not started
  ansible.builtin.service:
    name: apache2
    state: started
```

Listing 9.1 Example Playbook Instructions for Installing and Starting a Web Server with Ansible

The example playbook shown in Listing 9.1 defines two tasks. The first task uses the Ansible apt module to install the apache2 package. Or to be more precise, the aim is to ensure that the package is installed. This corresponds to the declarative nature of Ansible. The desired state is always described, and Ansible then ensures that the state is achieved.

The second task is to ensure that the service is started accordingly.

The playbook seems like a script, and that's exactly how it can be described on an abstract level: A playbook is a collection of YAML snippets that are easier to write and read compared to a shell script. The important thing is that Ansible processes the jobs linearly from top to bottom.

Ansible is very easy to use. Only Python needs to be installed on the controller and the target systems. Otherwise, only SSH access is required. This allows a system to be reached quickly in order to execute the Ansible playbooks.

Ansible works according to the push principle: SSH connections are established from the controller node to the target systems in order to execute the sequence of tasks. This keeps the infrastructure for using Ansible nice and lean.

However, Ansible also has a few disadvantages. First, it has disadvantages related to idempotency. Most tasks in Ansible are idempotent. For example, if you define that a configuration file is to be uploaded to the servers and execute it multiple times, the file should be there exactly once. However, because the tasks are always executed one after the other, different configuration files that overwrite each other could be uploaded several times. This is not a problem with simple setups; with a complex setup, however, the execution of the tasks defined in a playbook can become more complicated, so you have to look closely at how and under what conditions the file is uploaded.

Another disadvantage of Ansible is the speed at which tasks are executed. For each task, Ansible must check whether a change is necessary; only then is it executed. Parallelization is not easy to achieve since playbooks are executed from top to bottom.

Playbooks for many servers and with different services are complex to write, and they cause Ansible runs to take quite a while to complete. Waiting for an Ansible run to complete is often a good time to go and make coffee.

Also, if you are unlucky, you may encounter an error shortly before the end of an Ansible run that causes the run to abort. Such errors must be corrected so that they do not lead to a faulty system. However, you then have to run the program again with the corrected code, though you can skip a few steps by using tags.

Puppet

A widespread alternative to Ansible is Puppet, although Puppet's popularity has declined significantly in recent years. Puppet is written in Ruby, which means that many of its language components are closely based on Ruby.

In contrast to Ansible, Puppet works according to the pull principle. The Puppet agent must be installed on the servers to be managed; the agent picks up the configurations at regular intervals and rolls them out on the servers (see Figure 9.9). The agent must be provisioned beforehand, which requires additional tooling.

> **Note: Running Gag about Ansible and Puppet**
>
> There is a running gag among developers regarding which tool should be used: Ansible is practical for rolling out the Puppet agent, as Puppet itself cannot do this.
>
> Getting started with Ansible is pretty easy, which is not necessarily the case with Puppet.

9 Operating the Service

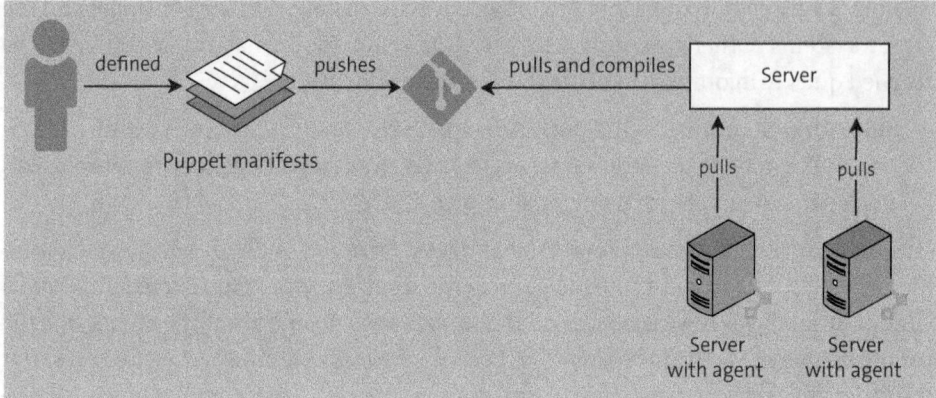

Figure 9.9 The Basic Structure of Puppet

Similar to Ansible playbooks, Puppet's *manifests* declaratively define the target state of the systems in a separate language, based on Ruby. The same applies to the language for templates, which has two dialects, Embedded Puppet (EPP) and Embedded Ruby (ERB).

```
# execute 'apt-get update'
exec { 'apt-update':
  command => '/usr/bin/apt-get update'
}

 # install apache2 package
package { 'apache2':
  require => Exec['apt-update'],
  ensure => installed,
}

 # ensure apache2 service is running
service { 'apache2':
  ensure => running,
}
```

Listing 9.2 An Example Manifest for Rolling out a Web Server with Puppet

The example manifest shown in Listing 9.2 does three things:

- The first block defines that `apt-get update` can be executed. Because Puppet manifests do not execute linearly from top to bottom, this condition is only defined and linked to the corresponding command.
- The second block then defines that the `apache2` package should be installed. However, this requires the package sources to be updated beforehand.
- The third block ensures that the `apache2` service is running.

The advantage of the language for defining manifests is that the entirety of a manifest usually works only if it has been written syntactically correctly. In the example in Listing 9.2, the dependencies of the resources are defined so that Puppet knows in which order the tasks must be executed.

While Ansible processes playbooks from top to bottom, Puppet checks each time whether a manifest are valid. If not, Puppet does not roll out any changes. One advantage of this behavior is that no half-baked steps are executed, as can be the case with Ansible. Puppet works faster than Ansible, especially with complex, multistage rollouts, and since a job cannot fail in the middle, you need significantly fewer attempts to successfully execute a Puppet manifest than to execute an Ansible playbook.

Another advantage is that there is a large ecosystem of modules that Puppet can use. Ansible has a similar ecosystem of modules, but with Puppet, there is a greater separation between the definition of a manifest and the data with which the configurations are rolled out.

If you want to use Puppet, you need a monitoring tool. Since Puppet runs on the pull principle, you need to ensure that manifests are rolled out properly everywhere. If you do not use a monitoring tool to check the rollouts, it is quite possible that you will not notice errors in changes to configurations.

The advantages sound good, but there are a few reasons why Puppet has lost popularity. While you can get started with Ansible with almost no prerequisites, there are quite a few hurdles to overcome before you can get Puppet automation up and running. These hurdles, along with the fact that you need to learn a separate language to write manifests, introduce additional effort and complexity in using Puppet.

I would therefore suggest using Ansible to get started with configuration management and to carry out simpler tasks. If your infrastructure has grown to such an extent that Ansible playbooks become too confusing, you can always consider switching to Puppet.

Conclusion

In addition to Ansible and Puppet, there are many other tools for configuration management. The best-known alternatives are CFEngine, Saltstack, and Chef. Any of these tools can give you the same end result, and all have their advantages and disadvantages; the choice is also a matter of taste.

The important thing is that if you use a tool for configuration management, use it as consistently as possible. The more uniform and standardized the configurations are, the better. And if everything is available as code, the typical developer workflows can be applied to the infrastructure. Changes and updates are checked and secured by code reviews. If an error occurs, the old configuration can be restored by means of a rollback.

> **Note: Reflection**
> Both Puppet and Ansible allow you to define the configuration of systems and applications as code. Ansible is particularly useful if you want to quickly automate recurring tasks on systems. It also works quite well if you do not yet rely on complete automation of the infrastructure in the cloud.
>
> One way to check whether configuration management has been implemented properly is to use SSH. While it was common for a classic system administrator to log in to a system via SSH to adjust a configuration, it is now more likely to make such changes in the Ansible code in the repository. The less SSH is used to look up or adjust configurations, the closer you get to full automation.

9.6 Chaos Engineering

If your team is already working according to DevOps principles and you want to test how good your workflows are, then you should take a look at *chaos engineering*, an approach to testing a system to see how it will react to problems.

Basically, the aim is for software to always run correctly, reliably, and securely, even under difficult conditions. But what you often see instead is the motto, "It will work somehow. After all, we've survived this far."

When teams work by this motto, it means they do nothing more than hope everything is working as it should, rather than actually ensuring it is. Over the course of this book, I have already discussed many steps in the development process that are designed to improve code quality. These steps include code reviews during programming (as discussed in Chapter 5) and use of the different types of tests described in Chapter 7. Deployments to review and staging environments also uncover errors at an early stage.

9.6.1 Making Systems Fail

All these important steps of the development process are good and necessary. Nevertheless, you can go one step further, because many systems are very complex and it is difficult to really understand all the components and all the possible problems.

In particular, you want to know whether you can ensure that the service will essentially continue to function if a subsystem fails. Systems can become quite complex, especially when microservices are used, so it is important to ensure that the system will still work if one or more subsystems fail (for whatever reason).

Many companies set up their systems with high availability but hardly ever test their setups. When failures do occur, you always hear, "That shouldn't have happened!"

Chaos engineering involves deliberately creating "chaos" in order to bring down parts of the system, such as creating a high load and a lot of stress on production systems to see how the system behaves. These experiments can reveal opportunities to make the entire system more robust so that there are fewer problems in the event of actual chaos. Your system should be built in such a way that it still works.

Netflix started using this approach early on in its development. Netflix developed the Chaos Monkey tool (*https://github.com/Netflix/chaosmonkey/*). One of its core functions is that it randomly plays with virtual cables and switches off servers or throws virtual machines and containers out of the production environment.

In addition to Chaos Monkey, there is also the Chaos Toolkit (*https://chaostoolkit.org/*) and, in the cloud-native area, Chaos Mesh (*https://chaos-mesh.org/*) and Litmus Chaos (*https://litmuschaos.io/*). You can find more information on this website: *https://o11y.love/topics/chaos-engineering/*.

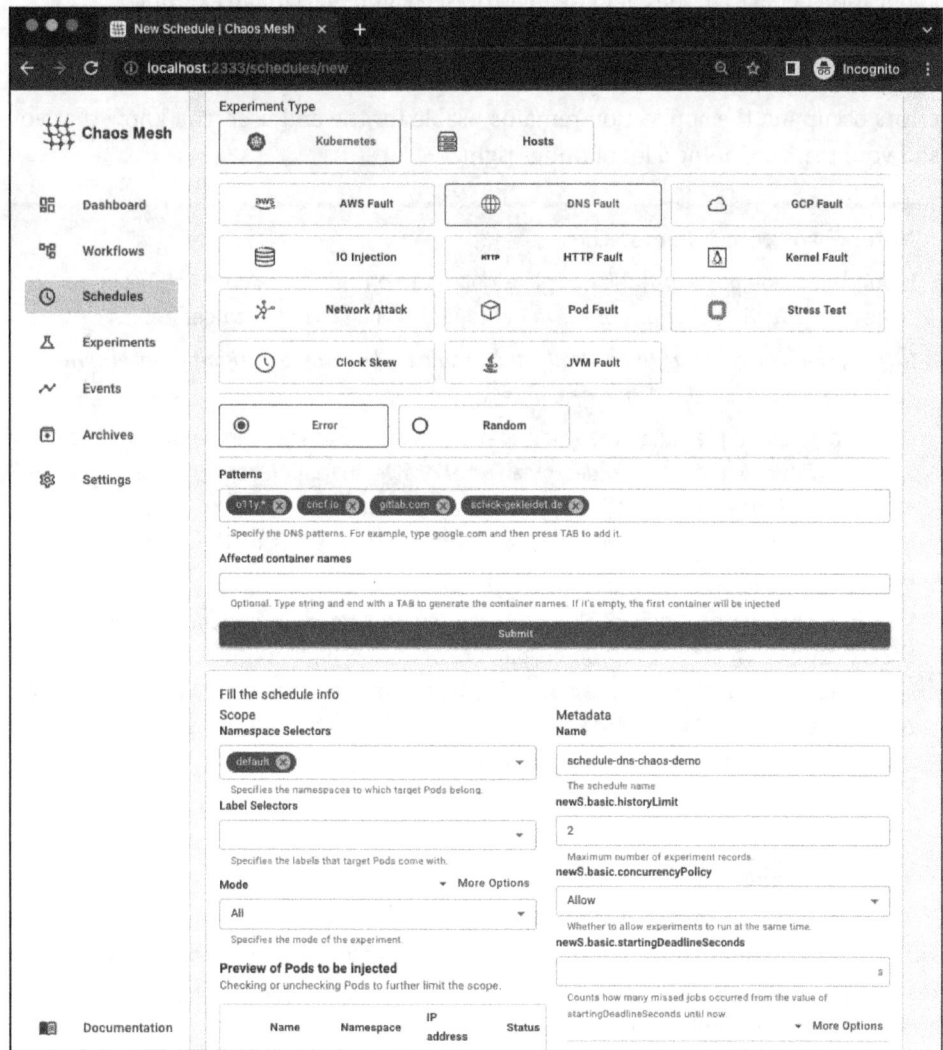

Figure 9.10 Recurring Experiments in the Chaos Mesh UI on DNS from Specific Domains

Although the term *chaos engineering* contains the word *chaos*, the engineering part is anything but chaotic. It has to be planned properly so that the whole process is carried out in a controlled manner. This is the only way to learn anything new from these experiments. Chaos engineering can be used to test a system at various levels: the infrastructure, the network, and the application. Tests are carried out at each level; during an infrastructure test, for example, servers are randomly switched off, or a high load is generated to see how the system reacts.

During a network test, the network bandwidth between various systems is throttled, or random packets are dropped. Both of these cases can occur frequently, so you should test them in a controlled manner in the production environment. In extended chaos experiments, DNS and HTTP responses can be overwritten in order to evaluate how the application behaves in the event of an error (see Figure 9.10 for an example).

Finally, during an application test, chaos experiments test how the application behaves under a heavy load or what happens when nonsensical requests are sent to it. If no errors occur and the application remains usable for the end user, you know that you and your team are doing a lot of things right.

> **Note: Chaos with Pokémon Go**
>
> I found this blog post from the Google Cloud team on Niantic's launch of Pokémon Go to be an interesting example of the advantages of using chaos engineering:
>
> *https://cloud.google.com/blog/products/containers-kubernetes/bringing-pokemon-go-to-life-on-google-cloud?hl=en/.*
>
> The Google Cloud team had assumed that the worst-case scenario would be that the planned traffic would be exceeded by a factor of five. In reality, however, Pokémon Go received 50 times more traffic than expected. A lot of things didn't work right at launch, so they had to deactivate some load-intensive parts of the game on short notice.
>
> I don't know whether the Google Cloud team were already using chaos engineering back then and whether they would have tested such high traffic. But the example shows how helpful such tests can be. Of course, you can't test everything—that's just the way it is—but introducing a bit of chaos before launch can give you more confidence in the launch and in your own product.

9.6.2 Chaos Engineering without Chaos, but with a Plan

In order to avoid actual chaos when carrying out chaos engineering, you should work with a concrete plan so that you can proceed in a targeted manner and can measure the results. The Principles of Chaos website (*https://principlesofchaos.org/*) specifies four steps for practical applications (see Figure 9.11):

1. **Definition of the status**
   Formulate a definition of the stable system in which the normal behavior is expected.
2. **Definition of the hypothesis**
   Formulate a hypothesis that the stable state will continue to exist in both the control group and the experimental group.
3. **Adaptation of variables**
   Implement real events that upset the state of the system. Inject errors such as failing servers and hard disks, network connections that no longer function properly, and completely overloaded subsystems.
4. **Disproving the hypothesis**
   Compare how the system has changed and evaluate the system's behavior using metrics.

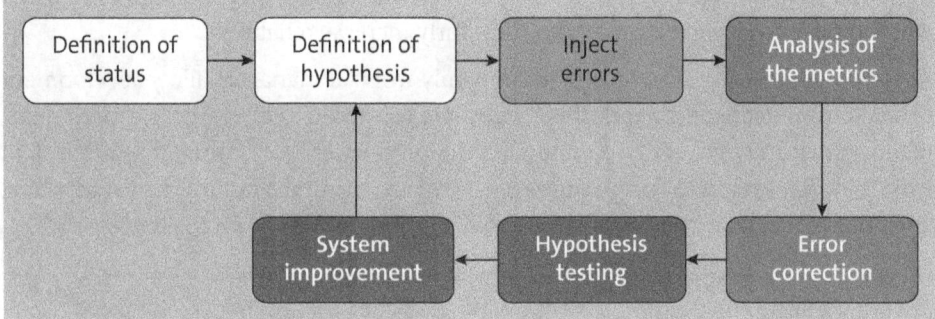

Figure 9.11 Chaos Engineering Step by Step

After evaluating the system's behavior, you can then determine the next steps. The principle of chaos engineering assumes that the testing is carried out on production environments. You may not carry out the testing on production environments at first; instead, you can do this initially in staging environments.

Before you carry out chaos testing, you should fully understand the entire system. You can then divide problems into these four categories:

1. Known knowns: These are problems that are known and understood in the organization and among employees.
2. Known unknowns: These are problems that employees are known to not fully understand.
3. Unknown knowns: These are parts of the system that the organization understands but that contain problems the organization isn't aware of.
4. Unknown unknowns: These are problems that the organization is neither aware of nor fully understands.

Regardless of chaos engineering, as much as possible should fall under "known knowns." It is important to minimize the other three categories as much as possible. Chaos engineering can help with this.

One advantage of chaos engineering is that it can increase the robustness of the entire system. However, you also need to have the confidence to run it on production systems. Even with the right monitoring and the right metrics, you should consider choosing a time to run chaos engineering when there is little load on the systems anyway in order to minimize the risk of failure. What is often ignored is the fact that with controlled failures, as is the case with chaos engineering, the errors can be corrected much better, faster, and cheaper than if errors occur under high load with only real users.

Chaos engineering gives teams a much better understanding of how the entire system works, where there may be problems, and how they can be corrected to actually find problems in good time, especially problems that are not quite so obvious.

Once you have carried out these tests once or several times, you can also automate the whole process and run it regularly to constantly increase reliability.

As is so often the case, an implementation only works if managers allow development teams to conduct experiments. If managers are too afraid to allow these experiments, it shows that they generally do not trust the production environment. But a lack of trust is a good reason to run these tests—if you can cause the failures you're afraid of, you can address their root causes and avoid major problems down the line!

> **Note: Reflection**
>
> Chaos engineering behaves like blue-green deployments and feature flags in the operation of an application and infrastructure, which we discussed in Chapter 8, Section 8.4; they are ways of working that you can carry out only if you have full confidence in your team and your application.
>
> As a first step, you should experiment with chaos engineering methods on staging environments before moving on to production environments. This alone will provide you with many insights, even if your organization still considers running such testing on production environments to be too dangerous.

9.7 Reliability Engineering

Reliability engineering is another role in the DevOps environment, although it is only vaguely defined and is split into different specialties. Probably the best-known role is the site reliability engineer, which was largely developed and shaped by Google.

9.7.1 Site Reliability Engineering

Google never provided a very clear definition of the site reliability engineer, as the role looked different depending on the given team and product. However, as this type of role has become widespread outside of Google over the years, the main principles can be summarized in general terms.

Site reliability engineers are engineers who develop and support large distributed systems. They take care of many aspects of both software engineering and systems engineering, with a focus on the reliability of the application and the entire system.

This may sound like a fancy term for the traditional role of system administrator, but the details are exciting: Site reliability engineers work closely with the development teams to strengthen systems in the areas of error susceptibility and scalability. They also support the teams in the implementation of monitoring, testing, and deployments, particularly with regard to operational stability.

Whether you use the term *site reliability engineer* or not, Google has succeeded in focusing on the availability of the application, which is the key point for users. For users, it doesn't matter whether Google Maps doesn't work because a developer has deployed a faulty feature, or whether Gmail doesn't show any emails because an admin has done something wrong. The only thing that matters to users is *that* it works, and everyone is responsible for making that happen.

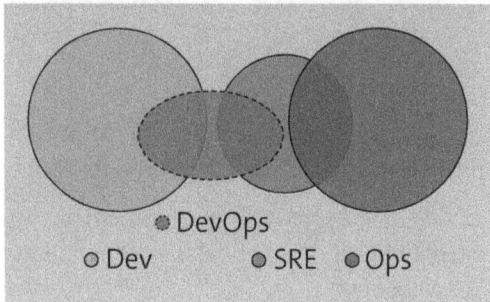

Figure 9.12 The Role of Site Reliability Engineers Compared with Development and Operations

The activities of site reliability engineers therefore lie between development and operations (see Figure 9.12). Overlap exists primarily between site reliability engineers and the operations team. In this type of collaboration, site reliability engineers take on more responsibility for the operation of the application and have a much stronger development background than traditional administrators on the operations team. As a result, aspects such as testing, which have always been important in development, have been reorganized under new terms such as fault injection and chaos engineering (see Section 9.6).

This type of collaboration means that the site reliability engineers also take over the on-call service for the application. Because of the site reliability engineer role's close proximity to software engineering with a focus on reliability, the developers do not necessarily have to be available on call.

In the event of incidents, site reliability engineers are responsible for debugging and troubleshooting, generally ensuring that the incidents are resolved; and they are the central communication bridge between the various teams in order to introduce and deploy the necessary corrections. Site reliability engineers should therefore bridge the gap between development and operations by having a strong software engineering focus—that is, they can program sometimes, but they are mainly concerned with the stability and scalability of the application.

> **Note: Books on Site Reliability Engineering**
>
> At *https://sre.google/books/*, you will find various books by Google that describe how the site reliability engineers work there. They contain many interesting insights.
>
> But don't forget that Google is a very big company that has to scale all the services it operates to a very high degree, almost without exception. The likelihood is that your company will not face the same challenges. Nevertheless, many of Google's ways of working can be adapted to your situation, and many of Google's ideas can also be helpful in smaller environments.

9.7.2 Database Reliability Engineering

Database reliability engineers work according to a similar principle, only that their work concerns only the database for the application and not the application itself.

Close cooperation between database reliability engineers and development is also required to ensure that performance problems identified by the database reliability engineers are corrected accordingly by the developers.

Good software engineering skills are also required so that database reliability engineers can efficiently bridge the gap between the database queries written by the developers and the database itself.

Unsurprisingly, database reliability engineers also need to know the application they are helping to maintain on a software engineering level. They may not be developing the application, but they provide valuable support to the development team to ensure that the application runs reliably with the database and can scale accordingly.

So again, collaboration and culture are important! What the technical implementation looks like is basically of secondary importance.

9.8 Summary

This chapter looked the benefits of bringing the development and operations team together and at the importance of their collaboration in the operation of an application.

Closely related to the operation of the application is the topic of monitoring the system landscape, which we deal with in Chapter 10.

Chapter 10
From Monitoring to Observability

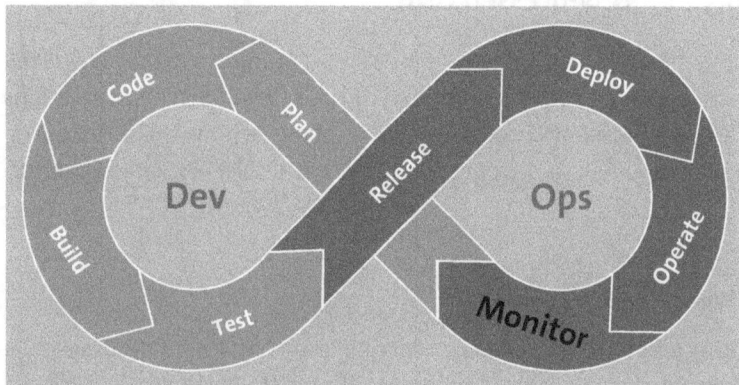

Figure 10.1 Monitor Phase of the DevOps Software Development Lifecycle

The last step in the DevOps loop is monitoring. After monitoring, it's straight back to planning, as the findings from the monitoring step are then used to guide ongoing development.

In terms of the Three Ways, monitoring is one of the core elements of the Second and Third Way: the flow from operations to development and therefore from the end user to the business. Essentially, monitoring allows us to learn from the end users and their use of the system with each iteration in order to continuously implement and roll out new improvements.

Regarding the CALMS model, the focus in this chapter is primarily on the metrics value. But as is so often the case in DevOps, many values and principles play a role, and so there is also a focus here on the Sharing value—on sharing what has been learned from monitoring and observability. This is also quite important. And at this point in the book, you should not be surprised that automation also has a strong influence here.

Monitoring is a topic that most people working in traditional software development have not heard much about, as it is a responsibility of the operations team in traditional software development workflows. In these environments, the performance of the application and the problems that become visible via monitoring remain hidden from the development team, as monitoring is not under this team's purview.

In a DevOps environment, monitoring is an important link between development and operations. Logging events, statuses, and other useful data about the services are relevant to evaluating a system's availability and performance. However, we want to go beyond this classic view of monitoring and take a look at *observability* as well.

This chapter concerns not only at the monitoring of an application and its infrastructure. It also concerns what DevOps has achieved in the area of monitoring and what metrics and analysis options exist for it.

10.1 No Visibility at nicely-dressed.com

The previous chapters described how the *nicely-dressed.com* online store was developed, tested, and later deployed over a long period of time. As soon as the operations team took over the newly developed code for deployment, responsibility for operations was transferred completely to the operations team. This also included monitoring.

10.1.1 Service Outages Happen Every Day

But what is monitoring actually for? There are several reasons for implementing monitoring, and all of them concern the continuous operation of the system.

First of all, it is important to identify problems early. Fortunately, not all minor issues lead directly to system failure, but problems should still be identified as early as possible so that they can be rectified before major issues arise. It is important to have a constant overview of the status of the entire system.

Reactive Symptom Treatment

However, a common challenge with classic monitoring is that only symptoms of problems become visible—that is, problems are noticed only once they have already occurred and the symptoms are detected. These can be major or minor issues, and it can be difficult to deduce the true extent of an issue or its cause from the size of its impact; for example, even a wrong line of code or a small error in deployment can cause a total system failure.

Essentially we need to ask the question, What exactly is a problem? Every problem in a system can have a different impact. Some problems are simply warnings that need to be corrected soon, while others have an impact on end users.

The following statuses have become the norm in classic service monitoring systems:

- OK: The system works as expected.
- CRITICAL: It doesn't work.
- WARNING: It works, but not as expected.
- UNKNOWN: The status is unknown.

As a rule of thumb, it has been established that problems that affect end users—that is, complete or partial failures—fall under CRITICAL and should be rectified as quickly as possible. In addition, problems should be prevented from becoming critical in the first place.

The first task of monitoring is therefore to immediately point out critical problems so that they can be rectified. You should not discover that a service is not working only through error messages from users. Problems that fall under the WARNING status should also not be ignored: If this status persists, the problem could be become more severe and fall under the CRITICAL status later.

> **Note: Any Error Messages from End Users Are Still Too Many**
>
> How many reported problems from end users are acceptable? This is difficult to answer, but remember that only very few people actually report bugs. If you get a handful of bug reports that something is wrong with your system, it's very likely that this problem is affecting many more customers who haven't reported it. You may have to stop what you're doing immediately and start looking for the root cause of the problem.

Monitoring should help not only to identify problems but also to troubleshoot them. It should provide material for error analysis so that the fault can be described in as much detail as possible.

And this is where *nicely-dressed.com* really struggled. The operations team used a rudimentary monitoring system that indicated whether the service was running. However, of the alert was a simple ping, limited to whether the web store was available. The monitoring system could not provide information about the exact status of the store or the consequences the current load had for general operations.

The operations team's primary focus when monitoring *nicely-dressed.com* was only to check whether the various services were running. This was indicated simply through binary messages: Either the service was working or it wasn't. And if not, an alarm was raised. There were also common checks that looked less at the services themselves and more at the underlying operating system. Typical checks of this kind are file system checks, such as how full the file system is, whether critical and noncritical package updates need to be installed, or whether TLS certificates have expired.

Only the operations team had access to the monitoring system and received alerts for messages in the system; the development team did not have access to it and did not receive any alerts.

The rudimentary nature of the monitoring system meant that every warning in the monitoring system required manual troubleshooting to find out the root of the problem. Of course, it would have been better if problems had not just been fixed temporarily, but if a real error analysis had been carried out to get to the root cause of the problems.

However, without the right information and without knowing how the individual parts of the system are connected, in most cases admins were only able to fix the symptoms instead of tackling the root of the problems. This naturally meant that the same problems would crop up over and over again—a problem of silos between teams.

Lack of Visibility

A typical problem that affects admins, the other teams working on the project, and the monitoring setup itself, is the lack of visibility in all directions. The developers have no access to the monitoring system for the most part, and the admins have no insight into the software development.

Due to the lack of visibility and the way in which monitoring systems are used, action is inevitably always reactive. Problems are often only noticed when something is already going wrong. By then it is far too late.

These issues happen regularly on the *nicely-dressed.com* team—the online store was regularly down. Although it was rare for the store to go completely offline, subsystems failed every now and then. These failures meant that some functions did not work properly, restricting the actions customers could take in the store.

Sometimes the billing system failed; this had no effect on some of the customers, as they had only looked at the items but did not intend on buying anything. For those who wanted to make a purchase, however, these failures meant that they could not complete their purchases, and the warehouses remained full. From a business perspective, this is the worst possible situation that needs to be avoided.

By developing individual benchmarks and targets that the monitoring system must monitor, such situations can be prevented. Don't be lulled into a false sense of security by a volume of monitoring data if a large part of that data is not relevant to you or is of interest only for further troubleshooting. Instead, you must clearly define which systems, API endpoints, websites, and so on are central. Think about how you can test and monitor their function. The operations team responsible for keeping things running smoothly will have the most experience to contribute here, but a lot of subtle monitoring criteria will come from the business departments and developers.

If you have collected and prioritized a number of important systems in this way, you can carry out monitoring in a much more targeted and efficient manner and achieve real visibility.

Alert after Deployment

At *nicely-dressed.com*, the most intensive time for all of the teams was the time around deployments, which happened only twice per year. Most of the problems occurred after changes were made, and a large number of them also popped up in the monitoring system.

However, some errors became apparent only when customers reported them; these reports often supplemented the monitoring system, so to speak. So many errors naturally lead to abandoned purchases and lost customers who are not prepared to tolerate bugs during a purchase.

In the previous chapters, we used the problem of the full file system as an example. The problem was only ever recognized after it had already occurred and various threshold values had been reached.

The admin team always chased down the problems at the surface level and extinguished any fires that occurred in the system, rather than finding the root of the problem. It obviously makes more sense to fix the problem at its root, especially when the file system will inevitably continue filling up.

> **Note: Avoid Shortcuts**
>
> An alternative solution to dealing with this file system issue is to enlarge the file system. However, this may only postpone the problem for a few hours or days. Depending on the environment, this can also lead to additional costs.
>
> But how would the team actually accomplish enlarging the file system? How are the permissions and freedoms actually regulated? With the lack of understanding and insight into the application on the operations team, this solution could be hard to implement.
>
> But even this solution would really just be treating the symptoms. It would be better to analyze where the large space requirement comes from and which system is responsible for the large amount of data that is written to the hard disks.
>
> Maybe it's a mistake that so much data is being produced; maybe it could be better organized. But the admins who are responsible for taking out the garbage would have to know where this pile of data comes from in the first place.

Missing Feedback Loop

The strict separation of the development team and the operations team at *nicely-dressed.com* meant that potential improvements and problems were not discussed. Admins implemented workarounds for problems instead of actually fixing them in the software.

In the example of the fill system that kept filling up, an admin would simply expand the memory as a workaround, and the reason for the problem was not analyzed: The application was writing too much log data in the form of long error stack traces.

This meant that the development team did not receive any information from operations. Any problems with the software remained hidden, as the development team had no access to the monitoring system or tickets from the production system.

Communication between the various teams took place only via emails and a shared ticket system, which should not be confused with the teams' own ticket systems, to which the other teams had no access. Error analysis was slow because all the information was gathered in bits and pieces in question-and-answer correspondence or was often misinterpreted.

Unprocessed Alerts

Another problem with *nicely-dressed.com* was that although the monitoring system reported many problems, no one intended to fix them. For example, every time security-relevant updates were available for a dependency or the operation system, a corresponding message appeared in the monitoring system. However, the operations team deliberately ignored these messages, as they already had enough more serious problems to deal with that were causing downtime. The admins certainly didn't need additional changes through updates.

The operations team was actually constantly on fire duty; there was always a fire blazing somewhere that had to be extinguished. This meant the team had no time to deal with security updates, which in turn flooded the monitoring system. This made it increasingly difficult to keep track of the other problems, and other alerts simply went unnoticed.

The problem could actually be solved relatively easily by automating the import of updates in a controlled manner. But there wasn't enough time for that—it was taken up by the many fire department operations.

10.1.2 Performance, Performance!

Traditional service status monitoring offers little insight into the actual performance of the system or service—or into its historical development or future trend. This gap has led to the development of *metrics monitoring*, which stores metrics as *time series*. The name of the metric, the value, and additional attributes (*tags*) are saved and can be visualized in graphs in a timeline.

When we talk about monitoring tools these days, metrics are always included or are the standard, as in the monitoring tool Prometheus. There are different types of metrics that are mainly relevant in the operation of the application, such as *counters*, *gauges*, *histograms*, and *summaries*. These types of metrics are explained in Section 10.3.2, in which we take a closer look at metrics using practical examples with Prometheus.

Even the operations team at *nicely-dressed.com* realized that a monitoring system that reports only whether a service is running is sometimes useless. The team also collected all kinds of system metrics to see the overall the performance of the systems. There were dashboards for each server that showed how high the CPU, memory, and network utilization was.

This made it easy to see that there were performance problems after a deployment and that the load on the web store was unpleasantly high. The monitoring system repeatedly reported these metrics and even triggered alerts. The problem for the operations team, however, was that they had no idea whether the high load was due to the latest changes or whether there was simply more traffic in the online store (which would have been a positive case).

So when load on the system was high, were there simply more visitors to the store? Perhaps a marketing campaign was also running in parallel with the latest deployment. Or could it be that the store's performance had declined rapidly? Or was it a mix of both? That is, did a higher volume of user traffic cause problems that could be ignored at a lower utilization rate? There are a number of questions here.

The operations team could actually analyze such a situation well if it had historical data with which to compare the current bottlenecks. However, the monitoring data was not kept for very long, and in order to have a point of comparison with a previous version, the team would have had to look back to the last major deployment. However, because deployments happened only every six months, it is completely inconceivable that meaningful findings could still be derived from this jumble of data!

This meant that the operations team could rely only on memory and experience to at least roughly estimate whether the current traffic was already causing too much load in the past. It would be important here to be able to look at the various metrics in isolation from each other in order to make fact-based decisions and implement changes.

> **Note: Latency**
> However, sometimes problems are caused by completely separate issues. Latency, for example, is a problem whose cause is not necessarily immediately apparent with classic monitoring tools. The problem with latency is that everything seems to work, but long response times can delay the entire process and cause problems in completely different areas.

The development team, which was largely responsible for the application's performance, had no idea what the problem might be. Without access to monitoring and information about the load on the production environment, their attempts to improve the website's performance were doomed to failure despite their best efforts.

As is so often the case, the problem also concerns the transparency of the available information. The development team could help solve the problem if they had access to the relevant information. An important step in improving monitoring is to increase collaboration by making access to the systems standard, simply and without high hurdles.

10.1.3 Logs

Closely related to monitoring is *logging*. States and events are written from the application to the terminal or a log file with a time stamp and textual information. For example, the start and end times of a program, the results of database connection checks, and unexpected error messages that the admins should look at can all be logged. Logs also often have debug information, such as SQL statements and HTTP queries, which make it easier for developers to work on solutions to problems.

Logs are valuable because they allow teams to investigate errors more effectively, especially with more detailed logs. They are an elementary component when debugging problems, and they are part of good monitoring.

However, too much of a good thing can be considered harmful. Take our example of the full system filling up at *nicely-dressed.com*: the logs caused the file system to fill up, which in turn caused the monitoring system to sound an alarm. Logs that consist only of messages that nobody looks at are of no use.

A major advantage of logs is that they are fairly easy to write. The simplest is plaintext written to the log system, sorted according to different log levels: `Debug`, `Info`, `Warn`, `Error`, and `Fatal`. The log system adds a timestamp to logs so that all entries can be viewed in a timeline.

This simple filter option with log levels ensures that sometimes more and sometimes fewer logs are output depending on the level set. How these levels are defined, however, is important, and it was a major problem at *nicely-dressed.com*.

At *nicely-dressed.com*, there was an issue in which the web application was logging far too many messages, which were then deliberately ignored by the admins because it took a lot of effort to find a needle in this haystack. Other parts of the application largely logged nothing, and what it did was unhelpful at best: `System Error`. `System failed`.

Here, too, the solution is closer cooperation between development and operations. If there are failures and there is no meaningful output in the log, this must be communicated accordingly so that more information can be available the next time. Or, even better, the relevant development team must be involved in troubleshooting during operation. The developer who wrote the code involved in the error will understand correlations much more quickly, which will simplify debugging. This experience from production also ensures that it becomes clear which information needs to be logged at which stage—and what leads only to performance losses.

In order for the various teams to have access to the log messages from the production system, a corresponding tool is required that archives the messages, makes them searchable, and processes them.

> **Note: Reflection**
>
> You must also consider the problems presented here in the context of the problems and challenges presented in the previous chapters.
>
> See which problems do or do not apply to your environment. If you are on the operations team, offer the development team access to the monitoring system. Conversely, if you are on the development team, inform the operations team of features that should be closely monitored to so that appropriate monitoring can be set up.
>
> Many problems can be solved or at least reduced through contact with other teams and discussions via official channels, even if the official structure in the company does not provide for this. Ideally, you should ensure that these adjustments, such as access to the monitoring system for all relevant people, are implemented directly on a regular basis.

10.2 With Insight Comes Foresight

In order to improve the situation at *nicely-dressed.com*, the teams needed to transform monitoring from a purely reactive process into a process that could be used for farsighted planning and action. It will come as no surprise that the first step in improving monitoring is to open up the monitoring tool to everyone in the organization. This gives the development team insight into the operation of the project and ensures that these insights can be used to improve the software.

It's not necessary to completely turn the basic principles of monitoring upside down. In terms of monitoring, the DevOps environment works in a similar way to the traditional approach. The focus is still on visibility into the running environment, including all possible problems.

The aim of a joint DevOps-focused team is to ensure that changes are delivered to end users quickly. Regarding monitoring in the context of the Three Ways, the feedback loop emphasized in the Second Way and the continuous improvement emphasized in the Third Way are relevant.

In a traditional environment, there is no clear process for feeding observations from the monitoring system back into future development. Due to traditional silos between the teams, this is not particularly surprising.

At *nicely-dressed.com*, new load peaks were reached after almost every major deployment, and the impact that the performance of the latest changes had on this problem was unclear. This is exactly where observability comes into play. Instead of just checking the pure availability and performance of the application, teams should be able to look much deeper into each request to find problems.

The aim is to use a combination of logs, metrics, and traces to look deep enough into the running application to identify potential problems in detail so that they can be eliminated and improved. Let's take a look at what this looks like in concrete terms.

10.2.1 Observability Engineering

Nowadays, the generic term *observability* is used; in the past, the term *application performance monitoring* (APM) was used. By and large, the two terms are similar but have some differences. Both approaches provide a much deeper look into the running application.

While classic monitoring is still needed to be able to react to problems, observability takes a different approach. This approach is not reactive; instead, it looks to the future—that is, it works proactively. The aim is to find any problems that arise as quickly as possible. Once the alarm has been raised, we look at what the exact problem is.

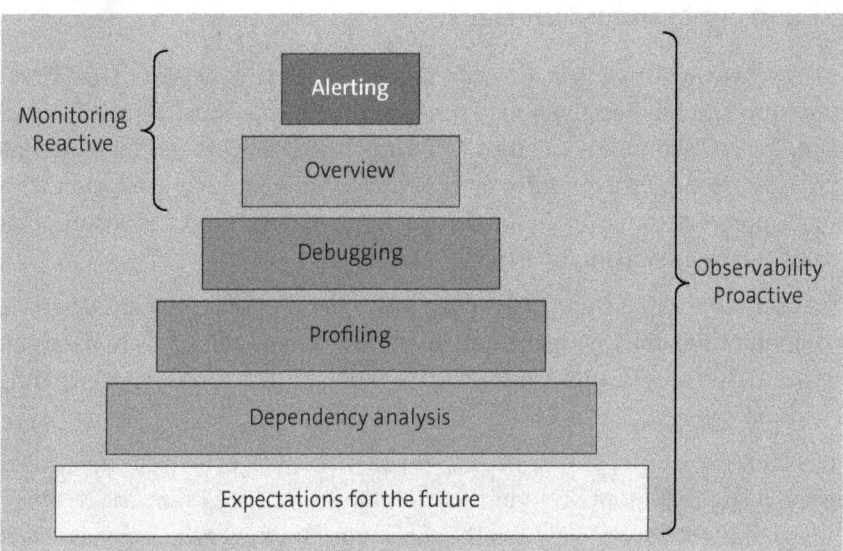

Figure 10.2 Monitoring versus Observability

As you can see in Figure 10.2, monitoring ends with an overview of the system. Observability, on the other hand, can and should also be used when there is no problem. It provides data for debugging, profiling, and analyzing dependencies.

Observability is used to be able to formulate expectations for future performance more elegantly. The aim is to be able to ask questions about the current status, progress, and future of all services in the environment at any time. It can also provide correlations between and insights into previously unknown problems (so-called unknown unknowns).

Observability was initially developed three pillars for data collection: logs, metrics, and traces. Both logs and metrics originate from the classic monitoring approach, so they are not rocket science. *Traces* allow you to completely trace the entire process of a request. I'll come back to traces in a moment.

In addition, further observability data has been developed, such as in the areas of profiling, end-to-end testing, and error tracking. The community is divided as to whether the three pillars are still legitimate; more information on this topic can be found at *https://o11y.love/#what-is-observability/*. To get started with observability, I recommend focusing on logs, metrics, and traces.

If you combine logs, metrics, and traces, you end up with observability practices. But what does this mean in practice? Let's take an example from *nicely-dressed.com*: A customer reported that he cannot complete his purchase because he received a 500 Internal Server Error message.

In this case, someone from the operations team first had to take over the ticket and check it. The first step of fixing an error is to find out where the error came from and which system is responsible. Typically, this means looking at the logs and hoping that stack traces can be found there that match the error description.

At *nicely-dressed.com*, for the admin and any developers called in, this meant quite a lot of work, as the (sparse) information from the customer and the large amount of information from the logs had to be interpreted in context: Why did the error occur at this particular time with this particular customer?

> **Note: Bug Reports versus Observability**
>
> In this example, by the way, there is a good chance that the reactive monitoring system would not have raised an alarm if this problem really only occurred with a single customer. If it is a major problem that affects all customers, the error *must* already be noticeable in a different way. In this case, you cannot rely on bug reports from customers. In this example, however, we assume that this is an *edge case*, or an unexpected and atypical problem, that is occurring only for this one user.

The error message should be understood as a work order for the development team. Of course, it is first and foremost the task of the admins to ensure that the online store is operational again so that the customer can complete his purchase. However, it is much more important that the ticket is not closed, but passed on to the development team.

For a development team to process an error, it needs to have an insight into logs, metrics, and traces. There must be no silos that prevent collaboration. Instead, cross-functional teams should discuss such problems in detail and from different perspectives. From a DevOps perspective, this means above all that problems and challenges can be addressed more quickly and proactively to prevent similar problems in the future.

Another sticking point is that a large number of different systems play a role in troubleshooting and their information must be brought together. Different tools usually exist for monitoring, logging, metrics, and traces, with their own authorizations, all of which have to be configured. The purpose of observability tools, which I describe in Section 10.3 is to combine these three pillars of observability so that it is clear exactly what has happened and that you can reconstruct the complete call in the system, including the complete trace, the log, and the metrics that belong to it.

This different information is simply necessary so that errors can be traced with as little effort as possible and so that it is clear why and in exactly what context the error occurred. Especially in production systems, it is hopeless to reproduce the exact initial situation that led to an error—and a "can't reproduce" response to a ticket doesn't help anyone either.

Observability engineering can also be used independently of specific error messages sent by end users. The data actively used by an observability management tool makes it clear where potential bottlenecks are hidden. All available metrics are used to see which parts of the system have a particularly high load, which errors are associated with this issue, and what could be the cause of a bottleneck.

10.2.2 Insights into Processes with Tracing

To understand errors in an application, log files and metrics are helpful but not sufficient—as anyone who has ever chased after an error that occurred only under very specific conditions and couldn't be isolated can confirm.

The exact time of an error in an application must be compared with the timestamps in the log data. In distributed systems and microservices, these log entries often occur at the same time, and when reading them you are staring into a colorful chaos full of possible program sequences from different host systems. What would be practical here is a concatenation of all log events.

In these cases, tracing is absolutely essential. Basically, a *trace* is a complete flow of a request or transaction.

By using tracing, you can gain deep insights into the individual function calls. This allows you to see how long individual calls take to complete in order to uncover any bottlenecks in the application, which in turn would otherwise lead to restrictions in performance and, therefore, functionality.

When examining a complete request (see Figure 10.3), you can gain insight into which method was called, which arguments were passed, what values these arguments had, and how long each call took.

Without this approach, problems have to be laboriously recreated, which is rarely possible in production systems.

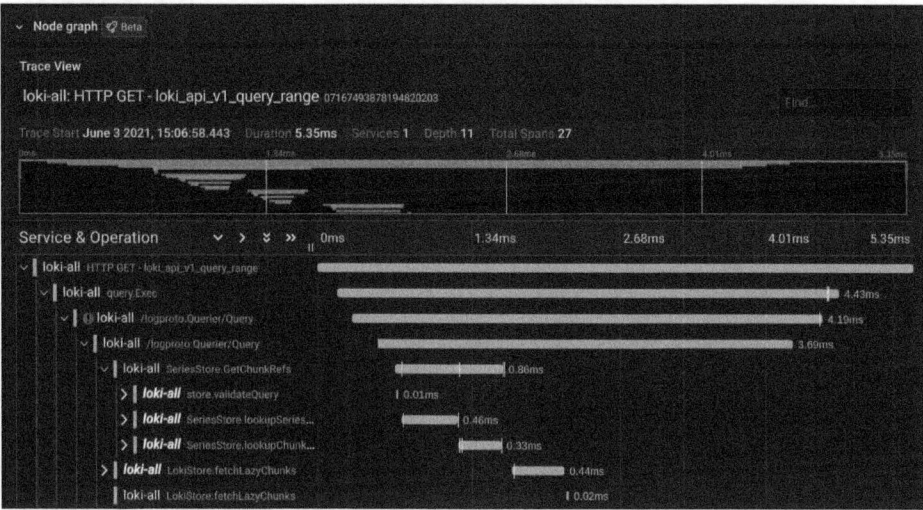

Figure 10.3 A View of a Trace in Grafana

All this information is important so that you have the time and the data to find opportunities for improvement faster and more effectively—and with data from productive operation. If you don't have real data and real traffic, it is rather difficult to find bottlenecks.

> **Note: Instrumentation of Telemetry Data**
>
> Of course, observability tools work only if the corresponding data also ends up in the system: There is no observability without data. Your own applications must therefore be instrumented so that they provide the data with which the observability tool can then work. OpenTelemetry (*https://opentelemetry.io/*), for example, is a collection and specification of APIs, SDKs, and tools for instrumenting, collecting, and exporting metrics, logs, and traces for the observability tool. More on this follows in Section 10.3.

One tool that has many similarities to a tracing tool is the profiler available in the browser. Modern browsers such as Firefox or Chrome have a profiler in the web console that shows which data is loaded and how long the JavaScript calls take. This is a common tool used in frontend development. In contrast to tracing, however, profiling repeatedly stops the states very briefly in order to record the data.

10.2.3 A/B Tests

If you set up and use the monitoring and observability tools correctly, then A/B tests are helpful for carrying out experiments. However, the prerequisite is that you have full confidence in the application and in continuous deployment; otherwise, you won't even get to start A/B testing.

A/B tests are tests in which two variants are implemented to check which variant delivers better results. At *nicely-dressed.com*, for example, two (or more) different variants can be implemented to advertise the offers of the day on the homepage. Depending on how the offer of the day is presented, more or fewer people may click on it.

Many factors can play a role in whether a customer clicks on an offer, such as the size of the advertisement, the text, and the color scheme. The *conversion rate*—the percentage of customers who click on the offer—can be used to measure the success of different designs (see Figure 10.4).

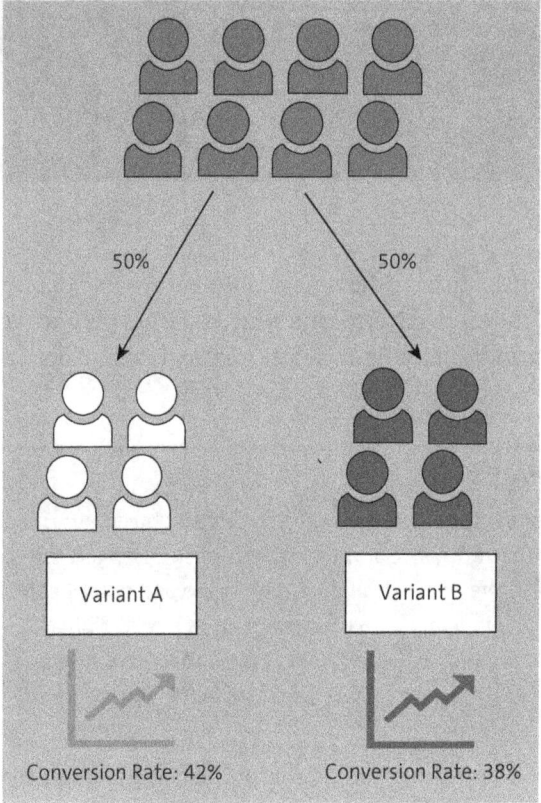

Figure 10.4 An Example of Conversion Rates

Depending on the type and method of implementation, two different versions of the otherwise identical software can be rolled out and the traffic can then be split between the two deployments via the load balancer. This is a practical and quick way to check which solution is better instead of spending time hypothesizing.

But A/B tests can be used for more than just the conversion rate. They can also be helpful if performance improvements have been implemented and need to be tested on a small proportion of users. In this scenario, you can use A/B tests to check which implementation performs better.

> **Note: Share the Success (or Failure) of A/B Tests**
>
> As part of the sharing value of the CALMS model, the successes and failures of A/B tests should be shared within the company so that other teams that may be facing similar challenges can learn from these findings.

The whole A/B testing framework should run as part of continuous improvement, which is one of the core DevOps principles. The findings from the A/B tests should therefore be constantly incorporated into the next development phases.

10.2.4 Business Monitoring

In addition to technical monitoring, the actual business should also be monitored. Up to this point, we have only talked about general load, usage, and performance of a software. In the end, however, it is also important how well or badly the business itself is running.

It was important that *nicely-dressed.com* sold as much merchandise as possible. Previously, reports on sales figures for the past 30 days were available only once a month. Data was pulled from the billing backend and prepared in Excel in a few manual steps.

Although this was better than nothing, it was only a purely reactive view of the situation: The data allowed the company to draw conclusions only about how well the past month went. There were no short-term evaluations or even tests of how specific adjustments could affect purchasing behavior.

It would have been better not only to automate the whole process but also to gain real-time insight into the business data. Several metrics could have been used by both business analysts and operations to gain such insight.

From a business perspective, it is first and foremost important to know how much turnover is currently being generated. For a business like *nicely-dressed.com*, this data is already available, as all sales are stored in the database. This data needs to be processed so that sales figures are visible to management. However, a number of related metrics, such as the following, can also be used:

- Number of sales
- Number of items sold
- Fill levels in the warehouse
- Conversion rates

The key point is that this data, which is already being collected anyway, is analyzed in real time so that it can be aggregated over different time periods, such as minutes, hours, days, or even weeks and months.

For example, it would be possible to directly evaluate how a marketing campaign that was run as part of the last major deployment actually performed by comparing the data with data with other time periods.

The changes to the monitoring system can therefore give the business team real-time insight into how well the store is currently running. And that's not all: If CI/CD and DevOps are implemented correctly, A/B tests and feature flags can also be used to effectively make changes during an ongoing marketing campaign to see how the conversion rate can be increased.

These findings are important for the operation of the infrastructure, as the detailed business figures and extended metrics can be used to assign a monetary value to the various hours of the day. With this data, any loss of revenue can be quantified more realistically.

In the context of an online store, this approach makes it easy to quantify system failures. It is obvious that an outage in the early evening, when most customers are using the system, is significantly more expensive than an outage in the middle of the night.

This approach can also show how strongly individual functions contribute to sales. In particular, you can carry out an A/B test on new convenience functions that result in a noticeably higher system load and correlate the result with sales or clicks. This could allow you to make a data-based decision as to whether the implementation is worthwhile without having to think about it for a long time.

Particularly regarding stability tests using chaos engineering, more experimental actions can then be taken, preferentially at times when less turnover is generated anyway. The associated downtime can thus be quantified more realistically in order to implement further improvements on the basis of these key figures.

10.3 Tools for Monitoring, Observability, and Tracing

In Section 10.1, we discussed the old world of reactive monitoring and the emerging world of observability. Classic monitoring tools are still just as relevant today as newer observability tools.

Regardless of which type of tool you use, the goal should be to make all relevant information usable and visible to everyone. This applies not only to the infrastructure on which the applications run, but also to the applications themselves. The deeper you can look into the system, the better.

You are certainly already using monitoring tools if you manage an application or infrastructure. But do you really see everything you need to at a glance in your setup? Do other people also have access to data that is important for their work? And are the checks and controls also helpful? Or does a flood of data convey the impression that everything is in view, even though important things are overlooked?

> **Note: Avoid SSH**
>
> In my experience, the use of SSH is a good indicator of whether monitoring has been implemented properly. Because if the monitoring of the entire tech stack is implemented according to best practices, SSH access to the systems is no longer required.
>
> If you use SSH to check if something in your system is still working, it is a good indicator that your monitoring solution is not providing the right information. Consider opening a ticket to note an improvement for the monitoring system.
>
> The same applies to log files: If you have to log on to a server and get log output manually, you should plan for a central log server that archives all important logs centrally and makes them searchable.
>
> This is especially true when working with virtual machines or containers—when you need the `kubectl logs` Kubernetes command. This is information that you actually already have in view and should not have to call up separately.
>
> (Important: This refers to production environments. For environments that are used for testing and development, such a comprehensive monitoring setup is usually not necessary.)

10.3.1 Monitor Systems with Icinga and Nagios

Classic monitoring systems include tools such as Nagios and Icinga. Icinga was originally a fork of Nagios. The newer Icinga2 version, however, is a complete rewrite of the monitoring system and provides a new configuration language and better support for distributed monitoring environments. However, tried and tested interfaces to plug-in scripts that are executed to record host/service statuses (OK, WARNING, CRITICAL, and UNKNOWN) can still be found. Metrics must be retrofitted with integrated external tools.

As already mentioned in Section 10.1, these tools are used reactively: Monitoring is configured, which is often either in the form of binary checks or triggered based on defined threshold values. If the monitoring tool triggers an alert, notifications are sent out via the configured alert methods.

However, classic monitoring systems have one major problem: The configured checks are usually relatively hardwired to the existing systems.

Especially if you want to work according to the "cattle, not pets" principle, working with monitoring solutions such as Icinga is not very convenient, as these tools do not come with autodiscovery by default but have to be configured for the services that are to be monitored.

With autodiscovery, the servers and services make themselves visible with little configuration effort so that the monitoring server can automatically integrate them. This is particularly useful in a cloud, as servers continuously appear and disappear. Of course, some actions can be automated using scripting, but this introduces greater complexity.

Another aspect of these tools is that they mainly monitor the infrastructure. By infrastructure, I mean primarily the servers themselves and the network. But the applications themselves are also monitored to a certain extent, although often only through binary checks, mentioned earlier (i.e., whether the application is running).

Many of these tools also include performance monitoring to a certain extent; however, the functions are limited, as those tools focus on service monitoring. Most tools with performance monitoring check only the health of the systems, such as how much disk space is still available.

Icinga2 and Nagios are not the only classic monitoring tools in widespread use. There are also tools such as Zabbix and Check_MK, which are dedicated to the challenges of cloud-native monitoring. For example, they both offer Kubernetes monitoring, and Check_MK has established itself as a pioneer in the field of service autodiscovery. In the cloud-native world, Prometheus has become the de facto monitoring standard. More on that tool in the next section.

Alerting and Incident Management

Monitoring is important, but you also need to receive reliable alerts if problems occur in the monitoring system itself. Email notifications are not a good choice for alerts about serious problems, as they often get lost in inboxes.

Tools should therefore be used for both alerting and incident management. Well-known commercial tools include PagerDuty and Opsgenie. The latter is another tool from Atlassian. A number of other tools, such as GitLab, also include alerting functions.

These tools also include functions for managing of who should be notified of detected issues, as well as when and how. For example, you can define in which rotation people must be notified. It is important that no critical warnings are missed and that the tools are managed accordingly.

10.3.2 Monitoring with Metrics and Time Series Databases

Relational database systems are generally not used to record and store metrics, as they are not designed to record and retrieve a large number of metrics with timestamps in a practical and high-performance manner. Instead, there are specialized time series databases in which time is always stored with the data as an additional dimension.

With time series databases, you can also automatically compress older data to save storage space. This means you can still analyze graphs from the last few years. For example, you could collect data points every five minutes in the long term. Collecting data points every half minute or so is useful for monitoring the current status, but it is not really needed to analyze data in the long term.

10.3 Tools for Monitoring, Observability, and Tracing

Prometheus and InfluxDB are widely used time series databases. There is a distinction between pure time series databases and monitoring solutions, but many solutions offer a mix of both, such as Prometheus.

Both Prometheus and InfluxDB have different concepts and approaches, which is not particularly surprising. Tools such as Graphite (*https://graphiteapp.org/*) can often still be found in older setups, although it comes with a number of architectural limitations, does not get much further development, and no longer has any significant relevance.

Although both Prometheus and InfluxDB contain their own visualization options, Grafana is often also used for visualization. Grafana is described in Section 10.3.3 in more detail.

Prometheus

Prometheus is both a monitoring tool and a time series database. It is a cloud-native application, so a container setup that can be created via Helm charts is recommended for installation. It is one of the first projects after Kubernetes to achieve the Cloud Native Computing Foundation's "graduated."

Similar to Kubernetes, Prometheus was created by former Googlers, though it does not come directly from Google. Prometheus is written in the Go programming language, so it can be deployed very easily on all possible systems—the compilation of the Go application is available as a single binary file that includes all dependencies.

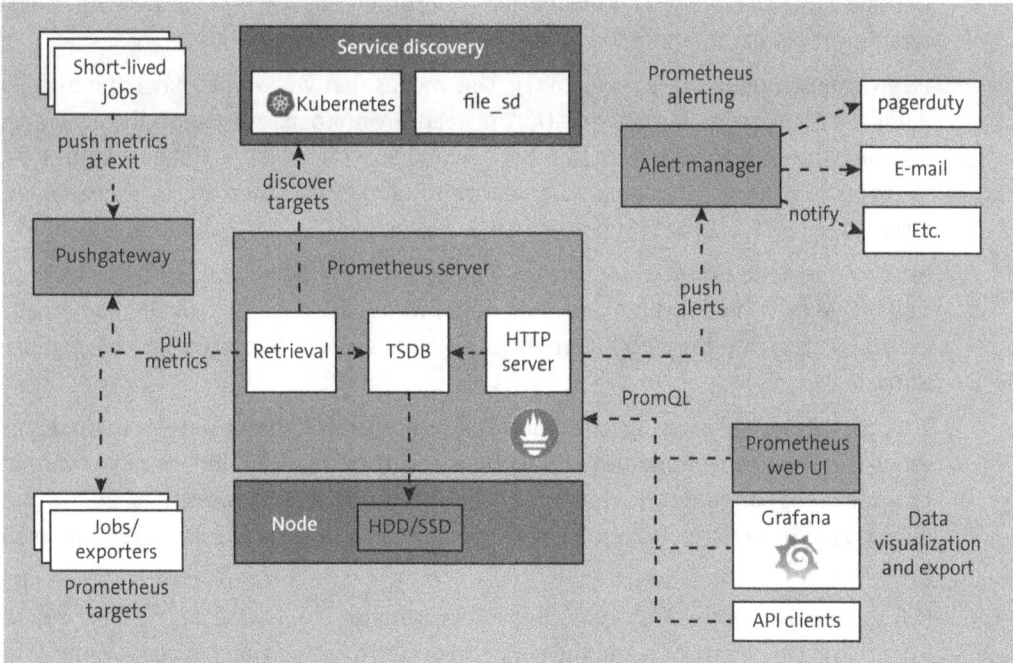

Figure 10.5 Overview of the Prometheus Architecture (Source: https://prometheus.io/docs/introduction/overview/)

However, the Prometheus project still consists of various components. The *Prometheus server* is the core element. It collects, stores, and, of course, outputs the data when queries are made. provides an overview of the architecture of the monitoring stack, which can, of course, look different depending on how it is used.

The First Metrics

When planning the architecture of a Prometheus implementation, it is important to understand that Prometheus collects data using the pull principle. It scans the configured servers and retrieves the data it needs to save. Prometheus is relatively simple to use, and you can use simple methods to check whether the metrics are provided by the services at all.

The servers and services that need to be monitored provide their metrics via HTTP. The Prometheus node exporter, for example, provides the node's metrics via HTTP. It can be accessed at http://192.168.2.3:9100/metrics and returns these lines, among others:

```
# HELP node_network_up Value is 1 if operstate is 'up', 0 otherwise.
# TYPE node_network_up gauge
node_network_up{device="eth0"} 1
node_network_up{device="lo"} 0
node_network_up{device="wlan0"} 0
```

The Prometheus node exporter then displays a list of metrics with their current data under the specified interface. The Prometheus server collects this data at the exact time it is called up and saves it. This means that the Prometheus server must always perform scraping at regular intervals; otherwise, there will be gaps in the data records.

Prometheus supports *service discovery*. This means that you do not have to manually enter all endpoints in the configuration; instead, Prometheus automatically recognizes and adds the endpoints. A large number of cloud providers (Azure, Google Cloud, AWS, Hetzner Cloud, etc.), container orchestrators (Kubernetes, Nomad, etc.) and system services (DNS, HTTP, etc.) are supported.

Initially, the node exporter delivers metrics "only" from the node itself. Some applications have their own Prometheus endpoints, which they use to provide ready-made metrics. This is the case with Kubernetes, as well as some network and storage applications.

Of course, it is more exciting to monitor self-written applications. Development teams can define their own endpoints with their own metrics. There are libraries for common programming languages that simplify the whole process. However, as the output format is only a relatively simple text format, this additional effort is often not even necessary.

Due to the way Prometheus works, the power to provide the desired metrics lies primarily with the development team; the team can decide for itself which metrics it wants to collect and provide. If the team needs to add metrics, it can simply extend the endpoint and reroll the application so that the data automatically ends up in Prometheus.

10.3 Tools for Monitoring, Observability, and Tracing

This is why good collaboration between the operations and site reliability engineer teams is important. Which metrics help with debugging the production environment? Which data is still relevant for developers in order to better analyze problems?

> **Note: Training**
> Julius Volz, the cofounder of Prometheus, offers training courses at *https://training.promlabs.com/* that anyone can work through at their own pace. Many training courses are subject to a fee, while others are available free of charge. Other shorter explanatory videos are also available on the associated YouTube channel PromLabs: *https://youtube.com/@PromLabs*.

Four Metrics Types

Prometheus has a total of four different types of metrics: counter, gauge, histogram, and summary.

Figure 10.6 Visualization of the Metric node_disk_read_bytes_total of Type Counter

The first type is the *counter*. Determining metrics using a counter is very simple: The value is simply incremented. As a rule, the counter is reset to zero each time the service is restarted. Typical use cases for the counter type include determining mouse clicks, errors that have occurred, and completed requests.

If you want to visualize the data with a counter, you must note that the counter really only determines metrics in the upward direction if it does not reset to zero, as you can see in Figure 10.6.

The second type, the *gauge*, is a numerical value that can go both up and down. A typical metric measured through a gauge is the utilization of the CPU or RAM, which can be higher or lower depending on the load Figure 10.7 shows what that could look like.

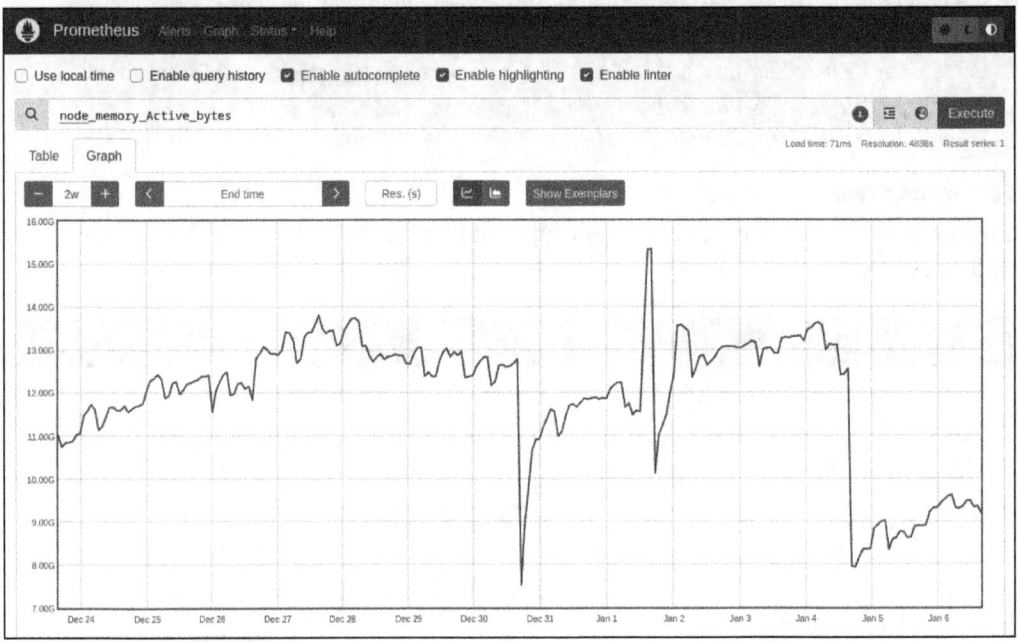

Figure 10.7 Visualization of the Gauge Metric node_memory_Active_bytes

A *histogram* is a little more complex, and it took me a while to understand exactly what it is and how it can be used. Many people outside the Prometheus world are probably familiar with histograms. To give a practical example, while you can use a counter only to measure the frequency of requests—as a counter can only increment values—you can use a histogram to measure the runtime of requests. How many requests have been answered in under 0.1 seconds? How many take well over 2 seconds? With such metrics and their visualization, it is easier to understand the distribution of the request durations and the number of requests for each duration.

The *summary* is similar to the histogram, but the window for the aggregations can be moved variably.

All metric types can also be calculated and output manually, but it will make your work easier to use the libraries provided by Prometheus for specific programming languages.

10.3 Tools for Monitoring, Observability, and Tracing

Labels

Metrics alone make no sense without labels in Prometheus. They play an important role if you want to create queries and visualize metrics. For example, the following lines are about network access:

```
node_network_up{device="eth0"} 1
node_network_up{device="lo"} 0
node_network_up{device="wlan0"} 0
```

Here we are measuring the metric `node_network_up` for three devices (indicated by the label `device`), called `eth0`, `lo`, and `wlan0`. However, only the network device `eth0` is currently running, indicated by the value 1.

Labels are key-value pairs that add metadata to a metric. If you want to export your own metrics, you should use the procedure shown above instead of using, for example, `node_network_up_eth0`, `node_network_up_lo`, and equivalent names.

There is a simple reason for this: It is much easier to search for a single metric like `node_network_up` in queries or for visualizations. Labels such as `device` are then supplied accordingly. Otherwise, the name of the network device cannot be filtered so easily.

Several labels can be assigned, which is often the case. This is particularly necessary for filtering and grouping.

The fact that many nodes or containers simply disappear and reappear plays a role here, as sometimes they are scaled up and sometimes they are scaled down. The use of labels means that data with the corresponding host names is sometimes available and sometimes not. This also makes it possible to deduce when a host was available and when it was not. Labels are therefore also used for identification.

Queries

The query language for data from the Prometheus database is not so easy to understand at the beginning and requires a little practice before you can get the hang of it.

Prometheus uses its own query language called Prometheus Query Language, or PromQL. It can be used to query data in real time.

The data that is queried can be either displayed as a simple table or visualized directly. The Prometheus server itself offers a simple query and visualization option. However, this option is intended more for ad hoc queries. As a rule, however, you should save and visualize various queries in panels and create dashboards in the longer term. Grafana is suitable for doing so (see Section 10.3.3).

A very simple query would look like this:

```
node_network_up
```

This would simply output all metrics unfiltered. Labels can also be used for filtering:

```
node_network_up{device='eth0'}
```

This query would return all network devices that output eth0 with the corresponding status.

But it can also be much more complicated, as the following example shows:

```
100 - ((node_filesystem_avail_bytes
{instance="192.168.2.2:9100",device!~'rootfs'} * 100) /
node_filesystem_size_bytes{instance="192.168.2.2:9100",device!~'rootfs'})
```

This command consists of only one line and contains some math and some result filtering. To be precise, it asks for the fill level of file systems by querying the total size and available size of memory for a specific instance, without including the rootfs.

> **Note: PromLens**
> PromLens (*https://promlens.com/*) is an extended query builder. It is now open source and was written by Prometheus cofounder Julius Volz. PromQL queries are much easier to build with PromLens. Not only is the tool practical for learning how to build queries, but it is also popular with experienced Prometheus users.

Scalability and Data Security

Prometheus is a relatively simple tool, which is evident when you look at how poorly it scales. As I have already mentioned, it is a cloud-native project, but it does not scale very well, at least not in the way you would initially expect.

Prometheus regularly retrieves and saves data from endpoints. As long as it does this continuously, no data is lost. Conversely, however, because Prometheus always saves the data that it has retrieved at the current time, no data can be imported retrospectively.

This also means that data cannot simply be deleted. For long-term storage, you don't necessarily need data records to be generated every 30 seconds; data that is generated every five minutes or even every hour is sufficient. This is also not possible in retrospect with Prometheus; instead there are separate projects, such as Cortex (*https://cortexmetrics.io/*), Thanos (*https://thanos.io/*), and Grafana Mimir, that enable the long-term storage of data from Prometheus. These projects also solve some of the scaling problems that stem from the simplicity of Prometheus.

However, if you don't want to integrate more tools, to keep the setup fairly simple, you could instead increase the storage period so that data is not deleted after a short time. This would naturally result in larger data sets. Alternatively, Prometheus servers can also scrape other Prometheus servers so that they retrieve the data at a different interval.

Prometheus in the DevOps Context

Now that you understand the concepts behind and setup of Prometheus, let's discuss the advantage of using Prometheus from a DevOps perspective. I have already touched on a few points: It is relatively easy to provide metrics for your own application, with or without the provided libraries. Through several iterations, new metrics can be determined and conveniently created and removed again without having to contact a separate team or ask for changes. The nature of scraping provides this capability automatically.

Most of the examples so far have been of a purely technical nature. However, the same features can also be used for business metrics. For *nicely-dressed.com*, for example, business-relevant metrics, such as the size of the shopping cart or purchases over certain periods of time, can be valuable.

A/B tests can also be used by adding labels and can be visualized accordingly. If Prometheus is used in a Kubernetes cluster, the corresponding pods only need to be provided with the corresponding Kubernetes labels for new Prometheus endpoints to be automatically recognized and retrieved via autodiscovery. Once you fully understand how to use Prometheus, you get a very high level of flexibility.

You can also receive alerts, like you would from a monitoring system, based on metrics. This is true for both technical and business metrics. The Prometheus Alertmanager exists for this purpose.

InfluxDB

While Prometheus is both a time series database and a monitoring tool, InfluxDB is a pure time series database. It is licensed under an open-source license and developed by InfluxData, Inc.

The special feature of Prometheus was that it retrieves data from the server itself—that is, it works according to the pull principle. This is different with Influx, as it writes data to the database externally. It therefore works according to the push principle, as is also the case with SQL servers. Metrics and events can also be saved in the database.

Therefore, monitoring with InfluxDB is fundamentally more complex than it is with Prometheus, but it can be scaled better than Prometheus. The option for horizontal scaling is available through the paid premium license.

Libraries for common programming languages can be used to write data to the database. This is particularly useful if you need to collect metrics from your own application. Alternatively, you could use Telegraf, an agent server that enables metrics to be collected from various sources and written to the InfluxDB database using simple means. For example, it can enable collection of sensor data via MQTT or data from Kubernetes or even Prometheus.

The data model is similarly flexible to Prometheus; the data can be written to the database with little effort and quite a lot of flexibility, so new metrics can be created again and again.

Previously, InfluxQL was the language used for writing queries for InfluxDB; queries written in InfluxQL were very similar to SQL queries. Version 2 of InfluxDB introduced the Flux query language, which is somewhat more powerful. Personally, when I first used Flux query language, I found it much more difficult to understand compared to PromQL.

With InfluxDB, you can clearly see that the focus is on the database, which is different with Prometheus. Data storage is also much more flexible; you can add data at a later date by transferring the corresponding timestamp, and you can delete data at a later date. These are capabilities that Prometheus was not designed for.

InfluxDB has its own visualization options. However, since many organizations deal with numerous different data sources, it makes more sense to rely on Grafana, which is discussed in more detail in the next section.

You may ask, Should I go for Prometheus or InfluxDB? I think the answer here is clear: "It depends!" Prometheus is simple, it's quick to get started using it, and metrics can be provided quickly and easily and are retrieved by autodiscovery. Getting started with PromQL is also relatively easy.

InfluxDB, on the other hand, focuses on significantly better data storage. Depending on which data you need to store and how, InfluxDB may be more suitable.

In the end, however, the important thing is what the team is familiar with and what gets you to your goal. This is possible with both tools. Of course, both tools can also be used in parallel, depending on the intended use. However, using both tools at the same time may require your team to build up even more expertise in maintenance and use.

> **Note: Reflection**
>
> If you want to use a modern metrics-based monitoring tool, there is no way around Prometheus. The tool is simple but still offers everything you need to monitor your own environment. See simplicity as an opportunity rather than a risk, because it is easy to set up a small Prometheus server and collect data, especially at the beginning. You can always scale up later.
>
> Instead, familiarize yourself with Prometheus. The server component can be quickly set up and can be rolled out without much effort. The challenge is learning PromQL, but little helper tools such as PromLens provide support.

10.3.3 Data Visualization with Grafana

Both Prometheus and InfluxDB focus on the time series database itself. A specialized tool should therefore be used to visualize the data: Grafana.

With Grafana, you can visualize data from different data sources. In addition to time series databases, you can also connect several other types of data sources, including logging and document databases, systems for tracing and profiling, and classic SQL databases.

Using Grafana, you can create various dashboards to visualize data in different ways. The quality and flexibility of visualizations created with Grafana are quite high. There are numerous possibilities, as you can see in Figure 10.8, such as time series, bar charts,

10.3 Tools for Monitoring, Observability, and Tracing

simple statistics, level charts, tables, pie charts, and even heat maps. There are many more; try out Grafana to see how you can best and most intuitively display your data.

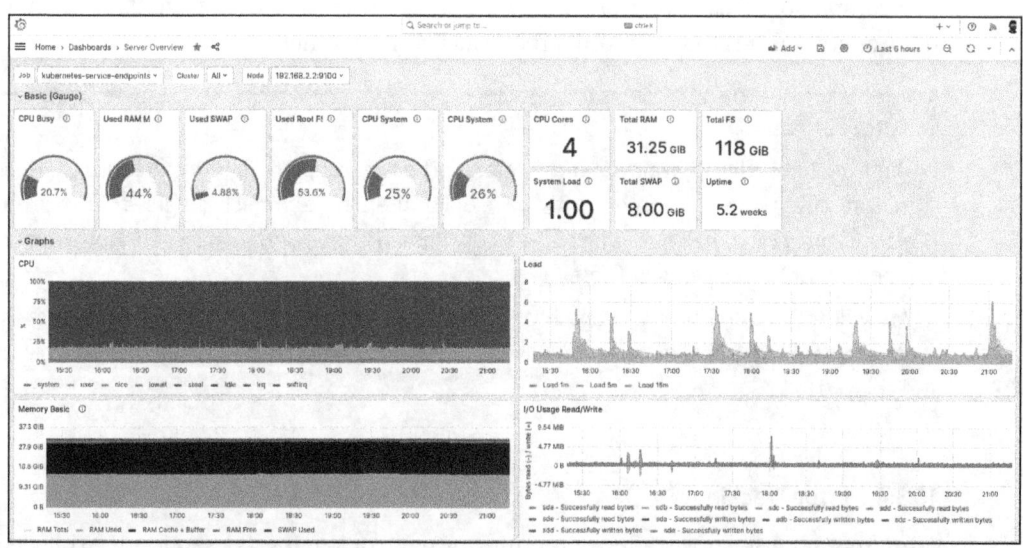

Figure 10.8 An Example Grafana Dashboard of a Server Monitored with Prometheus

The visualization options are available for the various data sources; the key is to choose the data that is really relevant and that needs to be kept in view at all times. A lot of time is required to get up to speed with Grafana, due to the wide range of configuration options. I can't help you decide which information is important in your environment, but I can give you some practical tips:

- You can find ready-made dashboards from the community at *https://grafana.com/grafana/dashboards/* for starting and working with standard software. These dashboards can be easily imported. Though the quality of the available dashboards varies greatly, you can usually find good dashboards on which to build your own visualization.

- Regardless of whether you are starting from scratch or using a community dashboard, once you have set up a dashboard, do not regard it as the final solution. Information requirements can change, and questions can be added or removed. You should collect feedback in order to adapt the presentation of your dashboard. And know that if your dashboard needs a lot of adjustment, it is often better to start from scratch with a new one.

  This is particularly relevant for scenarios in which you develop dashboards for your own application with your own metrics. Not only do the evaluated metrics have to be optimized again and again, but so does the visualization. Talk to everyone who uses the dashboards you build about how to best display your metrics.

- Don't overload the dashboards; rather, set up different interfaces for different target groups. The less information you have to display on one screen, the better.

The actual integration of metrics is now fairly easy using Grafana's Metrics Browser, which allows you to quickly click together metrics, labels, and other functions with autocompletion. Depending on the data source, however, the user guidance is not quite as good as it is today for other tools, such as Prometheus.

> **Note: Reflection**
> There is currently no way around Grafana. There are not really any practical and mature alternatives. A very recently developed alternative project is Perses (*https://github.com/perses/perses/*), which allows dashboards to be defined as code. However, practical use is still a long way off.
>
> You will soon realize that Grafana has many useful functions that make it easy to create diagrams. The challenge is actually deciding which information really needs to be displayed.

10.3.4 Error Tracking

Error tracking, also known as error monitoring or error logging, is a service that collects and logs errors that occur in the system at a central location. So if a user uses an application and encounters an error that throws an exception, this error must be recorded with a time stamp and other metadata. The central error list created through error tracking serves as a starting point for the development team when fixing bugs.

One of the best-known error tracking tools is Sentry (*https://sentry.io/for/error-monitoring/*). Clients and SDKs for error tracking with Sentry are available as open-source software and can be integrated quickly and easily.

However, Sentry can do more than just error tracking; it also offers functions for performance measurement and profiling—error tracking is helpful only if it is enriched with traces and logging. And this is exactly the way of observability platforms, where everything comes together to provide a complete overview.

Other providers of error tracking include the following:

- Airbrake (*https://www.airbrake.io/error-monitoring/*)
- Bugsnag (*https://www.bugsnag.com/error-monitoring/*)
- Datadog (*https://www.datadoghq.com/product/error-tracking/*)
- Integrated error tracking in GitLab (*https://docs.gitlab.com/ee/operations/error_tracking.html/*)
- HoneyBadger, *https://www.honeybadger.io/tour/error-tracking/*)
- LogRocket (*https://logrocket.com/features/error-tracking-issue-management*)
- Raygun (*https://raygun.com/platform/crash-reporting*)

However, most of these are commercial services that are not as easy to use as Sentry.

10.3 Tools for Monitoring, Observability, and Tracing

> **Note: Reflection**
>
> In my opinion, no tool is more important than error tracking. Learning from mistakes is central to the success of a DevOps environment.
>
> They must not be lost, swept under the carpet, or ignored. It is therefore essential that you organize a system that logs all problems centrally and helpfully.

10.3.5 Distributed Tracing

Tracing is about being able to understand the process of a request, including all the associated data. The concept of tracing is nothing new in principle, but it has become more complex over time. Due to the widespread use of containers and the associated microservice architectures, a large number of services now run in a distributed manner, so traditional tracing, which is limited to individual services, no longer gets you very far.

Instead, you need distributed tracing that can track a request across many different systems. You want to be able to follow these processes in order to find any bottlenecks and errors. In Figure 10.9 you can see what that looks like in practice.

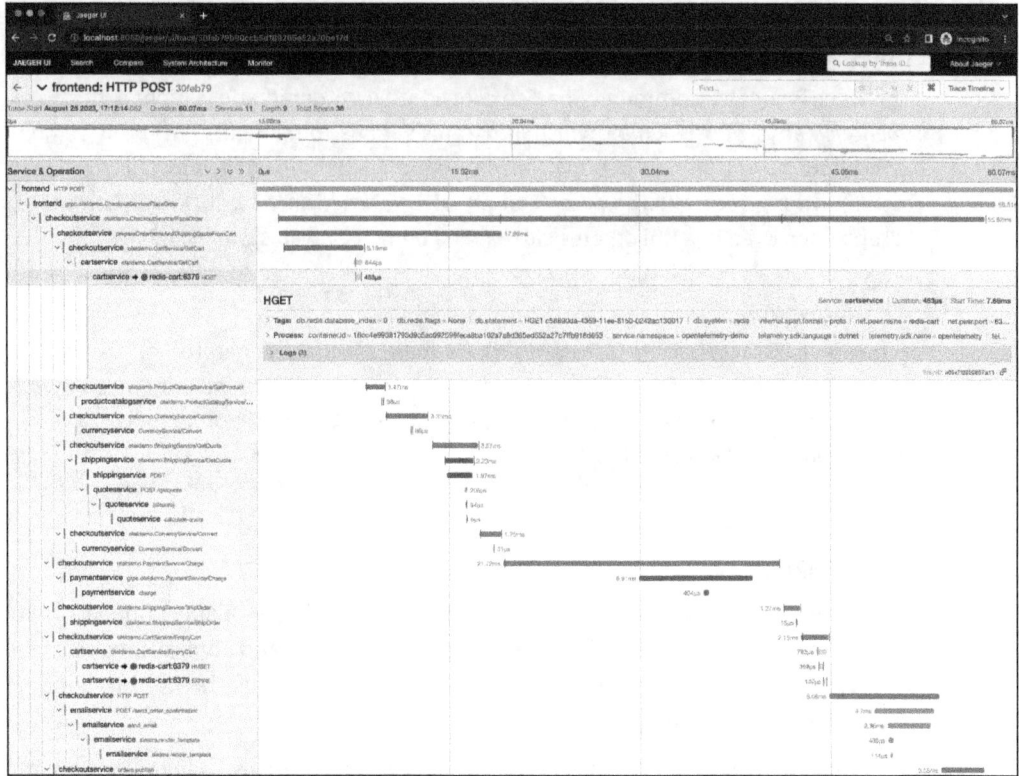

Figure 10.9 Visualization of Traces and Spans in Jaeger Using the OpenTelemetry Demo and the Checkout Service

To collect this data, you can use OpenTelemetry (*https://opentelemetry.io/*), an observability toolbox for generating and managing telemetry data, including not only traces, but also metrics and logs.

OpenTelemetry is *not* a single tool, but a framework and toolkit that is integrated into your own application as a library to provide telemetry data. In order to store, manage, and search the data, a backend that can work with this data is required.

If you cannot change the source code of the application to be monitored, OpenTelemetry offers a way to obtain telemetry data with *automatic instrumentation*. You can take a much more detailed approach if you can modify the source code and call the OpenTelemetry API in your code. You can find more information on this at *https://opentelemetry.io/docs/concepts/instrumentation/*.

One of the backends for tracing is Jaeger. Jaeger is a project of the Cloud Native Computing Foundation that has been given the "graduated" status, just like Prometheus and Kubernetes.

There are numerous other tracing backends, including the aforementioned Sentry, as well as Tempo, which comes from the Grafana project and is deeply integrated into Grafana.

> **Note: Start Tracing**
>
> The OpenTelemetry community has developed its own demo application at *https://opentelemetry.io/docs/demo/*, which integrates various components. With this demo, you can try out the process of tracing an application without having to install or instrument anything yourself. The demo can either be started in a local Docker Compose setup or deployed in a Kubernetes cluster. The demo is a web store including a checkout workflow and consists of several microservices that provide complex processes for tracing analysis.
>
> You can also install Jaeger to record and evaluate the data generated by the demo application. The traces can then be viewed via the Jaeger web interface. Alternatively, it is a good practice to integrate Jaeger as a data source in Grafana in order to have a central location for all information.

10.3.6 Logging

There are several log servers on the market. Many of them are available under an open-source license. A log server sends logs from various services and applications to a central service to store them there collectively. These logs can then be used to conduct analyses based on a large number of sources.

Log servers are now mostly as features in observability platforms, as logs alone are relatively boring and develop their true strength only when enriched with metrics and traces.

The more well-known and widespread log servers include Logstash and Graylog. Grafana offers its own, relatively new log server, called Loki. However, log servers are rarely found as standalone offerings these days. A connection to Elasticsearch is therefore often indispensable, at least if you look at Graylog and Logstash.

Logstash is part of the Elastic Stack, a group of open-source products from Elastic, also known as the ELK stack, for Elasticsearch, Logstash, and Kibana. Elasticsearch is where data is stored and searched; Logstash is used to collect and manage logs; and Kibana is responsible for the presentation of data. Beats, a tool that pushes the data to Logstash and Elasticsearch, is now also part of the Elastic Stack. Elastic now also uses agents as collectors and offers automated management, among many other offerings.

A widely used log server in the cloud-native environment is Fluentd, a project of the Cloud Native Computing Foundation. It is a logging layer between the system and application logs and data storage tools such as Elasticsearch, MongoDB, or Hadoop.

There are also a number of commercial solutions. The most common is probably the log management solution offered by Splunk (which offers other tools beyond its log management solution). Other well-known commercial log management providers include Elastic, Dynatrace, SolarWinds, Datadog, Logz.io, and New Relic.

10.3.7 Service Meshes

Container systems and software architectures based on microservices are undoubtedly useful, in that they ensure that services can be provided much more easily and quickly. But all these different services need to be managed and monitored. This includes routes, timeouts, encryption, authentication, and authorizations, as well as logs and metrics between the microservices. To enable the management and monitoring of these services, a *service mesh* within a container orchestration such as Kubernetes is used.

With a service mesh, each microservice works with a proxy that runs alongside the actual microservice. In addition, regardless of the implementation used, there is a control plane that can talk to Kubernetes APIs or other infrastructure services, for example.

Linkerd, Istio, and Cilium are the three main service mesh tools in use. Linkerd is optimized for usability and performance, while Istio comes with many more features and is much more adaptable. Cilium is actually an entire observability toolstack and implements service meshes using the extended Berkley Packet Filter (eBPF). All tools run under Kubernetes; Linkerd and Cilium even require a Kubernetes environment.

So what exactly does a service mesh like this do, and what is it good for in day-to-day work? For example, load balancing can be easily configured and tested using service meshes in order to route traffic from one microservice instance to another.

And that is just one example of the benefits of using a service mesh in terms of routing. Topics relating to resilience are also relevant, such as when a service does not respond in time and the request is executed again but is directed to another service of the same type.

Other benefits of using a service mesh relate to security and observability. The former is particularly relevant for authentication and encryption functions. In terms of observability, using a service mesh means that there is another dashboard on which you can see the communication between the microservices. This applies to visibility on routes as well as problems that may arise.

Before you jump straight in and plan a service mesh in your Kubernetes infrastructure, bear in mind that doing so will add another layer of complexity to the software architecture! It only makes sense to set up a service mesh if you are relying entirely on microservices. If you're working with a monolithic architecture, then you don't need a service mesh.

In addition, service meshes can introduce significant performance losses and a higher CPU load, as the administrative tasks also need to be completed.

> **Note: Reflection**
> Service meshes are a complex topic. I recommend using them only if you already have a good knowledge of the other topics in this chapter. In practice, a complete setup is rarely found.

10.3.8 Observability Platforms

Observability platforms are still relatively new. As previously mentioned, they involve the complete collection of observability data (metrics, traces, logs, etc.) and their analysis and queries in an overall view of the systems they monitor.

What is exciting is that many of the tools in this landscape are not new, but they have been expanded to include newer observability-related features. They previously fell under the term *application performance monitoring*, but today they are referred to as observability platforms, which expands on the performance monitoring concept.

When looking at observability platforms, one thing stands out first and foremost: There are a lot of tools from different providers. Some are quite sophisticated and very powerful but are not available under an open-source license; these tools are often available only as SaaS solutions. There are also open-source solutions, and some require more work to set up than others. It is not for nothing that most of the individual aspects of these tools have already been outlined in this chapter.

If you look at the purely commercial options, you may recognize a number of tools that were previously offered as application performance monitoring solutions, such as AppDynamics, New Relic, Datadog, Splunk, and Instana. Some of these providers were once independent companies but were swallowed up by larger companies over time. For example, AppDynamics is now owned by Cisco, Instana by is now owned by IBM, and New Relic has already been bought by a private equity company. This is particularly important to bear in mind if you trust some companies more than others—all of these tools are available only as SaaS solutions, so you're essentially handing over your data to another company.

A relatively new player in the observability platform space is the company Honeycomb.io with its SaaS-only tool, Honeycomb. The developers at Honeycomb.io have played a decisive role in shaping the area of observability. They also wrote the book *Observability Engineering*, published by O'Reilly (*https://www.oreilly.com/library/view/observability-engineering/9781492076438/*).

What has prevailed despite the prevalence of closed-source solutions, however, is OpenTelemetry. the players on the market, both large and small, now support OpenTelemetry, so you can at least rely on a free solution for instrumentation.

In the open-source corner, on the other hand, there are two companies that base their commercial products on original open-source software: Elastic and Grafana. Elastic has undergone a license change, as a result of which the tool is no longer considered open source. However, an open-source fork of Elastic called OpenSearch is available.

Both the Elastic Stack and the Grafana stack can be installed and managed on your own hardware and on their respective cloud offerings.

Grafana's stack is also known as the LGTM stack, which includes Loki, Grafana, Tempo, and Mimir.

While Loki is the log server, Grafana is the well-known tool for data visualization that combines the various backends under one user interface and makes them usable. Tempo is the solution for traces, and Mimir is the solution for metrics. Mimir is ultimately a long-term storage option from Prometheus that focuses more on horizontal scalability and high availability, which Prometheus cannot do by default.

Competition is known to stimulate business, and in the coming years we will see innovative features that make observability practices and DevOps workflows even more efficient. As a user, you are spoiled for choice between purchasing SaaS platforms or investing in self-hosted observability platforms. In any case, you should take the time to evaluate the options and choose one that works best for you.

Topics such as *AIOps*—AI-supported approaches to debugging problems—are also becoming increasingly common and could represent a decision criterion for DevOps teams. We take a closer look at AIOps in Chapter 14, Section 14.2.

> **Note: Reflection**
> It is exciting to see how many companies have rebuilt and relabeled their products to include observability features. This doesn't have to be a bad thing, because the basic idea—regardless of the marketing—is very good. You should decide on whether you want to host and manage such a setup yourself with open-source software or rely on a service from a SaaS provider.

> Of all the types of tools we cover in this book, observability platforms are the most recent. A lot will certainly happen in this space in the next few years. One driver of innovation is the eBPF, mentioned earlier.

10.3.9 Monitoring and Observability at nicely-dressed.com

As part of the company's efforts to modernize its tool stack, management wanted the teams at *nicely-dressed.com* to rely on modern and flexible tools. However, these tools had to be fully developed.

Because the company transitioned to a cloud strategy and the use of Kubernetes, relying directly on Prometheus was the obvious choice for monitoring. Prometheus made it possible for the teams to use the exporters from existing services and to quickly write their own exporters so that business metrics could be provided. Grafana was selected for creating visualizations.

For the observability platform, it was important to retain control over the data. The managers decided to use a self-hosted open-source solution, although they determined that the tools of the cloud environment already had sufficient features. In order to keep complexity to a minimum, the company therefore decided not to use its own Elastic Stack and LGTM stack until a real use case was identified. Service meshes were also ignored for the time being.

10.4 Availability

Closely linked to the issues of monitoring and observability is the concept of *availability*, which concerns availability of the entire system. How exactly is availability determined, and who is responsible for it?

Availability is often determined by the pure accessibility of the system. The number of *nines* is often used to measure and evaluate availability. For example, an availability of 99.99% means that a service allows a downtime of 52.60 minutes per year. However, this says nothing about how well the service is available at all. An available service that is so slow that it is unusable doesn't help anyone in the end, but its availability looks good on paper!

In order to define availability more precisely, *service levels* must be defined. Three terms in particular must be differentiated from one another:

- Service level agreement (SLA)
- Service level objective (SLO)
- Service level indicator (SLI)

10.4.1 Service Level Agreements

SLAs exist in other industries besides IT. Ultimately, such an agreement is not much more than a framework contract that defines how a service provider guarantees services to a client. In the case of a web service, an SLA is a contract between the company and the users; in the case of application software, it is a contract between the manufacturer and the company using the software.

An SLA defines various key figures, such as the availability in the form of a percentage and the response time, which determines how long a fault may last.

An SLA always includes contractual penalties, so it is an important tool for determining liability and the question of who is responsible for rectifying the damage in the event of a fault and when a service can be deemed not to have been provided. If it comes to that, SLAs are an important basis for settling legal disputes concerning compensation for damages or the failure of services.

However, SLAs do not exist only between different companies. There can also be SLAs within a company if one team—or another company in the corporate network—provides a service for other teams. In this case, no money or less money flows internally if the SLAs are not adhered to, but there is immense potential for conflict. SLAs can quickly cause responsibility for failures to be shifted back and forth. SLAs that are relevant only to internal obligations are also called *operating level agreements*. These are only internal and are not based on a contractual agreement.

If, for example, a server is overloaded by a clumsily programmed feature and the software is no longer accessible as a result, discussions about who is responsible could begin: Was it the fault of the developers? Or was it the fault of the network team, which was unable to meet the guaranteed availability?

Such discussions are rarely productive and should occur in the context of a blameless post-mortem. Rather than harping on the SLA, it is important to work together on solutions so that both sides can learn from the mistakes and improve their services.

In a DevOps environment, it must be questioned whether SLAs still make sense. An availability of 99.999% may sound excellent on paper, but it is not the most important thing. In the end, despite good availability, users may still not be satisfied with the product for other reasons, which cannot be mapped with SLAs, as they exist solely for lawyers.

Another argument against the use of SLAs is that the highest possible availability may not be desirable at all. To be more precise, is the highest level of availability may be desirable, but the costs of achieving it are too high. The more nines that exist in the measurement of the availability, the shorter the possibility of downtime. An availability of 99.999% means that we are talking about 5.26 minutes per year.

From an engineering point of view, such a level of availability is fairly difficult to achieve and could ultimately prevent developers from being able to implement improvements that would require downtime. Let's assume that access times for end users could be halved by an improvement to the infrastructure. However, this would require the entire service to be shut down for 30 minutes and put into maintenance mode. An SLA with five nines would make this update impossible, which is frustrating for everyone.

> **Note: Absolute Availability**
>
> There are, of course, examples of services that require absolute availability. Think of medical or emergency services, for example. Even the reception of the BBC is of fundamental importance for the survival of humanity: If British nuclear submarines can no longer receive the BBC 4 radio station and there is no other connection to home, the letters of last resort (*https://en.wikipedia.org/wiki/Letters_of_last_resort/*) must be opened.

It must be clear that a high level of availability comes at a high price. Each additional nine costs exponentially more money and effort. Keep this in mind the next time you need to negotiate SLAs.

However, there are also other examples that are not directly related to availability. For example, there are often agreements that state that certain errors must be fixed within a certain time—let's say 24 hours. Several problems can arise in such a scenario. First, the question arises as to when these 24 hours begin: At what point is the error reported? And what should the affected team do if it cannot reproduce the error at all because important information is missing and it takes more than 24 hours for this information to be supplied?

You can see from these examples that an SLA may not necessarily contribute to what's best for the end user. And to ensure the end user is considered in establishing performance targets, you have to look at the entire context, and that is exactly where SLOs come into play.

10.4.2 Service Level Objectives

While an SLA is more of a formal and legal construct, the *SLO* pursues a different goal— namely, a performance target that is set by the service provider itself. It can be understood more as an internal target that should be based on the needs of the users. For example, an SLO can specify a certain level of availability or performance of the system that the service provider is aiming for.

An SLO is explicitly *not* a legal construct, but a definition of the actual metrics. Therefore, while an SLA is binding and defines legal implications, an SLO is more flexible. The focus is not on legal protection in the event of a claim but on the question of how to define what brings the greatest benefit to the end user of the system.

Even with SLOs, there are challenges and typical mistakes that need to be avoided. Objectives can be difficult to measure, be formulated too vaguely, contradict each other, or block each other; for example, an "easy-to-use" but "secure" web store may be impossible to achieve.

It is therefore important that SLOs are clearly and simply defined in order to ensure measurability. And even if SLOs are not usually directly linked to sanctions, they should of course still be adhered to, as their values are almost always linked to SLAs.

10.4.3 Service Level Indicators

An *SLI* is the actual measurable value that indicates the performance and availability of the system or subsystems. The question here is not, as with SLAs, whether we are keeping our promises or, as with SLOs, what the targets are, but how we are doing.

For example, if we have defined in an SLO that the response times of the software should be less than 400 milliseconds, the corresponding SLI would be the current speed, which may be around 200 milliseconds. This would mean that the software is adhering to the SLO, which would in turn mean that the responsible team is complying with its SLA.

As you can see, these three points build on each other. An SLI is part of an SLO, which in turn is part of an SLA. So again, keep SLIs simple. Define a small but meaningful number of values that are useful for end users, because that's what matters in the end.

10.4.4 Error Budgets

In an ideal world, your software would have no outages, and you could guarantee 100% availability in the SLA. However, this is completely unrealistic. There will always be incidents that result in outages. So when you formulate an SLA with your contractual partners, be sure to agree on a realistic target.

A well-implemented monitoring system should give you an overview of how many failures and problems you can expect. Although historical data cannot easily be extrapolated into the future, it does provide a good indicator of realistic targets.

You can derive how much leeway the service has from the difference between the current status of the SLI and what the SLO and especially the SLA allow you to do. This is called the *error budget*. Figure 10.10 shows what an error budget can look like.

With this calculation, you have a quantifiable figure from which you can derive how much time you can plan each month for maintenance windows and routine work. Let's stick with the simplest example—the requirement for 99.9% availability. This SLA corresponds to a permitted downtime of 8 hours, 45 minutes, and 57 seconds per year. Converted to the amount of downtime for a month, that's 43 minutes and 50 seconds. That's not so little.

Exactly this time is the total error budget for the availability of the service.

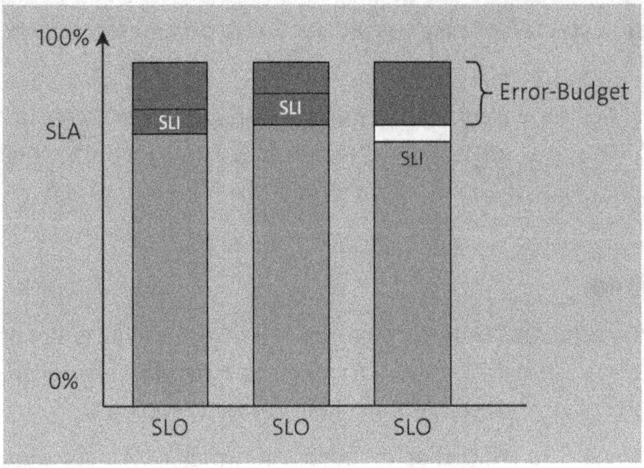

Figure 10.10 Determining the Error Budget

> **Note: Reflection**
>
> Errors, performance problems, and other issues that point to shortcomings in the software should always be viewed critically if you want to improve your team's culture. Nobody likes to admit to mistakes, as we discussed in the previous chapter in the section on blameless post-mortems.
>
> Technical implementation is, once again, only part of the solution. Although you can increase visibility with the right observability tools, this is only one building block for successful collaboration in a cross-functional team.
>
> Make sure that you collect useful metrics and make them visible to the team. This applies to both operational metrics and business-relevant metrics. If you can provide visibility into these metrics, then you should make sure that you specify and use error budgets.

10.5 Summary

This chapter dealt with the final phase of the DevOps lifecycle, monitoring. Findings from monitoring and observability serve as input for the first phase in next round of the DevOps lifecycle, planning.

As with all other phases, one thing is also important in monitoring: the ability of all teams to have visibility of the data and facts of the system. This is the only way to continue working efficiently and purposefully.

This information must be incorporated into the next iterations of the DevOps lifecycle in order to tackle problems and challenges at the root, together in a cross-functional team.

The choice of tools in the monitoring and observability field is exciting, as there is currently a lot going on in this area. You can't go wrong by using Prometheus and Grafana, but keep an eye on new developments so that you don't miss anything.

Chapter 11
Security and Compliance

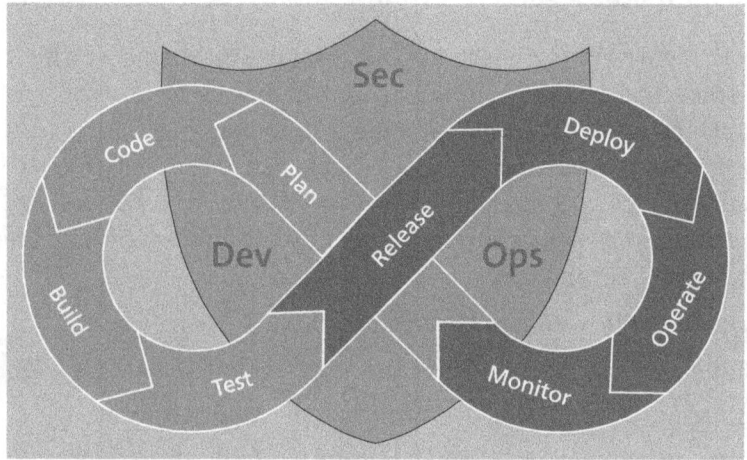

Figure 11.1 Security as an Aspect of DevOps

On a DevOps-focused team consisting of development, QA, and operations, one thing is still missing: the security team. The larger the company or organization, the more likely it is to have a security team that performs its work downstream and is not already part of the DevOps process. The smaller the company, the more likely it is to have subject matter experts in the area of security, although unfortunately it is all too common for this area to be either completely ignored or severely neglected.

This chapter looks at the various security mechanisms and concepts that are specifically formulated by the DevOps idea. These include the integration of the security team into the development department and the integration of security tools and reviews into the CI/CD pipeline.

In this chapter, we will use finding and fixing vulnerabilities as a prime example of the importance of security in a DevOps context. DevOps-focused companies that follow clever processes and use suitable tooling can quickly rectify these vulnerabilities, as they have a good overview of the dependencies and can roll out updates and bug fixes quickly.

The focus here is on *application security*. However, this part of DevSecOps is more than just looking at the security of the application—a look at *supply chain security* is also necessary. In addition, the practicable implementation of compliance guidelines in the company is also an important point in this chapter.

We will make repeated excursions into the security of the infrastructure, but this is not the focus of this chapter, as the planning and maintenance of the infrastructure is usually not the task of the software product team.

11.1 Security Disrupts the Agile Waterfall

At *nicely-dressed.com*, there was and is a security team that takes care of the security of the application and the infrastructure. Security is an important issue, because vulnerabilities that are exploited and lead to data leaks result in not only the loss of trust among customers, but also serious penalties, which should be avoided at all costs. For example, the General Data Protection Regulation (GDPR) stipulates severe fines if data is not handled carefully, and insurance coverage is possible only if the infrastructure complies with certain standards and rules.

To ensure security, the security team at *nicely-dressed.com* has three main tasks:

- Check the application itself to ensure that it is safe
- Check the infrastructure on which the application runs to ensure it is secure
- Check the infrastructure around the network to ensure it is secure

With these three tasks, the security team is not only very busy, but also often overloaded, and not all tasks can be solved satisfactorily because the staffing level is too low. As you can already see from the tasks the security team is responsible for, it has to work with almost everyone: the development team, the operations team, and the infrastructure and network team.

All three tasks are time-consuming and involve a lot of work that is never fully completed. It is necessary to continuously check whether new features and changes contain security vulnerabilities. The security team is also involved every time there are changes to the infrastructure or the network and has to check with the other teams to ensure everything is secure.

To this end, the team has drawn up various rules, particularly regarding the network structure and the isolation of the various networks from one another. These rules also explain which firewall rules exist and how the firewalls between the servers are to be set up.

Therefore, for the other teams, their work with the security team mainly consists of working through long checklists every time a change is made. And because not everything security-related is always fully documented, the other teams often have to make inquiries and give feedback to the security team, which only causes further blockages. These checklists then end up with the team that defines the compliance rules.

The workflow usually looks like this: The development team continues to develop its application as usual during the long development phase. As the development time in the agile waterfall model is around six months, a number of changes are made during this phase, which are then first deployed to a staging environment before being deployed to the production environment.

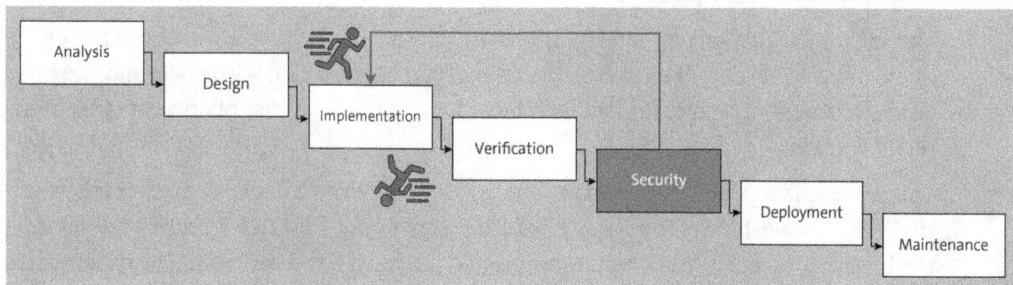

Figure 11.2 Implementation Blockers Due to Security Team in Waterfall Model

Before every major deployment of the online store, the security team is the last team in the process, as displayed in Figure 11.2. For each deployment, it checks whether it can find any security vulnerabilities. The team members use various techniques and methods for this. During this phase, they examine many questions, including the following:

- Can a simple penetration test find vulnerabilities? And can those vulnerabilities be exploited in practice?
- For example, is SQL injection possible in the forms?
- Do the dependencies in use have documented security problems that also affect the web store?

Some of the tests the team runs can be automated. However, the long six-month development phase means that everything has to be revised with every deployment. Just think about the process of checking the dependencies: Since the development team had previously neither used the package managers of the respective programming language nor documented the dependencies in a clean and up-to-date manner, the security team often has to manually gather and check the dependencies that are being used.

Since the timeframe of the development phase is so large and it is correspondingly difficult to get a working and reproducible build, the developers shy away from touching the dependencies: "Never touch a running system. The library has been doing what it should since 2013."

Such findings always cause sheer horror for the security team, as a lot of known and documented security problems find their way into the application. But even if the developers wanted to do everything right and updated everything at the beginning of a development phase, things usually look completely different six months later when the deployment ends up with the security people.

The security team therefore carries out an in-depth review before the deployment of new versions of the web store, at the end of which the team makes an assessment of the security status of the release. As security is not a priority for the development team, there are always a number of vulnerabilities and problems. Some of these problems are the use of outdated dependencies, but often they are simply errors in the source code that need to be corrected.

The security team is never able to approve a release on the first attempt, so the developers often have a lot of extra work, as they then have to deal with code that they had last seen six months ago. And all they have to help them is a list of criticisms from the security team.

This process is repeated more and more frequently before a deployment, and the security team always blocks the release because they either find new security vulnerabilities or that previously reported problems had not been fixed properly. The other teams are never on good terms with the security people, who are always blocking new features and urgently needed changes to the web store.

The pain is greatest in the waterfall model, but how can the team improve the situation through its transition into DevOps?

11.2 DevOps with a Separate Security Team

It is not uncommon for a company transitioning to DevOps to break down the silos between the development team, QA, and operations team but continue to have the security team live an isolated existence at the end of the value stream without having any influence on the cross-functional team.

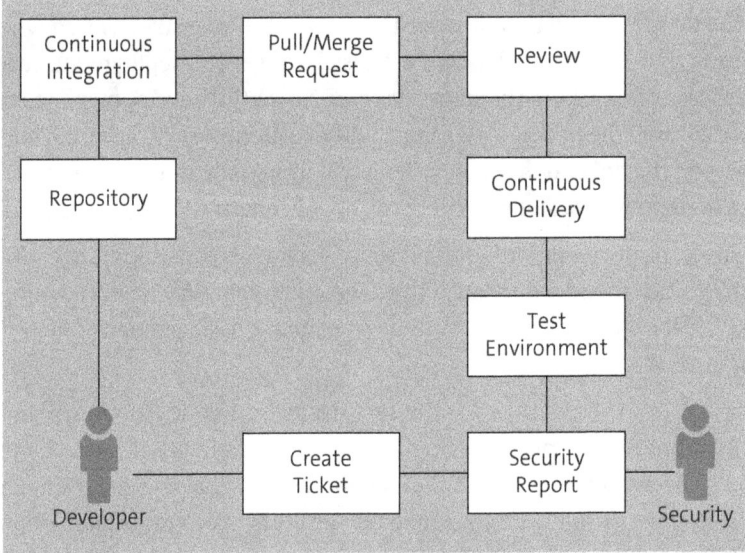

Figure 11.3 The Workflow with a Downstream Security Step

Such companies have taken the first steps toward reducing the lead time for development, but security is still an afterthought. Security is understood as an extra task after the work is done. It is therefore equated with a tick on the checklist that is common in a waterfall environment.

The workflow in such companies looks like this: Development continues apace, now according to the usual DevOps principles. The code is committed to the shared repository, the continuous integration pipeline starts after a pull or merge request, and, after a review, deployment to a test environment is triggered (see Figure 11.3). However, as the security team is still working downstream, the security report is still created manually and only at the end of the development phase. And this continues to be blocked, as the problems are only ever found when the actual work is already over.

> **Note: No Access to Security Reporting**
>
> Another common problem is that security reporting is not visible to the development team. This reinforces the silos between teams, because if the relevant people cannot access the same data, this data is not only quite cumbersome to use, but it is often simply ignored. From a development perspective, this is hardly surprising: You can't correct what you can't see. This is never ideal.

11.2.1 To Deploy or Not to Deploy?

At *nicely-dressed.com*, many of the security team's findings do not pose such a big problem. Sometimes a few lines of code need to be changed, such as to secure a field in a form against SQL injection or to place a check. This is annoying, but feasible. However, there are regularly nasty bombshells that fundamentally affect the design of the application or criticize the dependencies used—problems that should have been corrected much earlier in the development process. Now that an entire house of cards had already been built on this foundation, such drastic changes could only be made with great effort.

A simple case that occurs repeatedly is the use of outdated dependencies that not only have glaring security vulnerabilities but are no longer maintained at all. Essentially, these dependencies no longer receive security updates. The only solution is to replace the relevant package (see Figure 11.4).

However, such dependencies are often so deeply rooted in the online store's code that extensive refactoring is required. Shortly before the planned deployment, this is impossible if the teams want to meet the deadline, and they don't want to add any more untested changes to the web store's foundation on short notice.

In these situations, the security team and the management team at *nicely-dressed.com* have to weigh the options: Should we approve the release despite the serious security vulnerabilities? Or should the deployment be postponed until the vulnerabilities can be corrected? Should we hope that no issues related to these vulnerabilities will happen? Or should we start all over again with the planning and implementation of the new features?

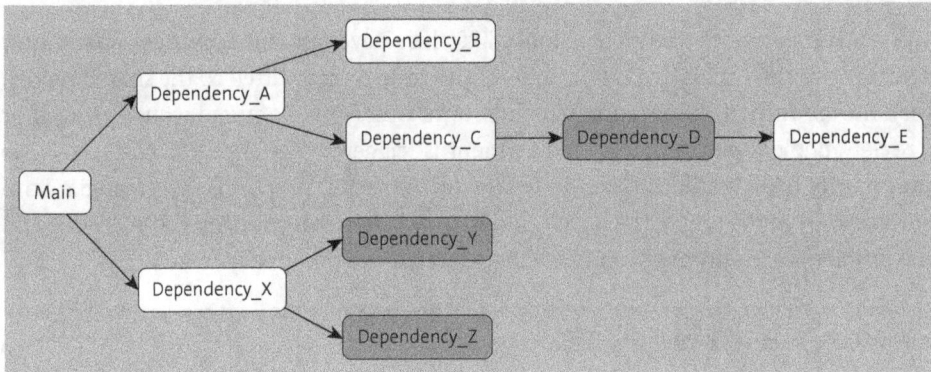

Figure 11.4 A Dependency Tree with Vulnerable Downstream Dependencies

Both options involve risks: Security vulnerabilities could be exploited, so it can be really expensive to leave them in the store. But postponing the development is also hardly possible, as the Christmas season is just around the corner and the new offers are likely to result in a considerable loss of sales if the store is not launched on time.

11.2.2 The Search for Undocumented Dependencies

In all the hustle and bustle, the security team often only checks the bare minimum. A proper security audit takes place far too rarely, and when it does, it takes forever before the development team half-heartedly fixes the vulnerabilities. Here, there are few concepts for tackling this strategically.

However, the security team is also in constant demand after deployment and during operation. The security team has subscribed to the usual security advisories and can therefore keep an eye on when new widespread security vulnerabilities are reported; when it learns of these vulnerabilities, the team passes the information on to the various internal teams. Some vulnerabilities even make it into the daily news, so that management becomes aware of them and naturally wants to know whether the vulnerabilities are also relevant for *nicely-dressed.com*.

For example, consider the log4shell vulnerability in the log4j library that emerged in December 2021 and made the evening news. Anyone who was responsible for IT systems inevitably asked themselves, "Are we actually affected by this too?" (I will come back to this in Section 11.5 when we discuss supply chain security.)

Security teams in many companies had to search all applications for use of this library. This vulnerability affected both applications developed in-house and these applications' third-party dependencies that used log4j. And if log4j was found, these companies had to make sure the vulnerable version wasn't in use.

And where could they even look? As a lot of shadow IT had crept in over time at these companies, the security people had to dig through numerous repositories by hand. Many departments had their own Git servers that could not be searched centrally; it was a lot of work to get an overview of what code was being used at any given time.

Some security teams found that their applications were using old versions of log4j or other libraries, so the applications weren't affected. But it isn't necessarily a good thing if your software is so old and outdated that it isn't affected by a recent security vulnerability.

Let's summarize: A major problem for security teams in traditional, non-DevOps contexts is that they have no up-to-date insight into the source code of individual development teams. Compartmentalization of development teams makes auditing incredibly time-consuming; timely decisions or even solutions are not possible.

This is not good news for management at these companies, who only want to know whether the vulnerabilities can be exploited. Even this simple question cannot often be answered, and plans to patch systems in a timely manner can often be a long way off.

11.2.3 Frustration and Blocking

The development team at *nicely-dressed.com* has hardly any security knowledge, as it has always relied on the security experts to ensure that everything is secure. On the other hand, the development team always feels frustrated when the security team hands it a long list of defects, sometimes months after the programming work had been completed. And since the security team doesn't know the code, many of the points on this list tend to be incomprehensible or simply impossible for the developers to implement.

The development team is therefore never on good terms with the security team and perceives the security experts as bureaucratic naggers and doubters. The security people, in turn, wonder why the developers don't just write clean and secure software: "It can't be that hard!" So each team always blames the other, and the finger-pointing starts all over again before every deployment.

> **Note: Reflection**
>
> Unfortunately, security is an issue that is all too often neglected. Development teams and managers often think that security issues won't happen. Together with the structural problems at *nicely-dressed.com*, this causes not only tangible security problems but also a lot of friction between the teams and inefficient processes.
>
> Of course, security vulnerabilities are bad and should be corrected as soon as possible. This is not only common sense but also required by various legal regulations.

> I can only recommend that technicians working in the traditional waterfall environments keep pointing out security risks to their team. Above all, however, it is important that managers also recognize these problems and understand that they need to invest in security.

11.3 DevSecOps: Building Security into DevOps

DevOps enriched with security is called *DevSecOps*. The basic idea is, you guessed it, that security becomes part of the entire software development lifecycle.

Before we dive into the depths of DevSecOps, there is one more conceptual question: Is DevSecOps a subset of DevOps, or is DevOps a subset of DevSecOps?

At first glance, you might think that DevOps is a subset of DevSecOps. The reasoning behind this is that DevSecOps is *more* than just DevOps, so DevOps must be a subset. The reality is actually quite the opposite: DevOps should be understood as the broader concept and DevSecOps as a specialization—it is DevOps enriched with security. There are also good reasons not to consider DevSecOps as the broader term encompassing DevOps as a specialization: The QA team and the platform team are not included in the DevOps name either—otherwise, term would become DevQASecPlatOps, or something similar.

> **Note: DevSecOps versus DevOps**
> When the term DevSecOps is used instead of DevOps, this is usually because the security part is being emphasized. If, on the other hand, the term DevOps is used, the topic is considered more generally, which should also include security.

So what exactly is DevSecOps? It concerns the integration of the security team into the complete development lifecycle at an early stage. The aim is not to completely disband and replace the security team, but to integrate its activities and knowledge into the development lifecycle as early as possible so that it becomes a natural part of the development work. The responsibility for developing a secure product, therefore, no longer lies in a process at the very end of the path, but is distributed among all teams involved. Security always plays a role, from the planning phase to implementation and deployment.

The security team should always be consulted during the architecture and planning phases and should be present when new features are implemented. This means that the security team can be consulted at an early stage during development without any major hurdles.

As is so often the case when a company is transitioning to DevOps, the first step is to greatly improve collaboration between the security team and the other teams. It is important to eliminate the typical view of security teams as a blocker that always causes more work for development teams.

11.3.1 The DevSecOps Team Structure

As is so often the case, there is not just one truth when building a cross-functional team. Figure 11.5 shows *one* possibility; in this example, the security team members are divided up among the cross-functional teams, so each new project team gets its own security expert.

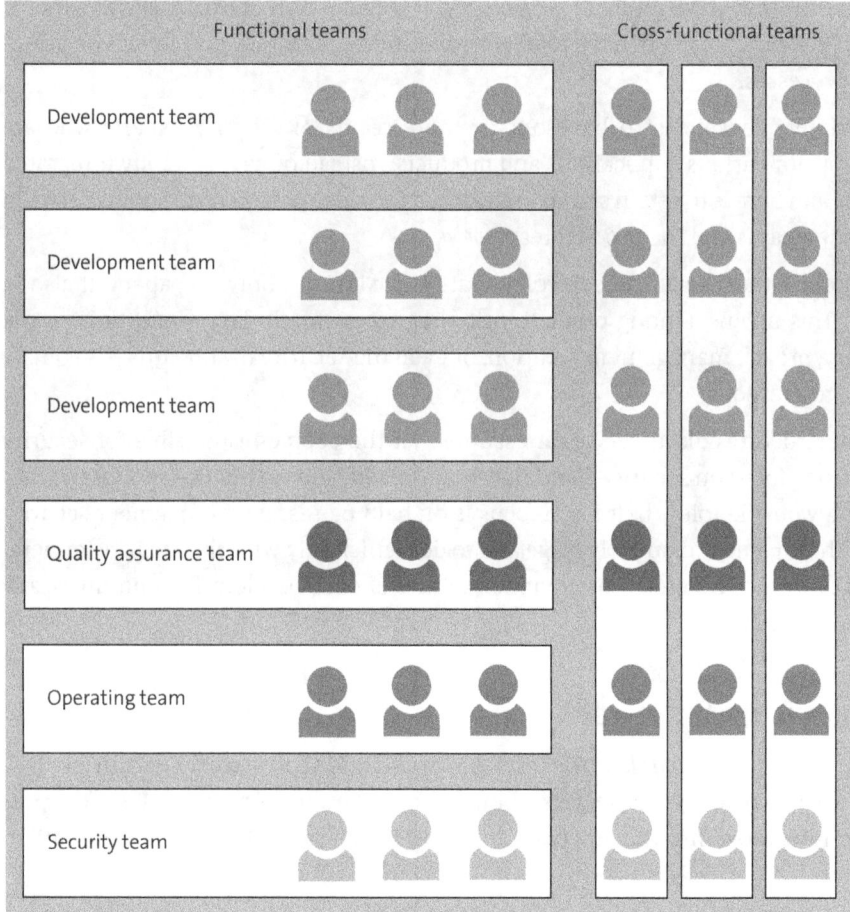

Figure 11.5 A Cross-Functional Team Supplemented by Security Members

However, it is worth taking a closer look here. Chapter 7, which dealt with QA, already described a dual team structure for QA: On the one hand, a person with a focus on QA is part of the cross-functional team for day-to-day work, but the standalone QA team continues to exist, which is responsible for the methodological improvements of the organization.

This concept can also applied security: A standalone security team remains in place to establish organization-wide guidelines on security issues, which are then to be implemented by the individual product teams. In addition, security experts work in all teams on day-to-day business and are involved in the development lifecycle from planning to operation.

Whether this approach works smoothly depends on various factors. One key influencing factor is the size of the company. Smaller companies often have only a few teams and usually hardly any dedicated security experts. A common problem in such companies is the limited number of security experts. The situation is often different in larger companies, as they have more employees and it is easier to fill positions with experts. However, they then usually have to deal with more bureaucracy and longer decision-making processes.

An alternative concept is to have one or more teams of security experts who are responsible for various applications and products. Instead of having a fully integrated team, the aim here is to retain standalone security teams but to keep the walls between the security teams and the product teams low.

It is important that the barriers to collaboration are low not only on paper but also in practice. This means, among other things, that access to all data—be it data in the repository, project management solution, or even the security dashboards—is equally available to everyone.

The team structures discussed so far assume that the people responsible for security have a strong focus on security—and rightly so. In addition to this, DevSecOps coaches can play a valuable role. Their main focus is on helping teams to implement security scans in the pipelines at an early stage and to deal efficiently with the security vulnerabilities that are discovered. These vulnerabilities should be identified automatically and closed promptly.

11.3.2 Shift Left: Find Errors Earlier

This approach is called *shift left*. The tests that previously took place far to the right (i.e., late in the software delivery lifecycle) should instead be shifted as far to the left as possible (i.e., integrated into the process earlier, as you can see in Figure 11.6). The earlier, the better!

If security vulnerabilities are found in the production environment, this is problematic; on the one hand, they can obviously be exploited, and on the other hand, it takes a while to implement, test, and roll out fixes for them. The more vulnerabilities are found in the production environment, the more time-consuming and, therefore, expensive the fixes are.

It is better to find vulnerabilities earlier. If they are found in a test or staging environment, then that is a little better. But here too, such fixes are expensive in the sense that the context of the development has been lost and all changes have to go through the review process again.

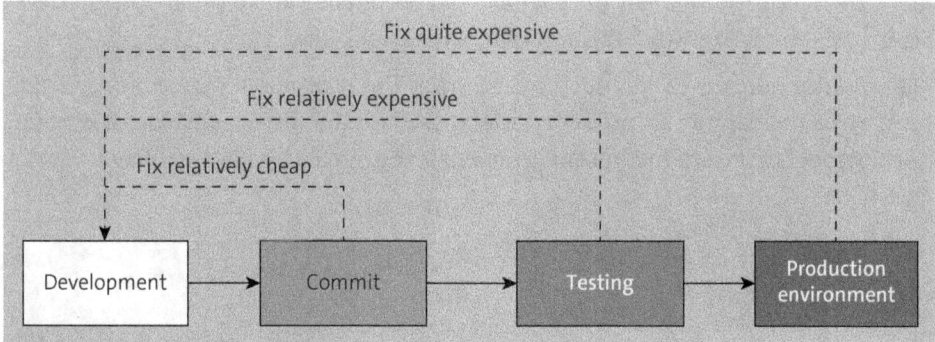

Figure 11.6 Shift Left Approach

Ideally, a security vulnerability is found at the latest during the commit—that is, when the pipeline is running, before a review takes place. The tooling should then support the process in such a way that the effects in the security context are shown for every change that you want to contribute. If no new vulnerabilities are added, then everything is fine. If they are, then countermeasures must be taken.

But it is also possible to go one step further: Some security checks can already be run in the development environment. Ideally, feedback is provided as soon as the lines of source code are typed.

11.3.3 Inner Sourcing Ensures Formal Security

The process of opening the repositories, described in Chapter 5, Section 5.5, and the joint project planning process for all teams within the organization also help the security team. The team can intervene in the transparent planning process at an early stage and provide feedback. This makes work easier for the developers, as they can quickly write to the security team and obtain assessments from the experts.

The prerequisite for this is, of course, that the entire team plays with open cards and that planning is transparent. The more problems and challenges are documented asynchronously in tickets or in a knowledge database, the easier it is to provide assistance.

Let's return to the example of the use of a dependency that has been discontinued by the upstream project and is no longer being maintained. By opening up the code and making it visible throughout the company, the security team members, who have become part of the previously standalone development teams, can directly see that the code depends on an outdated third-party tool. They can either veto its use directly or at least discuss within the team whether an outdated dependency can be used at this point in the project.

11.3.4 Security as an Integral Part of the Development Process

If DevOps principles are implemented well and there is a short lead time for development, then security fixes can and should be rolled out quickly to avoid potential risks. Figure 11.7 summarizes what this workflow should look like.

The workflow already looks much better: After the continuous integration pipeline runs, there is already a security report that is used for the current development before a deployment to the production environment is triggered. This has two obvious advantages:

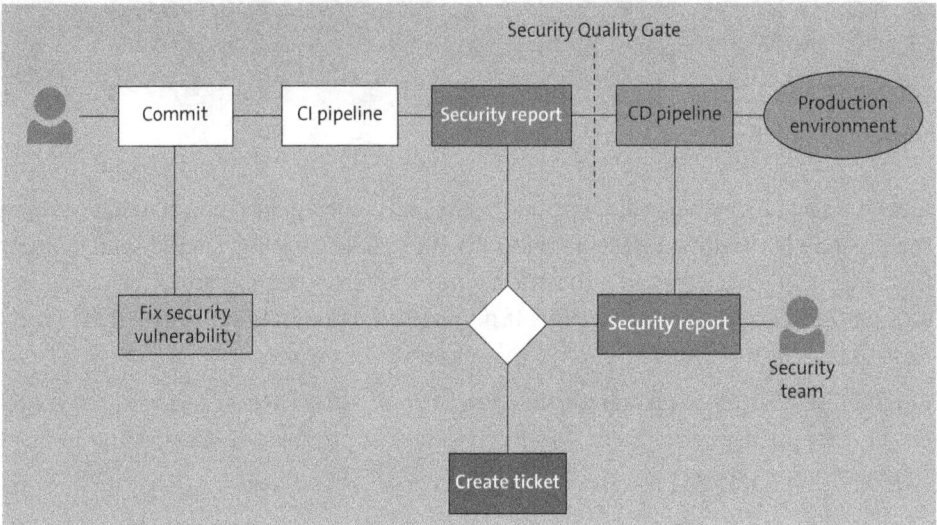

Figure 11.7 Security Quality Gates According to DevSecOps

- The developers responsible for the implementation receive feedback at an early stage. Right at the start of the work, it becomes clear that it would be a bad idea to implement the deployment as planned. Feedback about errors may be perceived a little negatively by developers—as if the developers are deliberately building in security vulnerabilities. In reality, however, most errors happen unintentionally, usually out of ignorance. The sooner errors are pointed out, the easier it is to avoid unnecessary work. With the right collaboration, everyone benefits from early intervention instead of intervention just before deployment.
- The security team can automate tasks and be much more agile. In a company composed of silos, security people are constantly overloaded because checking a deployment is like a single, huge, big bang release. Inevitably, they have to focus on what is most important, and many issues go unaddressed. However, by using security scanners early in the development process, the work is broken down into many small steps, making testing the release much quicker and easier. If everything is implemented optimally, the burden on the security team is reduced: The simple things

that can be automated no longer block their work but are solved by the development team with the help of the scanners. This is, of course, very helpful for chronically understaffed teams.

Simpler workflows also mean that manual work such as checking for outdated dependencies is no longer necessary. Subscription to and distribution of security advisories can instead be automated via the security scanner. If new vulnerabilities are found in the dependencies, a ticket could be created directly for the relevant team and prioritized if necessary so that the vulnerabilities are corrected as quickly as possible.

Security teams can therefore focus more on the production systems, which also still run separate scanners that regularly check the production environment.

11.3.5 Dealing with Mistakes

If you give security a higher priority in your workflow and integrate it early on in the development process, many errors will inevitably be found. Of course, nobody likes bugs and security issues, but it is important to find them. However, it raises the problem of how to deal with them effectively.

Quality Gates

Let's start at the beginning: What do you do with security vulnerabilities that are found before they are merged into the production branch? This is the ideal case, as errors can be corrected and rectified most easily when they are found early. There are *quality gates*, or defined checkpoints or criteria within the development process that are intended to ensure that the code meets certain security requirements before it moves on to the next stage of the development process. If the code does not pass a quality gate—for example, if a vulnerability is discovered during an automated security scan—then its development is stopped until the identified problem is rectified.

That sounds sensible in theory, of course, but in practice, the following questions arise:

- Which criteria should stop the development process?
- Who decides what a real showstopper is?
- How do you configure quality gates?
- What level of vulnerability stops the deployment?
- Who can approve changes, even if they lead to new vulnerabilities?

There are hardly any general answers to these questions; the answers will of course depend on your individual requirements. It will obviously look very different for the security of a nuclear power plant than for a company developing small online game. To begin with, it is important that the vulnerabilities become visible through quality gates in the first place. Ideally, once a merge or pull request is created, the people responsible will immediately recognize that changes need to be made during the development work.

> **Note: Compliance Policies**
> Compliance policies can be created in both GitLab and GitHub. Basically, the aim is to define policies that ensure that quality gates are passed. The typical compliance functionalities are included in the paid enterprise versions of both tools.

The second step is to create an overview of the existing vulnerabilities. This is where dashboards are necessary and helpful. Good scanners also provide useful information about vulnerabilities found and how they can be corrected.

Once this overview is available, changes are made to systematically eliminate the vulnerabilities found, which is the task of the development team.

The enforcement of approval rules when introducing new vulnerabilities should take place in a third step. This means that problems are accepted only under certain conditions—that is, exceptions under which known vulnerabilities or outdated dependencies may be deployed are defined. One reason for such approvals may be that fixing certain problems can be very time-consuming and new problems are likely to arise as a result. In such cases, it must be clarified whether the risk is acceptable. A transparent process should also record who is responsible for these approvals and what information was used to make the decision to allow these exceptions.

Obviously, this decision should never be made solely by the developers responsible for the code. Instead, the security team must be involved in this decision so that they can work with the developers on how to proceed. At the very least, it must be possible for security experts to intervene before changes are made and not just before the entire product is to be deployed.

In addition to serious decisions regarding the approval of code with known vulnerabilities, however, it happens much more often in practice that *false positives* have to be waved. False positives are alerts from scanners that do not reflect actual security risks. False positives can be problematic for development and security teams, as they require additional resources to review and fix perceived security issues that do not actually exist.

> **Note: False Positives**
> There will always be false positives. You will have even more false positives if you start all scans in a project from one go. *Tuning* the scanners is therefore important; it is best to proceed step by step when you put the scanners into operation and check whether the results make sense. Any false positives must be identified and eliminated.

Even more insidious than genuine false positives are reported security vulnerabilities that are dangerous only under certain conditions and pose no immediate risk in your system. You can live with such problems, but nobody knows what the future holds—

perhaps they will one day become actual risks in your system. The only action you can take in such cases is to document the vulnerabilities as transparently as possible so that your colleagues know immediately what to do if there is a need for action in the future.

11.4 Tools for Higher Security

So far, we have talked about security scanners, dashboards, and the like only in general terms without taking a closer look at how they work. As already mentioned, the security scanners are to be executed in the CI/CD pipeline, in which the code is checked early and automatically. The basic idea is that by running these scanners in the pipeline, feedback on security becomes a natural part of the development process.

The idea behind this is that a complete scan can be automated using the CI/CD tools. A complete security scan should provide an overall picture of the code that goes beyond individual analyses of specific products or the workflow, as compiling findings from across different analyses is so time-consuming and error prone that only a complete overview makes sense.

As always, the best tools are useless if their users cannot use them correctly. And since no one can fully grasp the complexity of software that has evolved over time, the various tools should take on precisely this task and clearly present and summarize security problems.

When implementing DevSecOps, it is important that the security tools are as easy to use and integrate as possible. It is the security team's duty to select the right tools with the right data and meaningful dashboards in order to effectively use the results of the scanners.

11.4.1 Dashboards and Reporting

Security dashboards and reports are the easiest way to share the findings of security scans with all teams. They visualize dependencies and highlight known security issues. Ideally, you can see at a glance whether a vulnerability affects only a small, insignificant part of your application or whether the entire monitor lights up red.

It is important for the successful implementation of DevSecOps that everyone can look at the same dashboards. A common problem that leads to silos is that only a few people from the security team have a view of the dashboards. This is not only a technical problem but also a structural problem.

If not all relevant people—developers as well as security experts and managers—can look at the same dashboards displaying the identified security vulnerabilities, there is, again, a risk that the vulnerabilities will not be fixed or will be discussed away, as everyone involved in such discussions are not basing their reasoning on the same standpoint. Conflicts arise again because not everyone can see everything and make decisions based on the real data.

When the security team sees the same dashboards as the DevOps team, the teams can take various actions based on the vulnerabilities listed there.

11.4.2 Pull and Merge Requests

When the security team identifies security problems, there are typically two basic reactions: For smaller problems, the security team could implement quick fixes directly in the source code by contributing the necessary changes via pull or merge requests, which then go through the normal reviews (which, of course, works only if there is easy access to the repositories). In the event of major problems that may even affect the architecture of the application, the team should open a ticket in the system so that fixes for them can be considered in the planning for the next sprint. But even if only minor changes are necessary, the feedback must be visible via pull or merge requests. Even without a review by another person, the developers can find out directly what needs to be changed in their code.

Once again, the intention is to integrate many changes as early and as promptly as possible. For example, when new dependencies are integrated, they should be scanned immediately by the security tools. And this is true whether changes need to be scanned automatically or reviewed by a human. If you wait too long and develop too much without running through the security checks or reviews, you still haven't won anything.

11.4.3 Security Scanners in Detail

So far, we have mainly looked at what the DevSecOps workflow looks like, which problems security teams look for, and when the workflow should take place. It should have been obvious that automated security scanners are very important here. They are part of the CI/CD pipeline and carry out security checks there. They can be roughly divided into the following task areas:

- Static application security testing (SAST)
- Dependency scanning
- Container image scanning
- Dynamic application security testing (DAST)
- Interactive application security testing (IAST)
- Secret detection
- Fuzz testing

Depending on the security scanner, security vulnerabilities are searched for at different levels. There are three basic levels: the source code, the binary, and the delivery mechanism. Figure 11.8 shows what most of the scanner task areas indicated in the list above focus on.

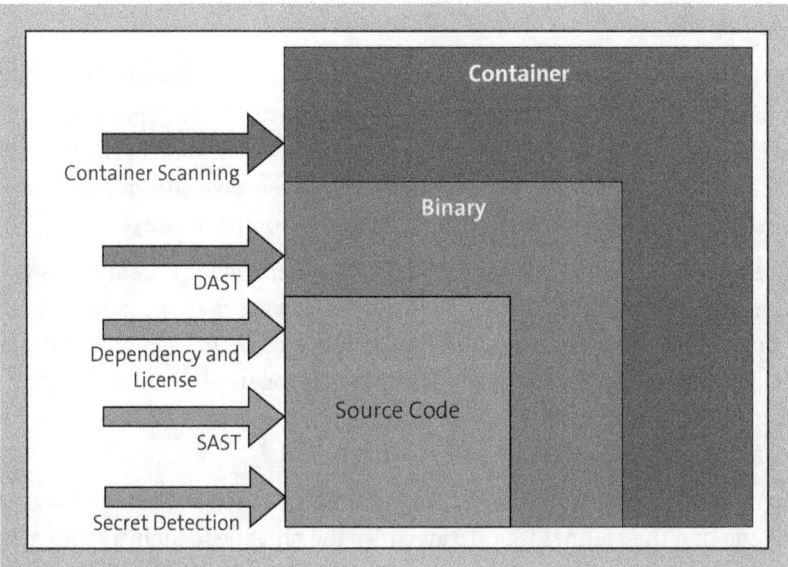

Figure 11.8 Security at all Levels

All security tools look for known security vulnerabilities. Every tool is backed by a commercial company with a commercial product, as managing and updating vulnerability databases requires a lot of work. Many of the commercial offerings also come with their own dashboards and reports. While these features are generally an advantage of these offerings, they are of no use if nobody looks at the results of the scans.

It is important that the integration is kept simple for the various teams and that the hurdle of looking into vulnerabilities is kept as low as possible. Companies often make the mistake of integrating too many tools with too many dashboards. A team member who just wants to do their job doesn't want to deal with constantly switching between tools and dashboards might start to ignore them.

There should be a central location for identified vulnerabilities. Ideally, you should use GitLab's or GitHub's built-in features directly, which are packaged in the commercial offerings. This is the simplest option, as it provides the best visibility and integration into the development workflow. Alternatively, there are various other tools for vulnerability management; in this case, however, the vulnerabilities found must also be reported there. DefectDojo (*https://www.defectdojo.org/*) is one such tool.

Dependency Scanning

Dependency scanning is our prime example of security scans in this chapter. It is very common for projects to be using a number of dependencies that are outdated, either because they the version in use hasn't been updated in the project or because the maintainers are no longer updating them. Dependency scanning does little more than search through the dependencies used in a project and build up the entire dependency tree (see Figure 11.4).

Building a dependency tree is important because not only can the directly integrated dependencies have vulnerabilities, but so can the dependencies of the dependencies of the dependencies. The scanner builds the entire dependency tree and checks the vulnerability database to see if any of the dependencies in the tree have known vulnerabilities. *Transitive dependency resolution* is the technical term for this process.

An essential prerequisite of being able to efficiently run dependency scanning is the use of a common, programming language–independent package management system. User-friendly dependency scanners should automatically recognize which programming language and which package management tool is being used. Having this information makes it easier for the scanner to build the dependency tree, look up security vulnerabilities, and then present them in the security report.

The development team must ensure that a package manager such as Maven in Java is used by default and that the scanner is integrated into the pipeline to identify the security vulnerabilities in the dependencies.

However, some dependency scanners go even further and have the ability to automatically update dependencies. They can either automatically create commits that update outdated dependencies or generate pull or merge requests. This feature minimizes the manual effort needed to update identified dependencies—only a merge needs to be performed to complete an update.

Updating dependencies by utilizing full automation may seem very helpful at first glance, but some forms of automation should be avoided—namely, automated merging of dependency updates. This type of automation can lead to unpleasant side effects, as incompatibilities with your own software could occur as a result of dependency updates. On the other hand, it makes a lot of sense to use automated creation of pull or merge requests, as it allows you to check whether the new version is still compatible or whether you need to make adjustments for the new version to be used.

Tools that have automation features include Dependabot, developed by GitHub (*https://github.com/dependabot/*), and Renovate Bot, developed by Mend (*https://github.com/renovatebot/renovate/*). While Dependabot works excellently with GitHub, Renovate Bot can also be easily connected and used on other platforms. A practical feature is that it can be integrated into the pipeline so that security vulnerabilities can be identified at an early stage. Both tools have automation features for generating pull or merge requests that your team can then review and accept.

Container Scanning

If you use containers for your deployments, these should also be scanned. This process is often referred to as *container scanning*. There are also several types of scanning mechanisms involving containers. One is *container image scanning*, which takes place in the pipeline, and another is *container runtime security*, which takes place when the container is running in the execution environment.

Container Image Scanning

The basic idea of containers is that they are based on images that contain not only the application but also the required dependencies and the base operating system.

In pure image scanning, the container image is scanned for possible vulnerabilities in the underlying base operating system. Security vulnerabilities that originate from the underlying Ubuntu image, for example, are often identified. In this case, the base operating system must be updated and a new image built.

In practice, however, the priority of these vulnerabilities is likely to fall behind those found by SAST and dependency scanning, as you have no influence over which images are used for your containers. If you cannot update the operating system and the basic dependencies or change the image, then you will have to live with these messages for better or worse.

> **Note: Scanning More Than Just the Image**
>
> Some container scanners do not only scan the underlying operating system but also perform dependency scanning and/or SAST. This means that some functions of different scanners may overlap.

The container image should always be scanned regularly in the pipeline so that any errors can be found early and regularly. Some container scanners, however, limit themselves to scanning the container registry in which the container images are stored. If this is not done correctly, vulnerabilities can be overlooked.

Container Runtime Security

In addition to checking the images, you also need to take care of *runtime security* by running the container in a test environment and to scan it for several types of attack vectors.

You always want to know whether containers are at risk of any known security vulnerabilities while they are running. After all, image scanning takes place only during the development phase, but what really matters is how the application behaves in the running environment.

Strictly speaking, checking runtime security is also pure image scanning. As an additional security mechanism, security vulnerabilities could be detected if malicious code is injected into the container from outside in order to adapt the behavior or access data.

Closely related to this is *drift detection*, which checks whether changes have been made to the container during runtime.

There are a large number of different open-source and commercial tools for scanning container images. One free tool is Trivy, developed by Aqua Security (*https://github.com/aquasecurity/trivy/*). As is the case with many of the tools we've mentioned, Trivy also has a commercial version with more features. An alternative is Grype, developed by Anchore (*https://github.com/anchore/grype/*).

It is also worth taking a look at the tool KICS, which is developed by Checkmarx (*https://kics.io/*). In addition to the configuration files of Terraform, Kubernetes, and Ansible, it can also read Dockerfiles and provide security information.

Static Application Security Testing

SAST is the static analysis of source code with a focus on finding vulnerabilities. With SAST, the source code is statically scanned once, and as a rule, the code is not executed. Depending on the programming language, some security issues can be found quickly and easily using SAST. In particular, problems with memory management, such as those that occur in C, as well as SQL injection vulnerabilities and similar problems can be detected quickly and automatically.

Because the fundamental question is whether the written code is secure, SAST scanners are heavily dependent on the programming language used. Some scanners focus on individual programming languages, while others focus on a broad selection. They should be run in every pipeline to detect early on whether new security vulnerabilities are being introduced.

SAST is a white-box test, as it looks directly into the code. This means that theoretically SAST scans can be executed even if the project cannot be built. So you can run SAST scans both in the pipeline and within the development environment in order to detect any security vulnerabilities very early on in the implementation.

However, SAST also has some minor disadvantages. For example, SAST scanners cannot directly validate whether vulnerabilities can actually be exploited. Intelligent SAST scanners could recognize that a vulnerability exists but that it is not being exploited and is therefore not critical. Nevertheless, it would also make sense to be able see the error in order to handle future calls securely.

Furthermore, a distinction must be made as to whether the affected code is actually used in production or whether it is just test code that is irrelevant for the real security assessment.

A free SAST scanner that supports many programming languages is Semgrep (*https://semgrep.dev/*). The basic version of Semgrep is available under an open-source license; more features are included in the commercial version. What's practical about Semgrep is that it is relatively lean, supports many programming languages, and is fast. It is also used as the standard for many programming languages in the GitLab security scanners.

A frequently used alternative is SonarQube. This tool was already mentioned in Chapter 7. SonarQube can be used to scan code for vulnerabilities and is available in both free and commercial versions.

Dynamic Application Security Testing

While SAST is about static execution, DAST is about dynamic execution. It is a blackbox test as, as it runs over a running application so that it can find actually executable vulnerabilities. In contrast to SAST, DAST is programming language–independent, as it does not look at the code directly but accesses it via its interfaces.

Execution of DAST in the pipeline looks slightly different compared to execution of SAST. The application is deployed on a review environment, after which DAST scanners are executed directly against the running application. While SAST looks at the code to see whether SQL injections, for example, could be possible in theory, DAST tests SQL injections directly.

Depending on the DAST tool used, the learning curve is relatively high, as you have to configure the tool extensively so that it can run against your particular application. For example, you may first have to log in to the application via the DAST tool so that various input errors can then be checked for security vulnerabilities.

A proper and complete execution of DAST can take several hours. These tests are usually not suitable for running in a CI/CD pipeline—neither in the main development branch nor in the feature branches. A DAST tool should be run automatically and regularly, but not as part of the standard development pipeline.

However, it may still be advisable to perform a quick DAST scanning run in each pipeline. Such runs must be reduced to simple and quickly detectable problems, such as problems with HTTP security headers.

One tool for DAST scanning is the OWASP Zed Attack Proxy, or ZAP for short (*https://www.zaproxy.org/*).

Interactive Application Security Testing

IAST is a mixture of SAST and DAST. Similar to DAST, IAST requires the application to be deployed on a target system so that the software can be examined.

The disadvantage of DAST is that there is no link between the running application and the source code. The disadvantage of SAST is that it can only find possible vulnerabilities and cannot determine whether they are exploitable.

IAST eliminates these two disadvantages. To stay with our previous example of SQL injection, while SAST would indicate that a SQL injection is possible at a certain line in a file, DAST would report accordingly which form field on which page is involved. IAST combines these two pieces of information and informs you that the relevant line in a file on the determined page can be used to carry out a SQL injection. Another feature of

IAST is that it produces a significantly lower number of false positives than SAST and DAST do, as it uses more realistic scenarios.

To use IAST, a certain amount of instrumentation in the code is required so that it can be verified that the identified security issues can be linked to the code.

In practice, this means that the application must be deployed to a test or staging environment in order to run IAST. During this time, regular, automated, and manual tests can be run. The IAST tool detects any security vulnerabilities during these tests. This is why IAST is called *interactive*—potential vulnerabilities are uncovered through interactivity.

Secret Detection

The purpose of secret detection is not to find vulnerabilities but to uncover credentials in the code. Secret credentials—including passwords, API keys, and other access keys—should never be committed directly to a repository.

> **Note: Passwords and Git**
>
> The reason is simple: A Git repository is a distributed version management system. If a password is pushed into a repository—regardless of the branch—then this password is assumed to be public, as at least everyone with access to the repository can see it. There is no point in deleting the password from the code, as the Git history saves it even after it has been removed. (It is theoretically possible to rewrite the Git history, but in practice, this is somewhere between "incredibly time-consuming" and "almost impossible.")
>
> Instead, passwords, API keys, and other credentials should not be managed in a repository but in a separate secrets management tool in which the credentials can be managed securely and with a fine-grained role concept. Use of a secrets management tool also comes with other advantages, such as the ability to completely rotate passwords much more easily.

When a secret detection scan is initially performed on a repository, it searches the entire history. Depending on the size of the repository, this can take some time. If secrets are found, they should be considered compromised and replaced accordingly. It does not make sense to rewrite the repository history, as the cloned passwords are lying around somewhere on the various computers.

Therefore, secret detection scans should also be run within a pipeline to find committed secrets at an early stage. Ideally, they should even be run when you push the code and the server then rejects it. Depending on the service, you can also specify that any keys found are automatically revoked (i.e., access is blocked immediately). This is useful for training team members accordingly: If their passwords to services in repositories are found early and are automatically revoked, new team members will learn not to do this.

Fuzz Testing

Another method of finding security vulnerabilities is *fuzz testing*. Fuzz testing involves throwing a number of input parameters at the application to see if they crash it. Such crashes indicate that there are hidden security vulnerabilities in the software, such as risks of memory overflow that can be exploited.

Two different types of fuzz testing are *coverage-guided fuzzing* and *behavioral fuzzing*. The former focuses on the source code and attempts to crash the application by generating various input parameters.

With behavioral fuzzing, the first step is to determine how the application should actually behave. Then, the fuzz test feeds the application with a number of input parameters. Any differences between target values and actual values could indicate potential security vulnerabilities.

> **Note: Reflection**
>
> As a technician, you have relatively low hurdles within your team to integrate initial security scanners into projects. Numerous open-source security scanners that you can integrate into your pipelines are available.
>
> It is essential to build an understanding of security both within your own team and on other teams to address vulnerabilities effectively. Although open-source scanners are already making good progress, there is no way around a commercial solution if your team wants to find and fix vulnerabilities quickly and easily. This applies to workflows and vulnerability management.
>
> It is also important to keep the lead time for changes as short as possible. Not only should security vulnerabilities should be fixed in the project, but the fixes should also be implemented for end users. In many organizations, the focus is often only on one aspect.
>
> It is also crucial to break down the barriers between the development teams and the security team. The security team should support the development teams in fixing security vulnerabilities to ensure a secure application. In this way, the likelihood that the security team blocks development is minimized.

11.5 Supply Chain Security

In business, the *supply chain* is a multistage part of the value chain for delivering a product. For example, is the supply chain for suits produced for the *nicely-dressed.com* web store consists of the various fabrics and buttons as well as the sewing of the suit before it arrives at the warehouses and to be sold from there.

There is also a supply chain in software development, as it involves working with many tools and dependencies that are required throughout the entire software development lifecycle.

Figure 11.9 shows the attack vectors that exist on the path from writing the code to execution. Errors made during coding are only one possible cause of security vulnerabilities. A project can be exposed to vulnerabilities through tools and dependencies as well. The question of whether the tools and dependencies used can be trusted is known as *build integrity*.

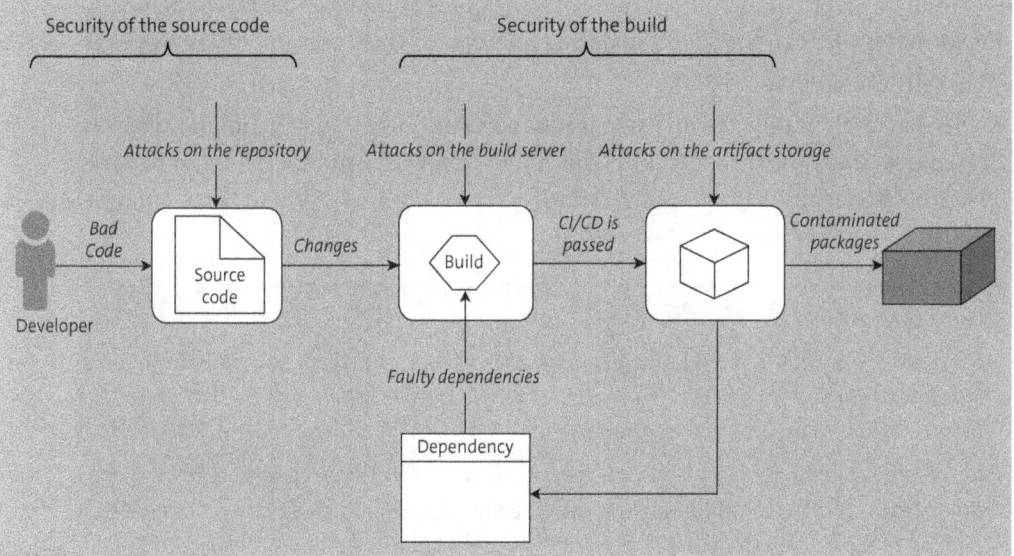

Figure 11.9 Attack Vectors throughout the Software Supply Chain

11.5.1 Attacks on the Supply Chain

Let's illustrate the importance of supply chain security all through two supply chain attacks that made headlines.

Solarwinds

Orion is software developed by Solarwinds that measures and monitors network performance in the IT stack. By definition, monitoring software must be able to look into every part of a system and, therefore, usually has very high access rights. Orion is used in many large and complex networks, not only by large companies, but also by public authorities and government organizations. This vector was extremely enticing for attackers—if an attacker could manage to implement a backdoor in this software, they could gain access to the networks of all customers, in this case almost the entire world.

And that's exactly what happened: Hackers did not infiltrate the networks of their actual targets, but the repositories of Solarwinds (which was very likely due to extremely sloppy security rules: *https://www.theregister.com/2020/12/16/solarwinds_github_password/*). There, the attackers manipulated the software. This access initially went unnoticed because the attackers deliberately did not immediately exploit the built-in backdoor—after all, they were not interested in the Solarwinds infrastructure, but in Solarwinds's customers. Solarwinds then released an infected new version, which customers diligently implemented.

In principle, installing updates quickly is a good, sensible, and important thing to do. This is because a number of security vulnerabilities and errors are usually corrected in updated versions.

In this case, however, updating the software was the beginning of disaster, as the Solarwinds update mechanism was used to deploy the backdoor on a large number of IT systems of major organizations. Among others, Microsoft was affected, whose systems were attacked via this vector: *https://blogs.microsoft.com/on-the-issues/2020/12/13/customers-protect-nation-state-cyberattacks/*.

log4j

Almost worse was the log4j security problem that I briefly mentioned at the beginning of the chapter. It concerned the widely used logging library log4j, which is used in virtually all Java projects. A vulnerability (CVE-2021-44228) could be exploited by attackers to inject malicious code into applications and thus gain access to the affected systems. After this vulnerability became known, IT specialists all over the world were busy checking whether log4j was used in their environments (which was almost always the case) and then installing the corresponding updates. Those who did not have a good overview of the software in use and were unable to update affected systems quickly had major problems.

Conclusion

In both cases, supposedly trustworthy software components were used as attack vectors: the project of the large, renowned technology company Solarwinds, which advertises its professional work, and an open-source project in which the code and thus the security problem was open for all the world to examine. Whether it was through the use of the commercial professional software or the open-source library, IT infrastructures became vulnerable, even if the operators had actually done everything right. So if parts of the infrastructure you rely on have a vulnerability, you are unfortunately in a bad position.

What does this mean in terms of DevOps? Of course, application security is important. But you have to take a look at the entire value stream to see whether parts of the supply chain are vulnerable.

We have looked at two types of security scanning: dependency scanning and container scanning. Both types of security scanning are relevant for supply chain security if external sources are used and implemented in your own project.

Note, however, that with dependency scanning, it considers only a partial aspect of dependency security. It looks only at which dependency is defined in which version and then checks the vulnerability database for security vulnerabilities.

11.5.2 Software Bill of Materials

The software bill of materials (SBOM) is also relevant to supply chain security,. The SBOM is a list of all the materials that were used to build a software. This list is essentially the result of dependency scanning, which notes the first-degree dependencies used in the software and the transitive dependencies that the first-degree dependencies follow. The version numbers and the corresponding licenses are also included.

Security vulnerabilities can be derived from SBOMs based on the listed dependencies and their version numbers. SBOMs ensure that all dependencies are known and that a rapid response can be made in the event of a possible attack on supply chain components. This is particularly important with regard to third-party applications such as monitoring software.

Even with smaller projects, it is almost impossible to manage dependencies by hand—that is, by relying on the development teams to document all used libraries and modules with their respective version numbers. In a DevOps environment with automated builds, this is not necessary, as there are always up-to-date lists of dependencies thanks to dependency scanning. It is important to manage and check these lists centrally so that that this information is easy to find and can be used sensibly. Automated Common Vulnerabilities and Exposures (CVE) scanners, which provide you with an overview of all security problems based on this information, can help with this.

Compliance rules, which are already mandatory in many industries, can also help. In this way, rules are laid down to ensure the secure development of software and to make it verifiable; this is the subject of Section 11.6.

In the absence of industry-specific compliance rules, legislators are increasingly stepping in and issuing guidelines that prescribe the careful management of dependencies:

- In 2021, for example, the US government issued Executive Order 14028 (*https://whitehouse.gov/briefing-room/presidential-actions/2021/05/12/executive-order-on-improving-the-nations-cybersecurity/*), which states that any software to be used in government agencies must provide an SBOM.

- Also, the EU is working on the Cyber Resilience Act (*https://digital-strategy.ec.europa.eu/en/library/cyber-resilience-act/*), which defines certain compliance rules.

11.5.3 Security of the Build and Deployment Server

If you use dependency scanning to return an overview of all your software's dependencies so that you can react quickly to any detected problems, you also need to make sure that the pipeline that builds and delivers source code to finished artifacts is secure. If you can't trust your infrastructure, you can never be sure that there are no backdoors hidden in the packages you deliver. This is pretty much sums up what happened with the Solarwinds server hack, and there are other examples, too.

It will come as no surprise to you that you should take some basic steps to ensure security in this area:

- Apply patches and security updates quickly.
- Have firewalls restrict network access on the build infrastructure (both at operating system level and possibly as a *web application firewall*).
- Use comprehensive monitoring to inform you about accesses.
- Make communication between the servers completely encrypted based on certificates or other tools.

These are all important tasks that require hard work and that should be a matter of course in a modern infrastructure.

11.5.4 Secure User Accounts

Some may be asking themselves, "Isn't there still a problem with access restrictions?"

On the one hand, you should restrict access to your servers—the fewer people who have access to the servers on which the artifacts are built, the better. On the other hand, I already wrote in Chapter 5, Section 5.1 that build servers should not be behind closed doors. To enable quick feedback and direct exchange within and between teams, it is much better if everyone can see the code and contribute changes.

With the right technical implementation, you can both restrict access to servers and avoid isolating them from your teams: The build infrastructure should be technically isolated to such an extent that it always builds the packages and executes the tests in a clean environment. At the same time, all teams should be able to continue adapting the parts of the CI/CD pipeline. The code review serves as a quality gate, ensuring that no errors are introduced.

Of course, you also need to prevent accounts from being taken over and reviews from being bypassed. User accounts and any required service accounts must also be secured by using a well-thought-out role concept that is restricted with strong authentication methods such as two-factor authentication (2FA). In practice, many companies actually use single sign-on, where multifactor authentication is used directly. This is also generally more advisable in order to automatically create and lock user accounts when new employees join the company or colleagues leave.

RBAC means that access to important resources is linked to membership of predefined groups. Access rights should not be assigned individually; once you start setting permissions individually, you will quickly lose track of who has access to what. Instead, the basic rule is to manage access rights via roles, and then individual persons or services can authenticate themselves via a central *identity management* (IDM) system and then receive the corresponding rights. This should always be done using the principle of least privilege principle—always grant as few access rights as possible. Visibility should be open, and changes should be made only after a review.

Access to Git repositories is also relevant. SSH should generally be used rather than HTTPS. However, there are often exceptions, especially in larger companies that need to follow more specific regulations.

> **Attention: Service Accounts**
>
> Service accounts or bot accounts are a difficult topic. Additional accounts are often needed to control automations. There must be governance rules to ensure that these bots do not run with too many rights. If these accounts fall into the wrong hands and this goes unnoticed, you have a problem.
>
> It is ideal if accesses by service accounts, which are usually available as access tokens anyway, are regularly rotated so that tokens are invalidated.

11.5.5 No Code Is Good Code

Last but not least, the fewest problems are caused by dependencies that do not exist. The fewer references to third-party software there are in your projects, the more calmly you can react to reports of security problems.

However, this recommendation is not quite so simple: It is often better to use libraries than to reinvent the wheel. For example, when it comes to cryptographic methods or other security-critical components, you should always use properly developed tools.

You will find the right balance only through experience—not necessarily through your own experience, but also through that of your colleagues and other teams. Therefore, take enough time in code reviews and pair programming to discuss whether it makes sense and is necessary to include a dependency. Can you solve the problem yourself with a few lines of code? Do your colleagues perhaps already have a function ready that can replace an external dependency (i.e., inner sourcing)? Or will you only bring more problems into the company with in-house development?

> **Note: Reflection**
>
> The software supply chain is a relatively new topic, even if the basics have been known for a relatively long time. The most important part of supply chain security is scanning the dependencies used.
>
> Some aspects of supply chain security (such as 2FA) require general training for all employees. Dependency scanning, on the other hand, focuses more on security. However, whether an SBOM is really needed depends on the legal situation. Regardless of your legal situation, SBOMs can be helpful, especially so that it is possible to quickly check whether your own environment is affected in the event of new serious vulnerabilities.
>
> Take a close look at the various aspects of supply chain security in your development environment and ensure that they are set up as securely as possible.

11.5.6 Security at nicely-dressed.com

At first glance, the selection of security tools for the various categories at *nicely-dressed.com* seem good: The pure security tools delivered outstanding results and excellent vulnerability management. However, although these tools run in the pipelines, they were never really integrated into the solutions. The security team uses a dashboard, which is important, but it hasn't been helpful because even simple tests show that the development teams hardly look at any dashboards. The integration is only half-baked, and time and again the developers simply ignore the results.

The development teams have therefore oriented themselves toward GitHub and GitLab. This makes it easy to integrate security scanning into the pipelines and to configure merge restrictions when attempting to introduce new vulnerabilities.

> **Note: Reflection**
>
> Security vulnerabilities must be detected as early as possible so that they can be corrected promptly before the affected code merged into a main development branch. It is therefore important to ensure that security scanners are integrated into the development process as early and regularly as possible. This is the only way to identify vulnerabilities during development.
>
> If you want to implement different scanners at the same time, it is better to introduce one scanner type at a time so as not to overwhelm the teams. Dependency scanning is a good and manageable way to get started, followed by SAST.
>
> The security team and the product teams should be looking at the same dashboards and reports to promote collaboration. Also, I recommend having your pipelines issue warnings rather than abort when security issues are found. This is because false positives will always occur, and aborting pipelines is an unattractive method and can be disruptive, regardless of the fact that the vulnerability management tool should support the process. I have seen many crude workarounds here that were more of a hindrance than a help.
>
> An important mantra to keep in mind is that the best security tool is useless if it is not used simply because it is inconvenient to use.

11.6 Compliance

In the context of software development, compliance is a topic that triggers even less enthusiasm than security. While security is about defending against hackers and securing your own application, compliance is "only" about ensuring that you're adhering to rules.

Is compliance still appropriate in a DevOps culture? It was not for nothing that Mark Zuckerberg once proclaimed the motto "Move fast and break things" for the development of Facebook: Do it first, and we'll deal with the consequences later.

Of course, this approach cannot apply to a professional team, and regardless of the motto, Facebook had and still has to deal with a lot of rules and regulations.

There are two main reasons for compliance requirements:

- The most important reason is that there are regulatory requirements. These are general laws, but also specific requirements for specific sectors, such as the automotive and financial sectors. There are also specific requirements such as those set out in the GDPR and those derived from accessibility regulations.
- The second reason relates to organization size. Large corporations in particular use many different tools and processes whose uses need to be standardized and who are subject to compliance requirements. Even if this is cumbersome in many cases, it is important to keep the big picture in mind: Good, well-thought-out regulations make collaboration easier and ensure a better product. For example, the best security scanners are useless if the teams can bypass them because they don't want to do the extra work. This is where the autonomy of the teams meets the obligation of the organization.

> **Note: Compliance and DevOps Are Not Contradictory**
>
> Critical but nonexpert voices often say that the DevOps culture cannot work in all industries, as each team would then simply do what it thinks is right. Some argue that the important autonomy of the teams is at odds with the regulations and laws that are enforced by a compliance process.
>
> This is, of course, an exaggeration and not true. What is true is that cumbersome compliance rules do not work well in a DevOps culture: If four different signatures have to be obtained from the management board by fax before a deployment, it is not possible to work in an agile way. However, this does not mean that there are no ways to implement rules in such a way that the result is faster, simpler, *and* more secure.

The basic aim is to enable secure deployment in the production environment while complying with regulations. Compliance with regulations is ensured by jobs in the pipeline that automatically enforce checks. After all, it is all well and good to have security and compliance rules, but if the teams are so autonomous that they can simply switch them off, then nobody is helped.

11.6.1 Define Compliance Guidelines

Where do the rules come from? Most companies have a team that defines the compliance rules for the entire organization and for specific projects. This team uses laws, general regulations, and industry-specific compliance rules and enriches them with the company's own internal rules.

It is important to focus on the *outcome* and not on the rules themselves. Unfortunately, you often hear people say, "We've always done it this way!" even though their way of doing things doesn't make much sense. It is better to focus on hard figures and metrics and to always check that the formulated requirement is in line with the value stream. This includes first defining the owners of the compliance requirements, and then considering the process and the *threat model* as a whole and in context: Which threat needs to be contained? Which guidelines are necessary and sensible?

The guidelines can be diverse and affect not only the development but also the operation of the application. The following compliance rules are frequently seen in software development:

- The test coverage must be at least 75%.
- The coding style is adhered to and checked by Linter.
- The security scanners must not find any security vulnerabilities that are classified as `critical` or `high`. Otherwise, the code must not be deployed.
- Changes may be merged only once another person has reviewed them. Code that ends up in the finished product or in the production environment must be checked by at least four, preferably more, eyes. Who was responsible for the review must be documented in the same way as the author of the code.

Compliance rules are also common in the company:

- All data connections between the services have to be encrypted.
- All HTTP endpoints must be encrypted using TLS.
- The running containers must be checked regularly for known security vulnerabilities.
- Vulnerabilities classified as `critical` and `high` must be patched within one week.

Of course, these are just a few examples. It is also important that the compliance team work closely with the other teams and never set up rules in a patronizing manner that cannot be implemented on a day-to-day basis.

> **Note: Introduce Policies with Caution**
>
> When defining compliance policies, consider the development status of the projects. If you introduce security rules in an old project that previous did not require security checks, the code and dependencies are likely to be teeming with security vulnerabilities.
>
> And introducing a compliance policy specifying that no security vulnerabilities may be included in the project would cause development to come to a standstill for the foreseeable future. Development cannot continue, and new deployments cannot be carried out. It may make more sense to concentrate initially on ensuring that no *new* security vulnerabilities are introduced. This can be a pragmatic solution, ensuring that

> development continues with security in mind and the mountain of problems found in the legacy code can be gradually removed.
>
> In addition to the number and level of security vulnerabilities, define in the policies who can still approve a merge or deployment if vulnerabilities are found. In this way, false positives or security vulnerabilities that cannot be exploited in the project can be waved through after careful examination so that development and deployment can proceed.

Compliance as a Process

Establishing guidelines is only the first step; it is much more important that compliance is understood as a process that affects the entire system. In addition to the compliance rules, a process that specifies how and when these rules are to be checked must be defined and documented. This process must also specify that the results of these checks must be recorded. Whether rules have been adhered to must always be verifiable—your legal department will thank you for it.

The actual handling of compliance rules in practice depends heavily on the type of implementation. In the traditional waterfall contexts, there is a lot of manual checking, whereas in the DevOps world, much more is automated. In the following sections, we take a closer look at both approaches, what such a process can look like, and what needs to be considered.

11.6.2 Manual Compliance

In waterfall environments, the compliance team provides a checklist that the development teams have to adhere to. As deployments take place only every six months, this list is typically completed by the developers only shortly before handover to the operations team. There is another checklist for the operations team shortly before deployment, which also has to be completed.

If the compliance team vetoes a deployment, improvements have to be made to fix the identified compliance issues. This creates potential for conflict, as compliance issues could arise at every transition between teams, which would mean that the whole process would have to be repeated from the beginning. This could greatly delay the development of a feature or its deployment and could mean that several months of work had been wasted if it is ultimately discovered that something was wrong.

The teams complete these checklists and indicate whether they had followed the rules, including any comments as to why a certain compliance rule could not be implemented. The compliance team reviews the changes and then approves them.

For smaller organizations, such a process (as shown in Figure 11.10) may be more feasible and cause less issues, as there are short distances to the compliance team and the list of compliance points to be processed tends to be manageable. However, it becomes problematic in larger organizations when it is impossible to process them promptly because the compliance team is working at full capacity.

Let's take a specific example. Your company imposes a compliance requirement specifying that only code that has gone through a review may end up in production, which is a sensible requirement. If you take this compliance rule seriously, then you would actually have to check all code changes since the last deployment and see whether there really was a review. This is obviously completely impossible. And even in a weakened form, such a compliance check could extend the lead time endlessly.

A manual process also scales poorly. The larger the organization and the more rules that need to be adhered to, the longer such processes take and the more cumbersome it becomes for the teams.

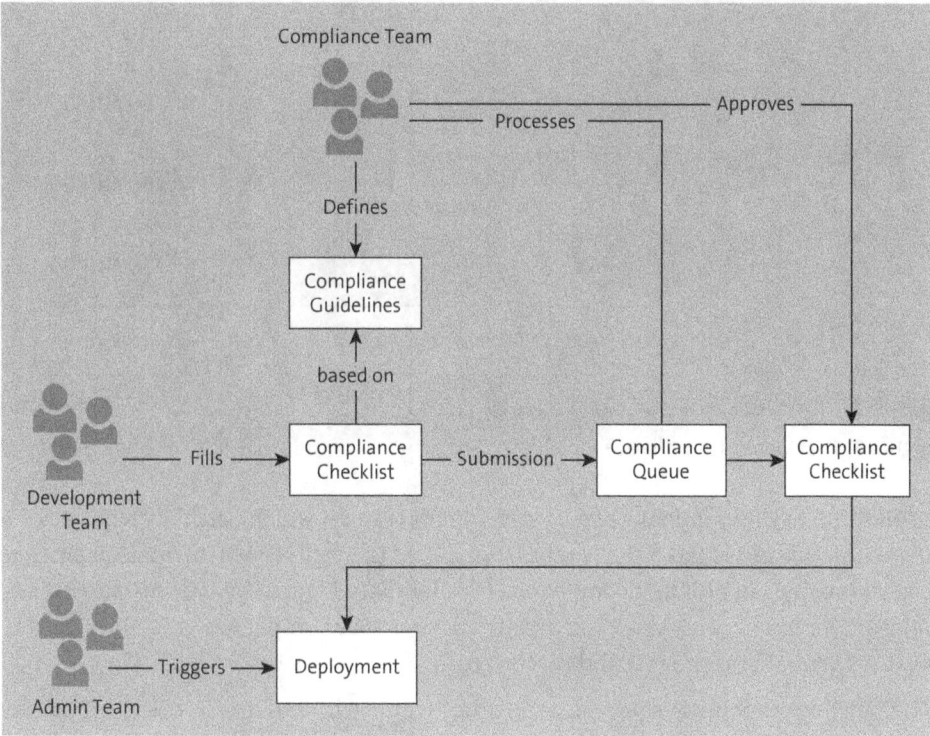

Figure 11.10 Manual Processing of Compliance Rules

11.6.3 Fully Automated Compliance

It is hardly surprising that the focus should be on possible fully automated compliance. As with the manual workflow, the compliance team continues to set the rules, but compliance is enforced and documented fully automatically.

11 Security and Compliance

Compliance Pipelines: Providing Freedom while Enforcing Rules

The first step is to define compliance pipelines. As with general security scans, it is important to follow a shift left approach: Compliance checks should also take place as early as possible in the software development lifecycle so that errors and violations are detected when they occur.

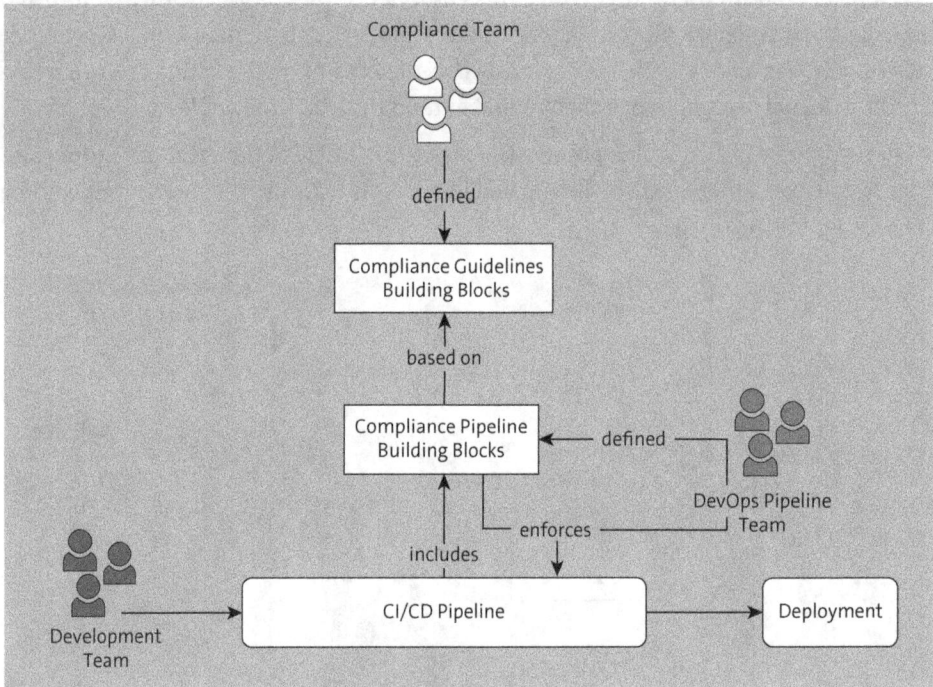

Figure 11.11 Building Compliance Pipelines

In addition to the compliance and development teams—which naturally work according to DevOps principles—the team that manages the continuous integration pipeline is also involved in building compliance pipelines (see Figure 11.11). This is particularly relevant for larger companies and organizations whose compliance guidelines need to be just as scalable as the rest of their systems.

The pipeline team always provides the other teams with building blocks for building pipelines in order to avoid multiple implementations. This applies not only to regular parts of the pipeline but also to the parts that are relevant for compliance.

The focus is on the security scanners, which must be executed with every run. There are two things to note here:

- It must be made as easy as possible for the teams to integrate these security scanners.
- The teams must not be able to simply switch off the security scanners.

The basic principle is clear: Teams should carry out compliance checks as early as possible in the software development lifecycle. And compliance checks are best done in the pipeline so that the they are carried out regularly in each branch and become part of the development process. The teams can then use the functions provided by the DevOps team. The compliance-relevant jobs are then automatically and inevitably injected by the pipeline. Complete automation takes work off everyone's hands.

The difficult part is enforcing the rules in the compliance pipeline. Choosing the right tool and the right pipeline architecture is essential here.

Some companies rely on a central pipeline that is managed by the pipeline team. However, this approach leads to many problems, as many different teams use different software stacks; the model is therefore almost impossible to maintain in the long term. Such a centralized approach will only lead to frustration, which is why it should be avoided.

It is better to prepare appropriate compliance modules for the pipeline. The individual teams then still have the option of writing and modifying their own parts of the pipeline themselves, while the right tool forces fixed, security-relevant modules to be run through. For example, each merge to the main branch checks whether new dependencies introduce security vulnerabilities. The compliance modules also serve as a quality gate shortly before deployment: The licenses of the dependencies used can be scanned automatically. If a problematic license is found, no deployment is carried out.

A good compromise looks like this: One part of the pipeline must be sealed off, while another part of the pipeline can be defined by the teams themselves. Figure 11.12 shows what this might look like.

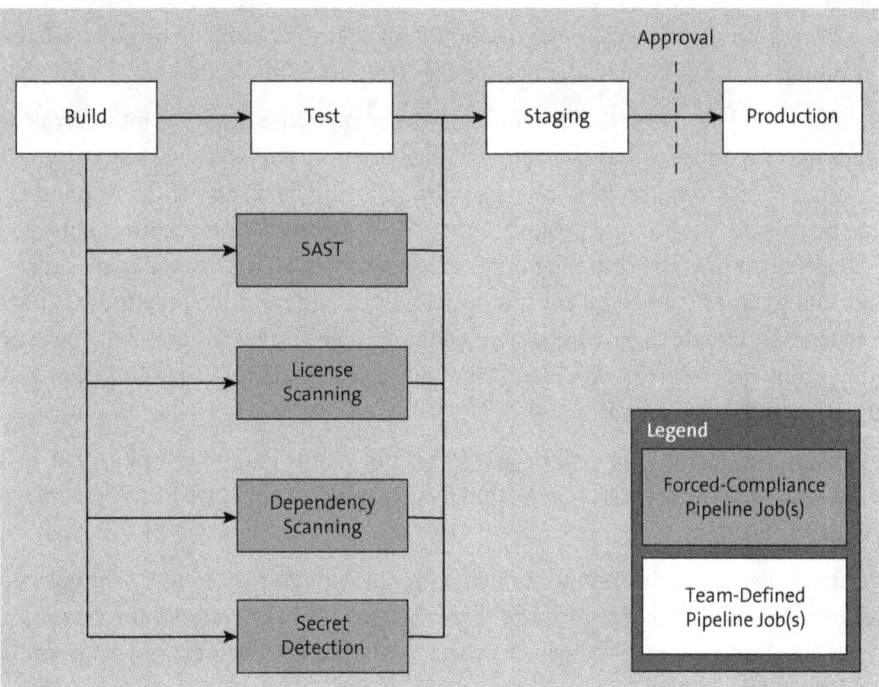

Figure 11.12 Pipeline with Forced Compliance Jobs alongside Team-Defined Jobs

Writing Compliance Pipelines

I have already mentioned that choosing the right tool is very important at this point. While there are usually different tool options that all do a given job well, in this case I can only recommend the pipelines from GitHub and GitLab for checking compliance rules.

> **Attention: Compliance with Jenkins**
>
> Unfortunately, Jenkins is not a good choice for writing compliance pipelines, as it is not a central platform from a single source. Jenkins allows a lot in the standard version and has a very modular structure. Isolating everything using Jenkins is somewhere between nontrivial and impossible, which is also due to the fact that source code management and continuous integration are handled separately in Jenkins.

Depending on the platform, the policies that are checked in the pipelines are available as configuration files. This makes it easy for team members to understand the content. The policies are usually stored in a separate repository and are, therefore, also versioned. The teams can propose changes themselves (in the spirit of inner sourcing) via merge or pull requests, which are checked by the decision-makers.

You can incorporate procedures such as static code analysis and security scanning, as well as custom scripts, to ensure that your code and development practices meet specific compliance requirements.

When writing compliance pipelines, you should make sure not only to focus on what is important but also look closely at *how* pipelines need to be implemented.

Many companies have several different compliance pipelines because not every project needs every scanner or has to comply with every rule. For example, a smartphone application will not require container scanning, but the backend for the application will. In the context of multiple projects with different compliance requirements, you need to either define the compliance pipeline so that it automatically recognizes whether the scanner is relevant for the project, or define multiple pipelines that are used depending on which feature set is needed for the respective project. However, note that problems could arise if several jobs have the same name—the CI/CD tool may not know which job is relevant.

Although automatic merging sounds good, it is not realistic in practice. Instead, it is more necessary to set up a hierarchy within the compliance pipelines by recreating the inheritance yourself.

In principle, however, the same applies here as for all other pipeline steps: The compliance definitions should be easy for the teams to read and understand. If the compliance part of the pipeline is longer than the actual build specifications, this could quickly lead to frustration.

> **Note: Not Suitable for All Projects**
>
> I advise you not to specify too much, because every constraint restricts the freedom of the teams, which at best could encourage the teams to implement workarounds for these restrictions and at worst could result in a dissatisfied, annoyed, and inefficient team.
>
> Unfortunately, many companies simply activate all compliance-relevant pipeline modules for all projects on the DevOps platform. This may ensure that no project is forgotten, but it creates a lot of unnecessary work. It is better to take a closer and more differentiated look at your projects and pipelines: Not all code is intended for production. Not every project is a software development project. Not every project needs to be strictly monitored.
>
> Enforcing unnecessary checks for projects only leads to the creation of a shadow IT system that is managed by the teams themselves because there are too many restrictions. This should be avoided at all costs.

Add Compliance Components

Compliance pipelines are a large and important part of the implementation of compliance guidelines in DevOps teams. However, not all problems can be solved with self-written rules in a compliance pipeline. Instead, some third-party tools must be integrated into the pipeline, such as the tools described in Section 11.4.3, which have their own vulnerability databases. Although a container image scanner may come from a third-party company, the compliance team could specify only that a scanner is to be used and enforce execution via the compliance pipeline.

However, using third-party scanners, vulnerability databases, and so on raises the question of who is responsible for them. And what happens if a serious vulnerability is found when, for example, a container image is scanned?

In smaller environments, it may make sense for the team that is closest to these tasks to take care of them. Therefore, the security team would be responsible for the scanners, while the development teams would be responsible for ensuring that up-to-date container images are always used. In larger environments, however, this approach could lead to a scaling problem, and it may make more sense to set up a separate team tasked with watching the security scanners and ensuring that the affected container images are rebuilt and replaced in the event of problems. That way, the development teams can concentrate fully on their work and the image does not have to be rebuilt by several teams. This component team must then ensure that the pipeline checks that only the approved base container images are used (see Figure 11.13). Such small, separate, and agile teams can, like the platform teams, help with special tasks that affect all other teams.

11 Security and Compliance

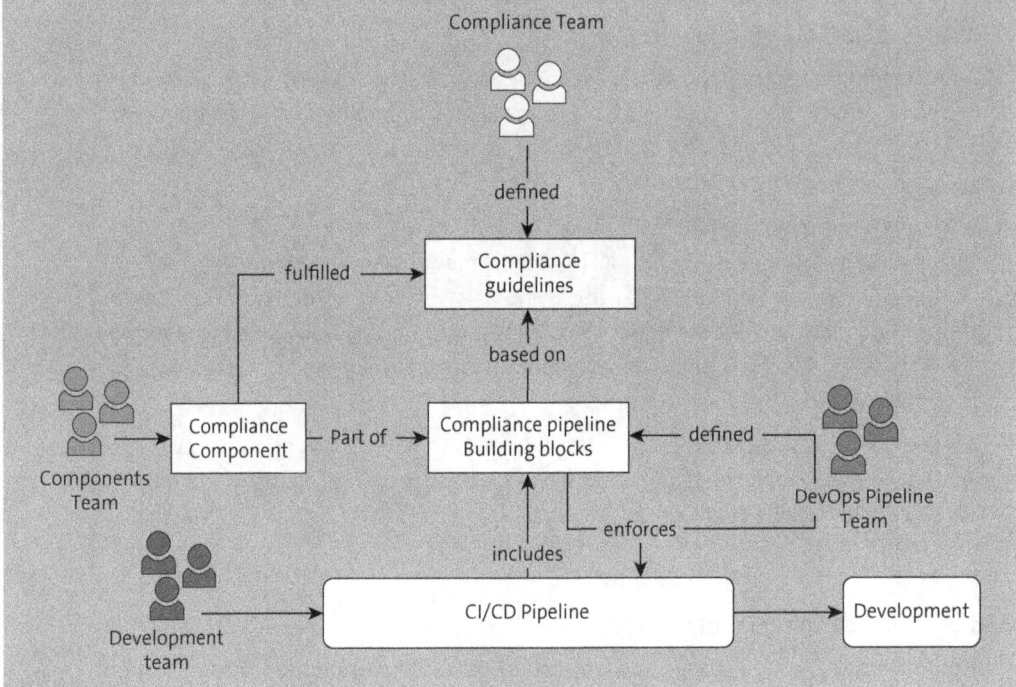

Figure 11.13 Team Structure Involving Separate Component Teams

Approvals and Audit Trails

By firmly integrating the compliance guidelines into the development process as pipeline modules, you have an easy way to manage approvals. Of course, no one can manually check whether every code change has been approved by a reviewer.

However, since all common source code management tools can check that merges are accepted only if they are approved by several people, you can fully automate this requirement. A separate check is no longer necessary because the process and the tools already monitor the main branch to ensure that no unchecked code enters it. This also applies to deployments. Approval is part of the deployment process and not a separate step beforehand, so deployments can take place automatically after approval.

It is important that approvals are logged by the corresponding tool—that is, that an audit trail is created to provide information on what was checked and when. The modules of the continuous integration pipeline are versioned for this purpose, and the individual runs through the pipeline are logged. For the code reviews, the source code management system records information about who made the code change and who accepted it directly in the Git history and in the associated merge or pull requests.

If the pipeline's security scanners raise an alarm, it is initially up to the developers to check the relevant location. If developers decide that the finding is not relevant, the security team must check the finding as well. If a joint approval is given (i.e., the code is merged into the main branch), this must also be documented directly in the source code management system. If problems do occur on production systems, the audit log provides information on how long the gap has existed and on the background to the implementation. This information should be evaluated and understood as part of a blameless post-mortem.

11.6.4 Compliance at nicely-dressed.com

Up to now, everyone at *nicely-dressed.com* has worked as they wanted: Somehow changes were made, but whether these changes were good and met the company's own requirements was never really monitored.

In selecting the tools for ensuring compliance at *nicely-dressed.com*, the aim is to ensure that the configuration can be carried out transparently and without security vulnerabilities. One compliance rule that *nicely-dressed.com* needs to be able to enforce is that newly discovered security vulnerabilities can be integrated only after the security team approves an exception for such vulnerabilities. However, many tools are not able to provide this capability, as they can usually prevent deployments only in production environments.

Therefore, once again, the company has decided to use GitHub and GitLab, as they enable the shift left approach to be implemented particularly consistently.

> **Note: Reflection**
>
> Compliance regulations, including both legal and internal company regulations, must not be neglected.
>
> The main focus of following compliance rules should be on the definition and execution of the compliance pipeline. In other words, you want to enforce that certain parts in the pipeline of the respective project must be executed. These are usually security scans. Make sure that the compliance guidelines are easy to adhere to and that the teams concerned are supported in doing so. Complete blockages due to unnecessary intermediate steps should be uncovered and removed.
>
> The simplest rules—such as the number of reviewers required for the review process or code quality and security requirements—best provide teams with healthy guidelines that improve the project but do not significantly restrict development.

11.7 Summary

Security is an enormously important part of the software development lifecycle. Unfortunately, security is not a high priority in many organizations. The reasons for this usually involve a lack of budget and unhealthy attitude toward changing long-standing procedures: "It has worked well so far without it."

People often say that security is becoming increasingly important. That's wrong, of course, because security has always been important! What can be said, however, is that the integration of security checks early in the software development lifecycle has never been easier than it is today.

If the culture between development, security, and operations is good and supportive, not only can security vulnerabilities be found early on, but they can also be corrected promptly so that patches can then be rolled out. And this is exactly what ultimately matters: Security must be an integral part of the software development lifecycle in order to enable secure software development.

Chapter 12
Successfully Implementing the DevOps Transformation

The previous chapters provided insight into the technical and nontechnical factors that play a role in working according to DevOps principles. However, we focused almost entirely on the individual phases of the software development lifecycle.

Dividing our discussion by the phases of the software development lifecycle was helpful for introducing the various DevOps principles, as it allowed us to look at the team structures and DevOps techniques in isolation. In many places, I repeatedly noted that the following always applies: People over processes over tools.

The best tools are ultimately not useful if the processes are not great. And the best processes don't help if the people in the company don't understand their purpose. And that is precisely the main problem.

This chapter focuses on three aspects: In the first part of the chapter, we take a deeper and more general look at the methodologies that can be used to introduce a DevOps culture into a company. This is important especially at the beginning of a company's transition to a DevOps environment, as the teams need to be made aware of the purpose and objectives of the DevOps concept. In the second part of the chapter, I show how the success of DevOps can be measured using the DORA metrics. Finally, in the third part, I discuss what value stream mapping is and how it can be used to optimize the value stream.

12.1 Introducing a DevOps Culture

Introducing a DevOps culture is not an easy task. Many teams transitioning from the waterfall model to agile software development have seen the same problem: People were not properly educated about *why* the team was making such a transition. It's not unusual to hear disgruntled team members ironically state: "We work agile now! We introduced Jira!"

Unfortunately, you hear similar statements regarding DevOps initiatives: "We are doing DevOps now! We have a CI/CD pipeline and Kubernetes!"

But teams who have this attitude may not even have a fully automated CI/CD pipeline, as no deployments are carried out on production environments.

12 Successfully Implementing the DevOps Transformation

There are many more points we could make, but it is time to take a conceptual look at the topic—as a guide for managers and as a source of ideas for pure technicians who could pass those ideas on to the upper management.

> **Note: The Journey Is the Reward**
>
> Building a DevOps culture is not easy. I have provided some examples that you can use. Not all of them may be relevant to you. There is also a good chance that you have experienced other paradigms in your company.
>
> If there are specific approaches that have worked or not worked for you, please feel free to tell me about your experiences by sending an email to *mail@svij.org*!

12.1.1 Bottom-Up or Top-Down?

The first fundamental question is whether you should take a bottom-up or a top-down approach when transitioning your team to DevOps (see Figure 12.1).

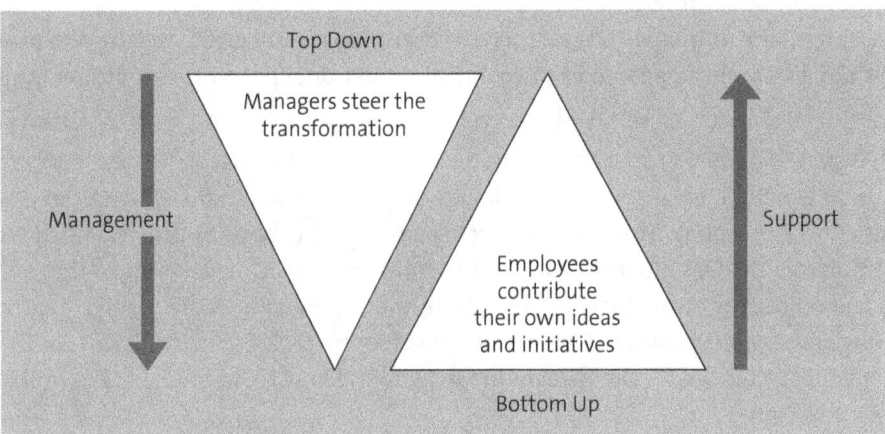

Figure 12.1 Top-Down and Bottom-Up DevOps Transformations

With a bottom-up approach, the initiative for the DevOps transformation mainly comes from individual contributors on the teams. These people are not responsible for management tasks.

With a top-down approach, on the other hand, the transformation is initiated and led by managers. The individual teams then have to follow suit. It is extremely important that as many people as possible in the organization are properly involved and that the planned transformation does not degenerate into coercion. Because in the end, that will not help anyone.

There is no clear answer to the question of whether bottom-up or top-down is the best approach. It depends on many different factors within the respective organization. Nonetheless, it is generally easier for smaller companies to pursue a bottom-up approach. However, because with this approach the employees have to take the initiative and convince their managers to make the transition to DevOps. the more managers who work with a "but we've always done it this way" mentality, the more difficult this process is.

The larger the company, the more essential it is to take a top-down approach. In larger organizations, individuals usually have a more difficult time introducing large initiatives, as there are already many structures and processes in place that would make such initiatives difficult to achieve. Often, initiatives are not even desired.

The biggest challenge is generally that managers do not support DevOps initiatives. It is difficult to say whether a DevOps transformation can succeed if the pressure comes only from the teams themselves. If managers don't see any problems with the status quo, then all the enthusiasm of the employees for a transformation won't help in the end.

In truth, neither an exclusively bottom-up or top-down approach will ultimately be possible, but *everyone* must pull together regardless of what combination of approaches is taken. Some employees and managers need to be convinced, while others will inevitably have to be pulled along if the transformation is to succeed. Ideally, ideas and initiatives should come from all sides.

I have already covered many points regarding the changes needed to make a DevOps transformation possible over the course of this book. The necessary changes to the organizational and team structure are probably the most obvious. However, these changes are actually the second step. The first step is for leaders to first fundamentally recognize the difficulties their teams are having with the traditional waterfall method and to want to act. If you are a manager who wants to change something with your team's software development processes, then it is excellent that you are reading this book.

However, if managers do not recognize the need for a transformation to DevOps, then they need to be informed of the difficulties their teams have been facing. But if there are few or no opportunities to pass on feedback to the top, then the outlook for a transformation does not look good—after all, an open feedback culture is also a basic building block of DevOps.

When feedback is given upward, it is always important to speak the same language as the managers. Unfortunately, the wrong approach is often taken—specifically, people often only highlight individual problems without communicating big picture issues.

I often hear the following statement from technical people: "We want more automation so that we can concentrate on the important work." This statement sounds good at first, because focusing on important work is not insignificant.

However, depending on how it is received from higher up in the hierarchy, this statement can be understood quite differently: "People are lazy and want to work less." Of course, the perception depends heavily on the company culture.

In fact, over time, I have learned that it is more effective for technicians to primarily emphasize how making DevOps-focused process changes could increase efficiency and ultimately bring projects to the finish line faster, which is ultimately also cheaper for the company.

However, it is also a fact that not everyone on a team may the same problems. People are creatures of habit, and many don't want change; they may feel that, after all, they've always worked this way and it has worked. Giving more responsibility to the teams could also sound risky at first (which I can understand, by the way). That's why it's important to get even the most hesitant people on board.

12.1.2 First Steps in the DevOps Transformation

The first step—once it has been determined that a transformation towards DevOps is to take place—is to get the relevant people in the company on board. At this point, it should come as no surprise that this doesn't just mean the development and operations team, but also members of the security and QA teams. And this doesn't happen overnight—it is a multistage process.

A DevOps transformation can essentially be summarized in the following five steps:

- Step 1: Define the DevOps vision and strategy
- Step 2: Create measures for acceptance
- Step 3: Create a DevOps transformation team
- Step 4: Start with the first value stream
- Step 5: Measure and iterate

And, very importantly, remember that every environment is different. Don't think of these five steps, which I discuss in more detail in the next sections, as an unchangeable, hard step-by-step guide! There are too many factors that influence the success or failure of a DevOps transformation, and these are different for many companies. So my advice is to use these steps as inspiration and to adapt them according to your own ideas, experiences, and circumstances.

Step 1: Define the DevOps Vision and Strategy

The first step is for managers to clearly define the exact DevOps vision and the general strategy for implementing it. Without a clear definition, a transformation to a DevOps environment cannot succeed in a meaningful way.

To successfully define a DevOps vision and strategy, it is essential to onboard the employees in the organization to it. Without a clear vision and strategy, employees will not understand why they should work differently and there will be little enthusiasm for change.

First of all, it should be made clear what DevOps is and is not, similar to what I have done here in this book. It is important to focus on the DevOps mentality and always explain the benefits for the organization's employees.

As a manager, the benefits of short lead times, frequent deployments, and faster delivery of changes to end users naturally sound good at first, as the work reaches the users faster, which they usually like.

However, when employees hear that speeding up work and deployments is the goal, many initially get scared. It's hardly surprising: They may not yet understand the new way of working, and it initially sounds like a lot more work and a lot more stress. And in the end, only the boss has more money in his pocket. Great! That would scare me too.

It is important that the aspects of collaboration, shared responsibility, and knowledge sharing are emphasized and that the team slowly moves in that direction. Without having empathy for your employees and their apprehensions, it will be difficult to convince them to go along with the changes. Nevertheless, you have to realize that nor everyone will follow you, because some people will ultimately have to be forced to be happy. The number of unwilling people must, of course, be kept to a minimum.

Defining the vision and strategy also includes communicating what the team structure should look like in the future. It is important to define the vision not only for development and operations but also for the QA and security teams.

You must also make clear *why* your teams are making this transition and *what* they should hope to gain from it. Efficiency gains and higher productivity are always important, but they're not very comprehensible for the employees carrying out the work, as they generally just want to get the job done.

Instead, in order to get your colleagues and employees excited about the idea, it is important to emphasize that the *software development and delivery experience* will improve considerably as a result of such a DevOps transformation, provided it is designed properly. Traditionally, the developer experience concerns how pleasant the working environment is for developers. However, as our approach is not limited to developers, but also includes the operations, QA, and security teams, I refer to this as the software development and delivery experience.

So if you manage to present the approach in such a way that working in general becomes easier thanks to a more open approach and fewer pointlessly restricted environments and tools, then many more people will be able to get on board.

Feedback Channel

In addition, your teams may bring up more points for consideration. That's why it's important to set up an easy-to-use feedback channel in order to regularly gather opinions: What do employees think? Do they think the measures are good?

Not only should you give the impression that you are listening, but you should also *actively* listen. Because you will hear some concerns and questions from your teams that should be taken seriously.

There are various ways to set up a feedback channel. Depending on how the company has worked in the past, not everyone may want to communicate openly. Not everyone may want to be recognizable by name. Therefore, while you can set up nonanonymous channels where everyone can discuss the team changes together, anonymous channels can also be helpful.

Another way to collect feedback is to task the managers of the respective teams with collecting opinions and evaluating the moods on their teams. This can make it much easier to see where things are stuck, and managers can also better assess possible points of criticism within their own team.

Surveys about the DevOps transition are also an option. They allow you to actively inquire about the teams' opinions and assess their mood. However, pure text-based feedback forms may not be particularly helpful for some people. In most cases, only those who have something to complain about will respond anyway.

You should conduct such surveys regularly during the transformation. That way, you always have an up-to-date insight into the teams' mood and can see whether the DevOps transformation is developing in the desired direction.

The feedback that comes in via the channels should therefore be treated like data. It is much easier to implement something data-driven if it is also measurable. This is not so easy, because a stable data situation must first be established.

Error Culture

An important point that should be made clear right at the beginning is that the overall team's error culture also needs to change. Blameless post-mortems, which are discussed in Chapter 9, Section 9.4, will not always be *blameless* at the beginning, and often people will still try to cover up mistakes.

One way to get off to a solid start in terms of establishing a healthy error culture is for managers to report on the mistakes they have made and how they have dealt with them.

Unfortunately, in many companies, you often see managers presenting themselves as if they never make mistakes. But of course they make mistakes and learn from them, just like everyone else—at least, I hope so. An open error culture should therefore apply not only to employees but also to managers, who can set a good example.

In this way, employees without management-related responsibilities on the teams can gain much better insight into how their own managers work, which is, unfortunately, often lacking. And with such an open culture, employees are more likely to report problems and challenges than if they have managers who seem unreachable.

To make it even clearer how to establish a good error culture and where it can lead, I would like to share a real-life example. A new employee starts at a company. In the first week, he goes to have lunch in the cafeteria with a colleague, where they meet the CTO. The colleague introduces the new employee to the CTO. The CTO welcomes him and directly asks, "Have you broken the production environment already?"

The new employee's reaction was hardly surprising: "Who, me? No, of course not!"

And what did the CTO say? "Why not?"

This example shows how things should work. The CTO makes it clear to the new employee right at the start of their work that it is perfectly okay for them to make mistakes. It also suggests that he should get his hands dirty quickly in order to get to know the infrastructure better. Of course, some details are still missing here, such as the fact that there is an error budget that the new colleague should use, so they should work with caution.

And that is exactly how everyone must work and act in a DevOps environment: It's okay to make mistakes if you work responsibly.

A lot can be said about the presentation of the DevOps vision and strategy. The points presented so far are just a small selection of examples consider. There are certainly other related methods that should not be ignored. But, as we all know, this is only the first step.

Conclusion

Before proceeding with the transformation after presenting the vision and strategy, a little time should be allowed to evaluate the feedback collected from the teams. Only then should the next step be taken.

> **Attention: Summary of the First Step**
> - Communicate the purpose: What is DevOps? What is it not?
> - Define and present the vision and goal for the organization
> - Exemplify changes to the team structure
> - Emphasize that the developer experience will improve
> - Make initial progress in improving the error culture
> - Create a feedback channel

Step 2: Create Measures for Acceptance

Once the vision and strategy have been communicated in the first step, it is time to start taking measures to break down the walls between the teams. However, you should *not* start restructuring the teams at this point! Such an approach would be counterproductive at this early stage, as your teams still have too little insight into the tasks and activities of the other teams.

The focus at this stage should therefore be on ensuring that all employees have as good an understanding as possible of how the other roles work. Only then does it make sense to break up the team structures.

You should carefully consider the measures you'll take in teaching your employees about their colleagues' roles so that this learning process is not a deterrent. You have to think about what you can do to get people talking to each other. There are many team-building measures that are not just aimed at DevOps and can certainly help in this endeavor.

In this section, I will focus on two measures that can be implemented from a more concrete DevOps perspective. The first measure is a kind of retrospective, and the second measure is a team exchange, in which individuals are swapped between teams that previously worked separately from each other over a certain period of time.

Cross-Team Retrospectives

Many people are familiar with the concept of retrospectives from the Scrum and agile project management methods. Retrospectives take place at the end of the sprint. The problem with retrospectives, however, is that they are often held only within the team itself and not between the teams. As we want to promote collaboration between the different teams and roles, a second retrospective can also help, during which all teams come together at the end of the sprint and present their part.

This assumes that the sprints within the organization end at the same time. If this is not the case, then this is a measure that should be brought forward and can be implemented without major upheaval.

The role of the moderator is essential for such retrospectives. This is because the points described below must be presented in a structured manner. Each team and each person should have enough time to present their results and experiences. Depending on the group, without a moderator, unproductive group dynamics may creep in.

In addition to presenting the team's progress of the respective teams, each team should also highlight any errors, problems, or shortcomings it has encountered. It is important to make sure that no accusations are made here. Unfortunately, making accusations while discussing a topic like errors and shortcomings is not uncommon, as it is a normal human behavior at times. However, this is an essential point at which the moderator of the retrospective must intervene, if necessary.

Ideally, presenting the team's problems and challenges can help the other teams to understand the whole system, which is part of the First Way of the Three Ways, which were discussed in detail in Chapter 1. It can also give the teams an opportunity to say that they want to try out one method or another in the future, regardless of whether they have a good knowledge of it. In scenarios like this, people from other teams may report that they have already looked into that method on their team, or privately. Therefore, these types of discussions can lead to synergy between teams.

This is a simple way to promote communication: Everyone involved not only gains better insight into how the other teams work but can also assess why some hurdles are built the way they are. And that helps enormously.

Temporary Team Exchange

The second measure is the team exchange. Other terms that could also describe this measure include inspection, work shadowing, and internship (but I don't particularly like these terms, for various reasons).

The idea is to have individuals temporarily swap roles between teams (see Figure 12.2). This is a way of strengthening communication between people on a team. This measure is already a big step forward compared to cross-team retrospectives.

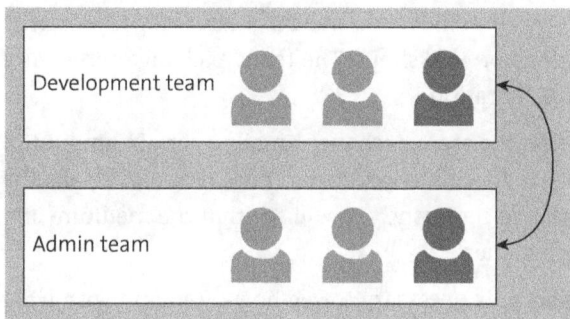

Figure 12.2 Team Exchange

The concrete idea of the team exchange is that a person from one team, such as development, is included in another, separate team, such as operations, in order to gain insight into the tasks and working methods on that team.

Just an hour or two would certainly not be enough for this exchange; a longer time commitment is required. A session can easily last four hours and should be repeated frequently over a certain period of time. The idea behind the team exchange is to allow developers to get a much better idea of how the operations team works over a longer period of time, and vice versa. The developer can look over the operations team's shoulder during the team's regular work but can also be present to give their own expertise when the application is being deployed.

The idea of looking over the team's shoulder is, of course, symbolic—there should be active communication during the team exchange process. The only problem with the team exchange process is that some people may not like the idea. It makes sense to explicitly select the first people and teams to participate in this process based on the results of a feedback survey in order to start with people who are willing to change.

A nice side benefit of this measure is that it implicitly introduces a feedback loop, as the developer repeatedly receives information from the operations team that they would otherwise not receive and that they can take with them into the next sprint. For example, through a team exchange, a person from the operations team could learn, for the first time, how and why the development team has built certain transitional solutions and how this relates to the work of the operations team. Of course, this is also often the case the other way around.

Team members often use the opportunity presented by the team exchange to talk directly about what bothers them the most. In this way, even without DevOps, you can quickly achieve big gains before you have properly started implementing DevOps. If the changes toward DevOps are then implemented relatively quickly, the team exchange is often a good first step toward breaking down silos. After all, the team exchange promotes direct collaboration, allowing team members to help those on other teams.

However, you don't have to rely on this measure for information on how the teams work to flow between the teams. Points that the teams need to address together can be discussed in the form of presentations or workshops. The team exchange serves as a simple trigger for discussion and collaboration.

However, the biggest challenge is that the team exchange entails a certain amount of loss of productivity in the short term. But it's important to note that if you take the whole DevOps transformation initiative seriously, you will see that the medium- and long-term productivity gain is much higher.

Here, too, a certain amount of iteration is necessary to see how often and with what intensity the team exchange should be implemented. After a few exchanges, each team should have much clearer view of the respective challenges of the other teams, so this initiative will have to be phased out after a while.

It is also important to measure the results of team exchanges in some way. If you simply leave measuring the results up to the teams themselves, such measurement may not actually happen. Therefore, department and team managers should keep a close eye on team exchanges and ensure that targets are set and tracked regularly; for example, by *objectives and key results* (OKRs) could be used to measure the results of team exchanges). Based on these measurements, it may be necessary to make adjustments.

> **Note: Objectives and Key Results**
>
> OKRs can be a good method for implementing a strategy in an agile way. The objective specifies what you want to achieve, and the key result specifies how you measure results or progress. These results can be used well for such implementations. There is separate literature on the effective use of OKRs.

Exchanges should take place with the various teams—that is, not just between development and operations, but across all other relevant teams, such as security and QA. As mentioned in the last chapter, the walls between the security team and the other teams are particularly thick.

> **Attention: Summary of the Second Step**
> - Conduct cross-team retrospectives
> - Carry out team exchanges

Step 3: Create a Team for the DevOps Transformation

At the start of the third step, the most important basics should already have been completed: The teams should understand and be well informed of the goals, tasks, and challenges of the various other teams and roles. A few months should pass before the teams have this level of understanding. A common mistake is moving through the steps of the DevOps transformation too quickly with too much ambition, which is not conducive to the initiative.

The third step is to establish the DevOps transformation team. This team has two tasks. The first task is to define and support the DevOps toolchain, and the second task is to help the other teams transform their cultures and workflows according to DevOps principles. This is not just about the technical aspects of the tools used but also about the processes and the culture surrounding them. Although the team's purpose is defined only here in the third step, its tasks start much earlier—especially the first task of defining and supporting the DevOps toolchain, which starts in the first step.

As a rule, such a team consists of people with a very technical background and with a strong focus on establishing a DevOps culture.

Ideally, some employees in the organization volunteer to work on this team, are motivated, and perhaps have even worked according to DevOps principles before. Such motivation and experience give these people a solid basis for initiating and supporting the DevOps transformation throughout the teams.

The selection of tools should be carefully considered so that a clean and lean DevOps toolchain can be created.

> **Note: DevOps Team Topology**
>
> The DevOps transformation team is one of the few teams that I would actually call a DevOps team. After all, this team is there to implement DevOps within the organization and introduce a DevOps culture.
>
> Many companies also call their teams—which may actually consist only of developers—DevOps teams, because at the end of the development process, containers fall out of a continuous integration pipeline that they don't even operate themselves. This is exactly what is not meant here!

Diverse DevOps Teams

When forming such DevOps teams—there can be several in one company—there should ideally be a certain degree of diversity. Because DevOps is first and foremost a culture and a way of working, and that's where people are important.

Diversity is not just about gender or nationality but also about age, experience levels, and job roles. A diverse professional background—such as some team members with deep experience in technical tasks and those with less experience in technical tasks—can also be helpful in general. A diverse, heterogeneous group makes it easier to understand different interpersonal problems than a homogeneous group.

With a view to the technical background, comprehensive knowledge is required in order to efficiently help the teams with the transformation. Training and education should therefore not be neglected. A great deal of knowledge is required here, as continuous integration, continuous delivery, infrastructure, and monitoring are major topics in themselves.

The team's goal is to abolish itself again. Only then will it have been successful. However, this does not mean that the team members are no longer needed afterwards: Their contact should remain close in the future in order to be able to address changes in the toolchain or problems.

The specific tasks of the DevOps team vary over time and are heavily dependent on how capable the teams to be supported ultimately are of putting the cultural and technical aspects into practice. Ideally, the DevOps transformation team teaches the teams as many of the points mentioned here as possible. To this end, cross-team training and courses are common, so ideally there should be a transfer of knowledge as well as assistance with taking initiative.

After training, the implementation should always be actively supported so that the members of the DevOps team can be available as competent contacts in the event of problems. Here, too, a certain amount of time is needed to build trust. The profile of the DevOps team should therefore not only be purely technical; there should also be a certain degree of interpersonal sensitivity in order to be able to deal constructively with conflicts.

Attention: Summary of the Third Step
- Introduce a DevOps transformation team
- Distribute the DevOps team to the various other teams to support them in implementing DevOps culture and technology
- Conduct training courses and seminars
- Provide support during the implementation of DevOps

Step 4: Start with the First Value Stream

Once the DevOps team has been formed and is ready to go, it is time to start the actual implementation of the DevOps transformation. It is important to start with an area of the transformation that the teams involved are passionate about. The more motivated the people involved are, the more likely it is that the transformation will be successful, as intrinsic motivation will get things moving. If you have carried out the previous steps sensibly, you have laid the foundation for increasing the intrinsic motivation of the team members.

But be aware that this can also backfire! An additional risk assessment in the area of *business continuity* must also be carried out. Motivated employees are essential, but if the transformation fails early on because a business-critical value stream is used, then the DevOps idea may not last long in the company.

This is the point at which the team structure is changed. I won't go into this in detail, as I have already discussed this topic repeatedly. Depending on the size of the company, it makes sense to set up a cross-functional team or, because there may be no other option, several teams at once.

However, you should not start all the different roles at the same time, as this does not necessarily make sense. I prefer carrying out the team merge in several iterations rather than all at once. The main reason is that the changeover is then much less disruptive and can be counteracted more quickly in the event of problems. In addition, day-to-day business has to continue, which is easier to achieve when the team structure is gradually transformed.

However, there is one more factor to consider when forming the first DevOps team. While motivated people are an important factor, you should also select a *value stream* that is suitable for the technical aspects of the DevOps transformation.

Young projects are usually predestined for this, as they often already rely on modern source code management and even have pipelines that are defined for continuous integration. Initial preparations may even have already been made to simplify deployment. For example, a web application that can be constantly updated is a good candidate.

Such projects are better suited for the initial transformation to DevOps than ancient applications that have to run on exotic hardware or industrial devices. If the project has special availability requirements or needs to adhere to complicated compliance rules, there is a high risk that even the first attempts will fail.

Once the first application has been selected and the team has been set up accordingly, you should slowly start to implement the various aspects of DevOps. It is important to establish all aspects of the CALMS model at roughly the same time. The most common mistake is to focus predominantly on the Automation value. It should be clear by now that DevOps is more than just a CI/CD pipeline.

Based on what you've read in the book so far, you may decide to straighten out the source code management as a first step. First of all, the repositories of the various teams should be opened and, ideally, merged. This has a direct impact on the automations that were previously used, as they are usually wired to the repositories, even if they may run separately. At the same time, you need to take care of the build process so that the workflows in Git match.

Because these parts go hand in hand, the introduction of code review only makes sense if there is also a continuous integration pipeline that builds the project once, even if no other tests are running. Code review without early feedback from a continuous integration system is possible but is somewhat more work. Don't forget that there are still people working here, and if you don't introduce code review wisely, they may not see any advantages at first and may wave everything they're reviewing through. This should be avoided at all costs by taking the right approach!

Once the build process is executable, the tests should also be adapted and automated to this extent. The same applies here: Introduce the tests continuously and not all at once. The tests should end up in the pipeline only once they have run successfully; otherwise, you will only cause further blockages. This also applies to the integration of security scanners.

In my experience, the operations team should be involved only once the previous stages of the DevOps lifecycle have been well implemented and established and the teams are collaborating well. This includes not only implementing the technical aspects but also establishing a more open error culture. If you involve operations too early before these aspects of DevOps are established, there will be too many construction sites open at the same time.

Introducing continuous delivery is not an easy step either—not only culturally, but also technically. The first step should be to set up a staging environment that can be deployed automatically.

It usually takes time to get there, as many technical foundations have to be laid. A safe step at the beginning is to start with the review environments and then move on to the staging environment, which you don't roll out every day at the beginning anyway. The intervals between deployments can then be gradually reduced. Then you can take the same approach with setting up the production environment for continuous delivery.

Creating trust is important for all steps of implementing DevOps, at all levels.

> **Attention: Summary of the Fourth Step**
> - Find a suitable value stream
> - Convert the team structure
> - Straighten out source code management
> - Introduce continuous integration with code review
> - Take the first steps toward continuous delivery in staging environments

Step 5: Measure and Iterate

The fourth and fifth steps must actually be considered together because iterating over processes is important not only within a project but also within an organization. The measures described in this book, such as retrospectives, post-mortems, and other opportunities for open exchange about what is going well and what is going badly, should be repeated regularly in order to learn from mistakes and implement improvements.

In the previous step, I already stated that continuous delivery should be introduced relatively late. This also applies to many other technical aspects described in Chapter 8, Section 8.4, such as feature flags and A/B tests. These can be implemented in a meaningful way only once the CI/CD pipeline has been successfully implemented. Therefore, you are more likely to find issues related to continuous delivery in a later iteration.

Security is also an important topic, but I have seen from experience that it will take a little longer to integrate the security team into the ongoing transformation in order to break up the team silos. During the process of transitioning to DevOps, it is primarily important not to block too much from the security team, as otherwise too much security-related will pile up again, leading to frustration. However, general security should not be neglected.

After a few iterations through the DevOps loop, all teams should be constantly learning new things. This learning culture and the working structure that will have been created should continue to exist in the long term so that silos are not built up again.

Attention: Summary of the Fifth Step

- Introduce continuous delivery throughout, where possible
- Have the ability and confidence to use feature flags and perform A/B tests
- Iterate over the entire lifecycle and constantly implement improvements
- Make success measurable with metrics

Note: Reflection

Successfully making a DevOps transformation is really not easy. Don't see the process as simply a list to be worked through, because all companies are different and may need to take different approaches and adjust the process to fit their needs.

Please note that many aspects of making a DevOps transformation are not steps that are "simply" completed at some point. They have to be lived in the long term. If something does not work as you expected, try to understand the problems and make adjustments.

12.2 Making DevOps Success Measurable with DORA Metrics

In this book, you are learning about many different aspects of DevOps. The question now is how to measure success in the organization, because evaluating the big picture is a challenge in DevOps transformation. There are various metrics for doing so.

The *DORA metrics* are important for measuring success. DORA stands for DevOps research and assessment. DORA is an organization that now belongs to Google. Every year, this organization presents the *State of DevOps Report*, which looks at current developments and research into DevOps in companies. It is definitely worth a look. As part of its research, DORA has developed four metrics for measuring the success of working with DevOps, which are therefore also referred to as DORA4 metrics:

1. Deployment frequency
2. Lead time
3. Change failure rate
4. Time to restore service

These metrics concern in particular the performance of software delivery. But looking at these metrics alone is not enough, because there is a lot of context that needs to be considered. It would be a misconception that these metrics alone can be used to assess how well DevOps—as a culture!—is implemented.

In addition to these four metrics, which I will discuss in more detail in the next sections, there is another metric that completes the picture with a view on *operational performance*: reliability

> **Note: Context**
>
> Whether these metrics can be used to measure the success of your DevOps transformation all depends on the context of the application, the teams, and the industry you work in. Not every company and not every team can deploy regularly and in very short periods of time—this approach is mainly suitable for software that runs on the web and where you have control over the platform. But in this scenario, you need to take a close look at how good the result is for the end users: Is a constant change to the application even desired? Is there feedback that demands updates, or are customers more interested in stability? Is availability for users perhaps even limited? Are perhaps only half-baked features being rolled out, which does not help users at all?
>
> The goal of the deployment is also essential to consider. Not every organization develops a web application that can simply be deployed in a cloud. If you are working very close to the hardware in an industrial company, (i.e., doing *embedded software development*), you can never make changes as often and as quickly as you can to an online store or an application. Different standards inevitably apply here. And it is obvious that projects that run in maintenance mode and no longer receive any major further development must also be evaluated differently.

> To a certain extent, however, these metrics can also help to evaluate work in traditional industries. But not every metric is meaningful and not every result is directly comparable. Depending on the industry, there are also specific regulations that must be adhered to.
>
> For example, you can regularly deploy software to cars—just look at Tesla—but without the exact context, it doesn't say much: Is it just a change to the infotainment system, or is the brake control unit being reprogrammed? The effects of errors are completely different, so the metrics cannot be applied in the same way.

12.2.1 DORA Metric 1: Deployment Frequency

The *deployment frequency* metric is about how often a deployment is executed. Deployments to the production environment are particularly important. The higher this value of this metric is, the better it is from a DevOps perspective. However, this metric should not be considered completely on its own, but always alongside the other three metrics.

The lower the frequency of deployments, the further away the organization is from a DevOps environment. The reason should be clear by now: The First Way of the Three Ways states that improvements should run steadily and regularly to end users, and this is possible only with a high frequency of deployments.

The higher the degree of automation and the leaner the entire development cycle, the easier it is to roll out a high number of deployments. A side effect of this is that with a high deployment frequency, the teams and the organization have a high level of trust in the software, the overall team, and the processes. Inevitably, the team often has the confidence to roll out changes if the quality is really right at all levels.

So that's the metric in theory, but what does it look like in practice? Which values are good, which are bad, and where is the middle ground? The figures, both for this metric and for the following metrics, are taken from the *State of DevOps Report 2023* (*https://cloud.google.com/devops/state-of-devops/*). Unsurprisingly, a frequency of one deployment per month to one deployment every six months is considered low. This is how often organizations that do not work according to DevOps principles typically deploy.

The medium is at least one deployment per week to one deployment per a month. Once you have reached this frequency, you are well on your way to a high deployment frequency, where you deploy as required, which can be several times a day.

It is important to roll out deployments every day. Figure 12.3 shows what the deployment frequency over the last 90 days can look like, in which there was an average deployment frequency of 2.8 days. However, the actual frequency of deployments on working days was much higher, as there were usually between four and eight deployments. There were no more deployments from Friday afternoon until Monday, as no work was done on the weekends.

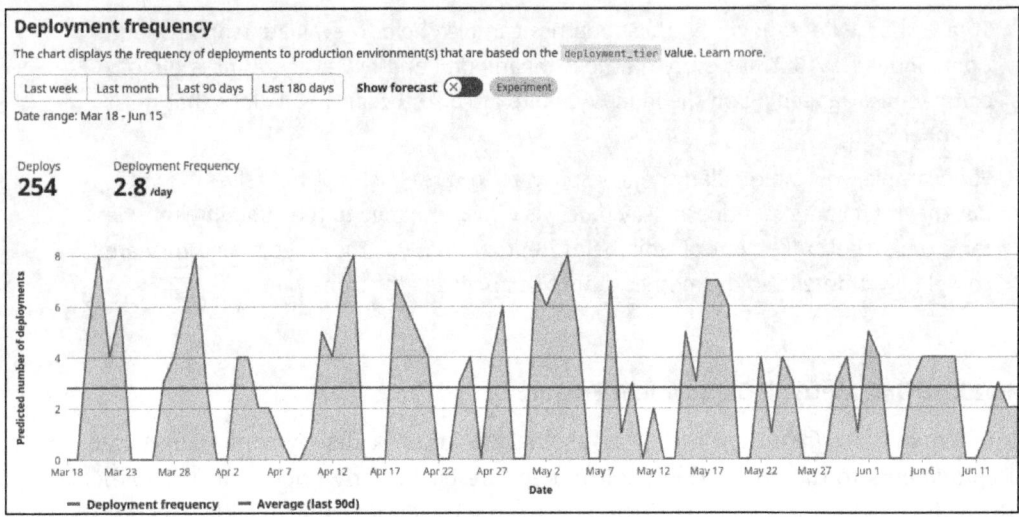

Figure 12.3 Deployment Frequency over the Last 90 Days

12.2.2 DORA Metric 2: Lead Time

The second metric goes hand in hand with the first: the lead time. As explained earlier in this book, lead time is the time it takes for a change to be rolled out. It begins with the creation of the commit (i.e., after the implementation of the change has been completed) and ends when the change is rolled out on production systems. Although there is a close relationship to the deployment frequency here, many deployments can also be made with many older changes.

The lead time should be as short as possible. This metric also relates to the first of the Three Ways: making the change available to end users in a timely manner.

A lead time between one month and six months is not a good value. A better lead time is between one week and one month. The lead time is excellent if it is between one day and one week, as this means that the changes really do arrive quickly.

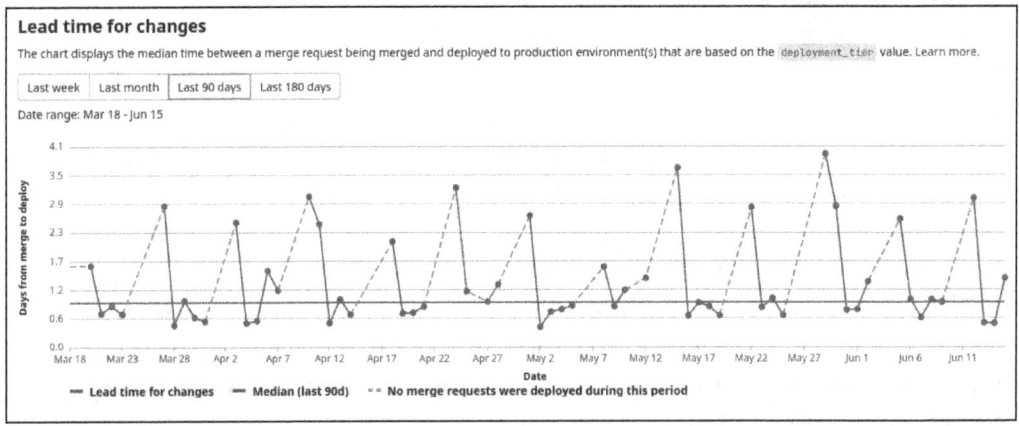

Figure 12.4 Lead Time over the Last 90 Days

As with the deployment frequency, the lead time is often significantly longer on weekends, as shown in Figure 12.4. The trend over a longer period and the median are also important here.

12.2.3 DORA Metric 3: Change Failure Rate

In contrast to the first two metrics, the third metric looks less at the frequency and speed of changes and more at how often errors occur. After all, rolling out changes quickly and frequently is not much use to users if errors are constantly occurring.

The *change failure rate* indicates how many errors occur per rolled-out change. The lower this value is, the better. This metric can also measure errors in the productive environment in order to assess the quality and productivity of the team.

If this value is between 0% and 5% this is considered elite, while anything up to 15% is already considered pretty good. The closer you get to 0%, the better. An average value is around 15%; anything avoe that can be considered a poor value. However, it is important to remember that the values only make sense in combination with the other metrics!

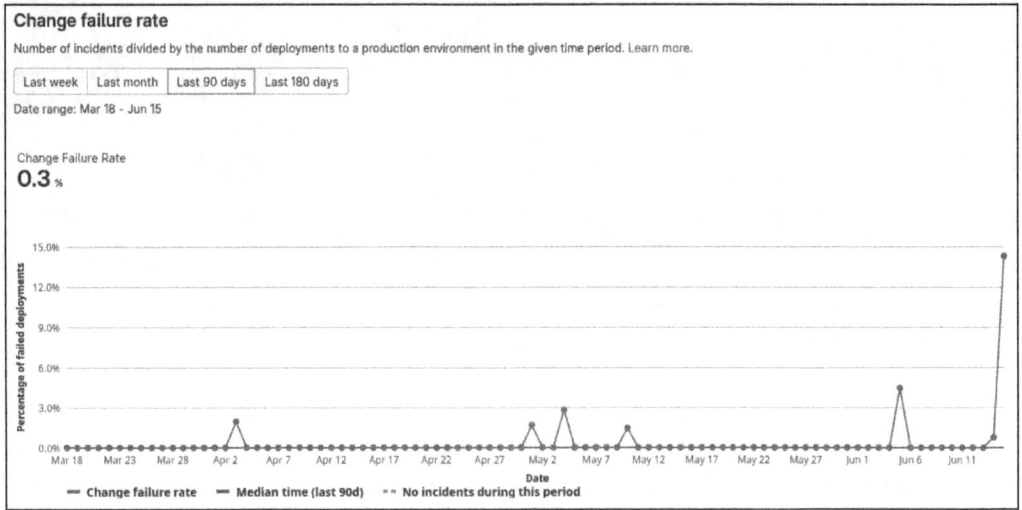

Figure 12.5 Change Failure Rate over the last 90 Days

In Figure 12.5, you can see what a change failure rate can look like. Most of the time, there are no errors or very few errors—until there is a major problem that drives up the failure rate. This metric therefore says relatively little about the actual quality, as a few larger problems can be worse than several smaller ones spread over time. So the same applies here: Considering trends is important so that you can take early countermeasures if problems accumulate.

12.2.4 DORA Metric 4: Time to Restore Service

Really bad errors lead to downtime. The interesting question is how quickly a system recovers from.

This is exactly what this fourth metric indicates: How long does it take for the service to be available again in the event of an outage? The lower the value, the better. This metric covers not only times when the entire service is offline but also performance losses and service issues affecting only small groups of users.

Ideally, the system should be fully operational again within a few hours (i.e., on the same day). This is considered a good value. Anything more than a day and up to a week is still acceptable. But it looks really bad if it takes longer than a week for services to be fully up and running again.

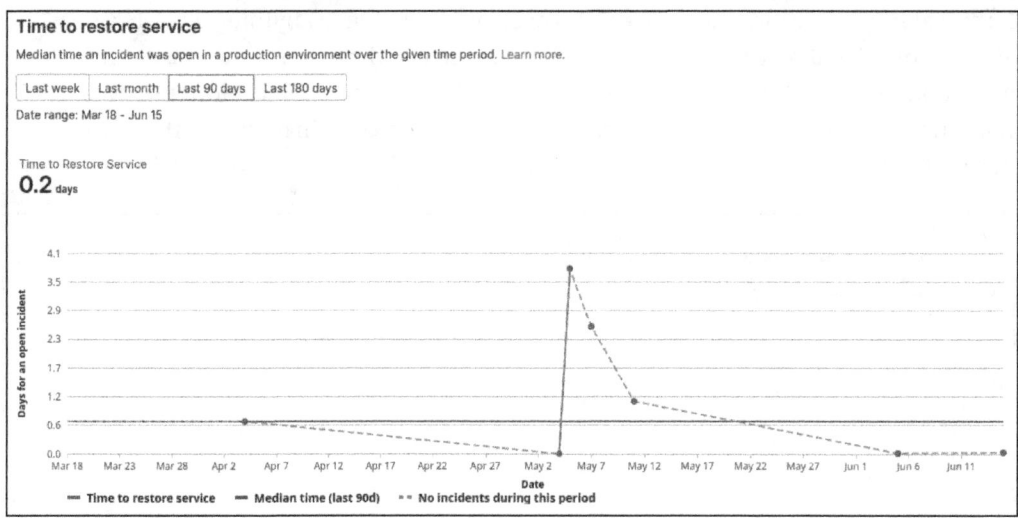

Figure 12.6 Time to Restore Service during the Last 90 Days

As with the previous examples, there are—hopefully—not so many errors that the service runs only to a limited extent (see Figure 12.6). Not every incident is resolved quickly, but every incident changes the median value of the time to restore service. The same applies here: Look at the long-term trends, learn from mistakes, and take action if necessary.

12.2.5 DORA Metric 5: Operational Performance and Reliability

The fifth metric is comparatively new and was introduced in 2021. Many sources therefore still refer to the DORA4 metrics, so don't be surprised to learn that there five metrics are recognized these days.

The fifth metric concerns *reliability*. about it measures how reliably a service works for users and whether it meets expectations. In the "State of DevOps Report 2023," the

DORA organization itself writes that the previous focus was primarily on availability, not on reliability. These two terms describe two different aspects of a service. Availability indicates only whether a service is running. If the availability of a service is 99.99%, an amount of downtime of 52.60 minutes per year is permitted.

However, this does not indicate whether the service works properly for the remaining 525,547.64 minutes of the year. A service that is always "there" but is unusably slow or produces errors with every second click is of no use to users. The evaluation and measurement here depend on defined performance standards.

Accordingly, this metric has nothing to do with software delivery itself, which is why it is considered separate from the four other metrics. Reliability is evaluated on the basis of availability, latency, performance, and scalability.

12.2.6 Findings from the State of DevOps Report

If you want to delve deeper into the facts and figures surrounding current DevOps developments, you should read the *State of DevOps Report* carefully. Once a year, the DORA team conducts large surveys and publishes the results in September.

The report is free to download, but not free in the classic sense, as you have to provide your own contact details, including your mobile phone number, employer, and job title in order to download the report. Conveniently, however, you can also find the report without having to input this information if you search Google for "DORA State of DevOps Report 2023/24/25 pdf".

The report always includes the latest developments, the latest figures, and changes in the figures from the last report. This provides a good and simple overview of how the figures have changed over the years and where the industry is heading—at least for those who took part in the survey. (You are also welcome to take part yourself.)

The sticking point, however, as is so often the case with such surveys, is the dispersion of data across countries and sectors. Unsurprisingly, there is a strong focus on the US, which is represented in the 2023 report by 28% of participants, followed by a similar amount in Europe. However the details of every European country is much more diverse. The report also tends to overvalue certain industries and company sizes.

Many points in the report are fairly obvious. However, they provide good, concise insight into how other companies work, although the report is not organized by individual industries. Many participants rely on the cloud and use tools to automate deployments. It can also be said that the more cloud and the more automation are involved, the more successful the implementation of DevOps is on a technical level.

However, other topics are also considered that should not be neglected. For example, the 2023 report states that a loosely coupled IT architecture contributes to team burnout during a DevOps transformation. The report provides more detailed background information on the study.

> **Note: Reflection on DORA**
>
> The DORA metrics provide a good—but not the only—overview of the success or failure of DevOps in an organization. These metrics can be used to measure and compare how well DevOps principles are being implemented.
>
> However, there is one thing you should not ignore: These are metrics that describe deployment frequency and errors. They do not provide any insight into the actual satisfaction of the team or the end users. However, this should be the main focus of the DevOps transformation: Focusing on the people, and only after that on the processes and tools. The DORA metrics deal with processes, as these are easy to measure.
>
> In order to record the DORA metrics for a team, a fairly complex tool landscape is required to collect the various values. This is often counterproductive because the more tools you use, the more difficult it becomes to reliably record all the data, as each transition between tools not only costs time but also makes the process more complex. DevOps platforms (see Chapter 13) can help to simplify the whole process. If you rely on GitLab, for example, and use its relevant functions, the DORA metrics are automatically recorded and displayed. However, you will need the GitLab Enterprise Edition with the Ultimate license for these functions.
>
> It makes sense to use look at the DORA metrics at them over a period of time to get a feel for how good your team's processes and technical implementation are. Try out various changes and see how they affect the metrics. Make adjustments if you are moving in the wrong direction. However, don't get too fixated on the metrics without looking at the context and background. A high deployment frequency is good and important; however, if nobody dares to roll out major changes in return, it doesn't fit again.

12.3 Value Stream Mapping

So far, this chapter has mainly focused on cultural adaptation and the use of meaningful metrics. What is still missing, however, is a look at a value stream. Ideally, a suitable value stream should be selected and value stream mapping should be carried out, especially at the start of the DevOps transformation.

But what exactly is value stream mapping? First of all, we need to look at what a value stream itself is. A *value stream* is a sequence of activities that must be carried out in an organization in order to bring a good or service to the end user.

This actually has nothing to do with DevOps and does not even have its origins in IT. The concept dates back to the 1990s and was popularized by the 1996 book *Lean Thinking* by James Womack and Daniel Jones.

A value stream is an entire chain of processes and people who are involved in delivering or providing a certain value or service. This is always cross-functional, as many different people with different roles are involved, as the following examples also show.

For the customer or user of the system, there is usually only one value stream, even if there are many different value streams internally that support the business.

> **Note: Processes Are Part of Value Stream Mapping**
> It is important to note that value stream mapping is not about visualizing a process. Processes are part of a value stream, so a value stream consists of many different processes.

In the next sections, we look at two examples of value streams, one almost completely outside of IT and the other a specific example from *nicely-dressed.com*. The easiest way to describe a value stream is to use an example that even nontechnical people can understand—here we use the example of a pizzeria.

12.3.1 The Value Stream of a Pizzeria

The value stream for a pizzeria customer is that they place an order and then receive a pizza. If you take a complete look at this value stream, it consists of many small steps for the pizza baker. The first step is the order. This can be done by phone, in person, or online.

The order is then processed: The pizza dough is topped with the desired toppings and then placed in the oven to bake. After baking, the pizza is either served directly on a plate or packed in a box for delivery or pickup.

Ultimately, three points are important for the customer: First, the ordering process should be simple, because nobody wants to order a pizza by fax, and the pizza should arrive quickly and hot for consumption and should taste great.

In a small pizzeria where the only employees are the owners, this is often not a problem, as the entire value stream is handled by one or two people, without bottlenecks or delays. Things get more complicated when the work has to be scaled: Significantly more orders come in at typical mealtimes, via online ordering and phone and also in person on site. There is a limited number of ovens, and the toppings for the pizza have to be prepared regularly so that waiting times are kept to a minimum. Vehicles and drivers for deliveries are also limited. Some steps in the process of completing orders take a very long time: It takes time for the pizza to bake, and then it takes time for a driver to be available to deliver the pizza, so it often arrives to the customer cold.

This is exactly where value stream mapping can be useful. It is not a tool or a process but a method for visualizing the previously defined and selected value stream. The aim is to identify the value stream of a system or process in order to subsequently optimize this entire process so that unnecessary steps can be removed and the process can be improved.

12.3.2 The Value Stream at nicely-dressed.com

At *nicely-dressed.com*, too, there is a key value stream for customers: Instead of a pizza, customers want order items of clothing that are to be delivered as quickly as possible. The start of the value stream is therefore the online order, and the end of the value stream is the delivery via a parcel service.

You may have already noticed that not every step in a value stream is completely in your own hands; for example, nobody can afford their own delivery service for their company, except perhaps Amazon. It is therefore important to visualize the value stream together with external dependencies.

For *nicely-dressed.com*, this means that the customer places an online order by placing the items in the shopping cart and completing the order process. Depending on the payment method, the customer pays immediately or after a few days by invoice. Therefore, there may be waiting times.

After the payment process is complete, someone in the warehouse goes and packs the goods, which are then handed over to the parcel service so that they can be delivered.

Waiting times and bottlenecks can occur with every transition: For example, the customer does not transfer the invoice on time, something goes wrong during the payment process, or the warehouse is understaffed. The goods can be collected from the warehouse, packed, and prepared for dispatch more quickly or more slowly depending on the given situation. As the parcel service collects the parcels only at certain times, pickups also need to be coordinated so that the handovers are correct.

Let's take a closer look at the example: The ordering system provides the warehouse with a list for each order, which the people in the warehouse use to pack the parcels. The problem is that an individual list is generated for each order, so each order is processed individually. In the context of value stream mapping, it would be more efficient if routes were optimized so that they could be calculated for several orders. This would significantly reduce the lead time for the packing process, as several orders would be viewed in parallel. More orders can be processed by the same number of people.

Is that obvious? Perhaps. In my experience, however, such ideas become clear only when everyone is sitting in the same room and explaining their working methods and challenges.

But what does this have to do with DevOps? The examples we use here aren't directly related—they serve to illustrate the lean method of value stream mapping, which can also be used in IT. Use this tool to visualize the value stream within your company and then optimize it.

One such value stream could be the delivery of a new feature for the *nicely-dressed.com* online store. I explain what the whole cumbersome process of delivering new features looks like at *nicely-dressed.com* at the beginning of each chapter. It is clear that collaboration between the various teams needs to be strengthened in order to deliver added value to the customer more efficiently and quickly.

This involves every transition, from customer feedback, to the requirements analysis, to the implementation by the development team, to the tests by the QA team, to the handover to the operations team, and finally to the handover to the security team, which can pull the handbrake just before deployment.

Value stream mapping is a tool for breaking through team silos and getting everyone around the same table to keep an eye on the entire product and service. An example can be seen in Figure 12.7. All information flows between the teams must be considered, which is why value stream mapping requires all stakeholders who are relevant to the value stream to participate.

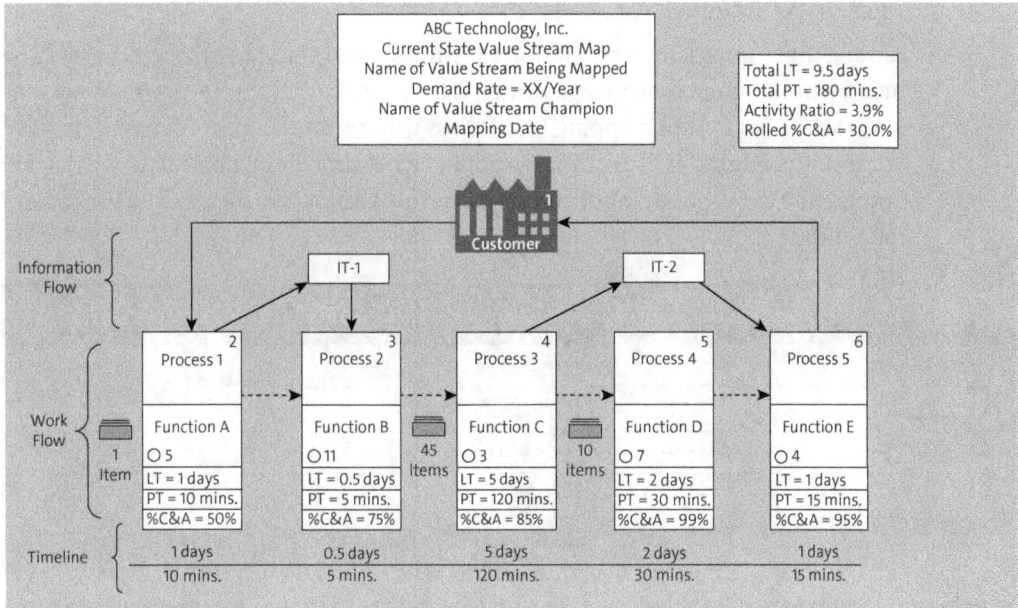

Figure 12.7 An Example of a Value Stream Map

For the larger pizzeria, these stakeholders are the owners, the people who take orders, the pizza bakers, and the delivery drivers.

Everyone is relevant here. When delivering new features for the *nicely-dressed.com* online store, it is not only the developers who have to make decisions, but also the operations team, the security team, and the managers. It is up to everyone to improve the entire value stream.

12.3.3 Implementation of Value Stream Mapping

Value stream mapping is not a task that can simply be done in between other tasks when there is time. It is a time-consuming process in itself that requires careful preparation, during which the necessary information is first gathered. You should plan for a few days to create the resulting map.

Value stream mapping is not a procedure that is carried out only once and delivers valid results forever. It is advisable to repeat the analysis either regularly or on certain occasions to check whether you are still working optimally.

In this section, I outline the steps for completing a value stream mapping exercise. A complete description would out of the scope of this book, but if you are interested in this topic, I recommend the book *Value Stream Mapping* by Karen Martin and Mike Osterling.

1. **Preparation**

   Before the actual map, a *value stream mapping charter* (VSM charter) is created, which notes the initial plans (see Figure 12.8).

   Among other things, the preparation meeting defines the value stream that is to be mapped, the first and last steps of the value stream, and the responsible persons involved. Value stream mapping is pointless without the support of managers; save yourself the effort if these people do not have the time or inclination to participate in this exercise. Without input and support from the top, nothing can be fundamentally changed.

Value Stream Mapping Charter						
Scope		**Accountable Parties**		**Logistics**		
Value Stream		Executive Sponsor		Event Dates & Times		
Specific Conditions		Value Stream Champion				
Demand Rate		Facilitator		Base-camp Location		
Trigger						
First Step		Logistics Coordinator		Meals Provided		
Last Step						
Boundaries & Limitations		Briefing Attendees ** required *optional		Briefing Dates &Times		
Improvement Timeframe						
Current State Problems & Business Needs		**Mapping Team**				
			Function	Name	Contact Information	
1		1				
2		2				
3		3				
4		4				
5						
Measurable Target Condition		5				
1		6				
2		7				
3		8				
4		9				
5		10				
Benefits to Customers & Business		**On-Call Support**				
			Function	Name	Contact Information	
1		1				
2		2				
3		3				
4		4				
5						
Relevant Data		**Agreement**				
			Executive Sponsor	Value Stream Champion	Facilitator	
1						
2						
3		Signature:		Signature:	Signature:	
4		Date:		Date:	Date:	

Figure 12.8 The Value Stream Mapping Charter (Source: https://tkmg.com/wp-content/files/Value-Stream-Mapping-Charter-KMMO-v14.xlsx/)

2. **VSM charter**

   The VSM charter must specify which current problems exist and which metrics can be used to measure them. For example, for mapping the value stream of the delivery of new features for the *nicely-dressed.com* online store, the VSM charter could state that lead time is the most important metric and that the time should be halved.

3. **Workshop**

   The actual mapping exercise should be carried out only once the VSM charter has been completed in full. The first step is to visualize the status quo. All relevant stakeholders may have to take several days to fully participate in and concentrate on a value stream mapping workshop.

   In the workshop, all people whose groups are involved in the value stream are interviewed in order to break down the individual processes. For each process, several pieces of information and metrics need to be determined. For our example, these are the pieces of information we need:

   - Lead time
   - Process time
   - Percentage complete and accurate
   - Persons involved

   The lead time starts as soon as work is available for a certain person in the development process and ends as soon as that person has completed their part and passes the task on to the next person in the process. Part of the lead time is the *process time*, which covers the time when the actual work is done. The lead time is therefore the time for the actual work plus possible waiting times. This is already a good metric for finding opportunities to optimize the process.

   Percentage complete and accurate (%C/A) refers to the quality of the work, which is specified as a percentage. If error-free work arrives at the start of the process, this corresponds to 100%.

   However, queries and corrections are often necessary, meaning that it is often necessary to go back a step. This increases the lead time and is another point at which significant improvements can be achieved, especially if several people are affected (i.e., if a bottleneck blocks several processes).

   These parameters are relevant for our example, but there are of course many other metrics and information that I have omitted here for the sake of clarity and simplicity. Finding out what is relevant for your value stream is a central task in this step.

4. **Visualization of the value stream and optimization of processes**

   Once you have visualized all the processes in the value stream and broken down the relevant metrics, the complete lead time, the process time, the total %C/A, and the activity ratio should be calculated. These values tell you the current status of the value stream.

12   Successfully Implementing the DevOps Transformation

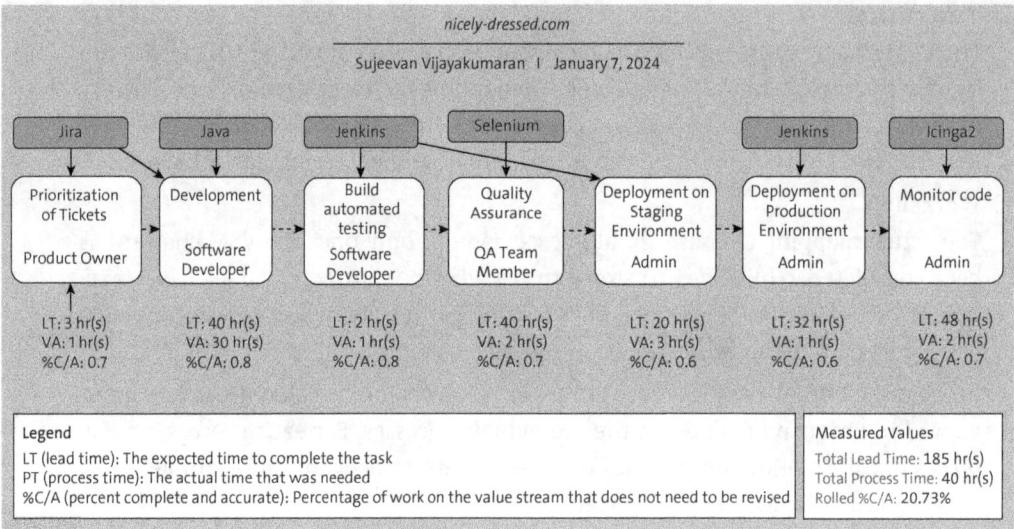

Figure 12.9 An Example Value Stream Map for the Software Development Lifecycle of nicely-dressed.com

A key advantage of this exercise is that all relevant people are involved across team boundaries and work together on improving the value stream. Without value stream mapping, only individual processes may be considered, which could result in the optimization only of individual process steps, regardless of the impact they might have on other processes.

Figure 12.9 shows another example, in this case for *nicely-dressed.com*. This should give you an idea of what a value stream map for the software development lifecycle could look like.

When mapping the value stream, you should not engage in finger-pointing, as this is not helpful. If you take your time with your colleagues, working together will help to clear up problems and clarify misunderstandings.

5. **Planning the future**

   The final step involves the question of what the process should look like in the future. To this end, it is essential to record the current status of the value stream, as this is the only way to identify opportunities for improvement.

   Once you have an overview of all parameters and metrics related to the value stream, it is relatively easy, at least in my experience, to optimize the entire value stream. Once a value stream has been properly described and modeled, everyone involved usually immediately notices where there is potential for optimization for the benefit of employees and customers.

6. **Understanding the whole system**

   If you can describe the central value stream(s) of your project, you are well on the way to understanding your entire system. This is very much in the spirit of the Three Ways.

In our example, the optimization of lead time and process time is the goal of the value stream mapping exercise, but only when you have an overview of the entire system can you say whether the changes really offer added value.

Many companies have complex approval structures in CI/CD pipelines that have to be passed through in order to carry out deployments. Only when all stakeholders are present can the question be asked: Do you need this? After all, the lead time is very high with a waiting time of a few days, while the process time is only a few seconds.

> **Note: Reflection**
>
> Value stream mapping is an excellent tool for uncovering inefficiencies. I myself have already carried out one or two value stream mappings with customers. By jointly noting and discussing the value stream with the various roles involved in it, inefficiencies and opportunities for improvement quickly become clear. It is therefore often helpful to bring in external help that already has experience in carrying out value stream mapping. A fresh look often (but not always) reveals rusty patterns of behavior.
>
> I can therefore highly recommend that you take a closer look at this topic. Value stream mapping helps to uncover problems of various kinds in the value stream, which can then be addressed in a separate plan. And only after you have carried out the value stream mapping should you look at the tools used. Because, again, "people over processes over tools" applies.

12.4 Summary

We began the chapter by taking a deeper look at the methodologies that can be used to introduce a DevOps culture. Next, the chapter walked through how to measure the success of DevOps using the DORA metrics. Finally, we discussed what value stream mapping is and how it can be used to optimize the value stream.

Chapter 13
DevOps Platforms

In recent years, it has become increasingly clear that the complexity of a tool stack leads to considerable problems in maintenance and long-term operation. For a long time, companies often put different tools together with a lot of hot glue, which can be seen especially, but not only, in historically grown Jenkins infrastructures that can be mastered only with a great deal of engineering effort.

In this chapter, we look at the history of DevOps tool landscapes that have evolved from various continuous integration tools, as well as the question of where DevOps platforms are heading.

> **Note: Stay Up to Date!**
> Note that the information in this chapter is the most prone to becoming rapidly outdated, as the DevOps tools market is constantly in flux. Some weaknesses in certain tools that were still relevant when this book was written, in the summer of 2024, may have been fixed by the time you're reading it.
>
> So evaluate specifically which platform is best suited to your requirements. Even if it is, of course, convenient to rely on other people's experiences, you know what works best for you.

The term *DevOps platform* is used by various manufacturers to promote their own products. The market research company Gartner, primarily known for its Magic Quadrant reports, also publishes the *Gartner Magic Quadrant for DevOps Platforms* report (*https://gartner.com/en/documents/4416199/*). Garner originally used the term "value stream delivery platform," which gradually moved toward the current term "DevOps platforms."

Forrester, another market research company, has also looked at the various DevOps platforms. This company refers to them as *integrated software delivery platforms*.

> **Note: Gartner and Forrester**
> Gartner's Magic Quadrant for DevOps Platforms is behind a paywall, so I will briefly summarize the results. Gartner divides the tools into four categories: leaders, visionaries, challengers, and niche players.

> The offerings from Atlassian, Microsoft, and GitLab are the only representatives of the leaders category in the 2023 report. The visionaries include offerings from Harness, JetBrains, and AWS; the challengers include Red Hat, VMWare, JFrog, CloudBees, and CircleCI. Codefresh, Bitrise, and the GCP can be found in the niche players category.
>
> The market research company Forrester has also published a buyer's guide on DevOps platforms (*https://forrester.com/report/the-forrester-wave-tm-integrated-software-delivery-platforms-q2-2023/RES178494/*).
>
> This guide uses the categories leaders, strong performers, contenders, and challengers. In the report from the second quarter of 2023, GitLab is the only leader. CloudBees, the company behind Jenkins, as well as Atlassian, Microsoft, AWS, Harness, and CodeFresh are strong performers. The rest of the providers included in the guide are in the contenders category.
>
> For both Gartner and Forrester, these are snapshots that are constantly changing.

13.1 Toolchain Complexity

If you look at the history of DevOps toolchains, you will recognize various patterns. Just as DevOps is always changing, so are toolchains.

The toolchain and its complexity depend on the size of the company and when the company first launches efforts toward automation. The more advanced the respective organization is in terms of DevOps culture and the DevOps toolchain, the more different patterns can be identified.

In the previous chapters, I have tried to delineate the aspects of DevOps by assigning them to individual stages. Of course, this simplifies the description of DevOps, but in many places you may have noticed that it is not possible to make clear distinctions. Neither the tasks nor the tools can be strictly assigned to the conceptual steps of the DevOps loop.

Some aspects of today's DevOps toolchain could already be seen in organizations that were already thinking in terms of automation when DevOps existed neither as a culture nor as a term. It simply started with the need to automate tasks. That is step 0.

13.1.1 Step 0: Toolchains Grow Historically

Step 0 is the "zeroth" step because it simply just appears. There is no plan regarding the toolchain, but the toolchain grows historically over time. You may be familiar with this scenario: A team is faced with a new challenge, so it introduces a new tool, and this happens continuously over time until the stack is barely maintainable.

In the past, source code was often stored in a Subversion or Concurrent Version System (CVS) repository; Git was gradually adopted, but the systems were always on their own.

Each team set up its own infrastructure, so there was no central management of the source code.

Jenkins, whose advantages and disadvantages have already been discussed in Chapter 6, Section 6.6.1, was often used for automation. Over time, Jenkins has become the automation hub for many companies.

Jenkins was often initially only connected to source code management systems (i.e., to various CVS, Subversion, or Git servers). For each connection, a plug-in had to be installed and access authorizations to the individual systems had to be configured and managed. Every person and every team that needed access to the systems needed access data that had to be created manually. So for most interested parties, there was simply no access, or standard passwords were used, which were extremely insecure. If you wanted to do it right, the configuration tasks were very time-consuming and multiplied for each team and for each tool that had to be connected.

The connection between the source code management system and the continuous integration server is already complex, but it becomes even more complicated when servers for storing artifacts also have to be integrated. Tools such as JFrog Artifactory, or Sonatype Nexus were often used for this purpose. The same challenges were faced here, too: The integrations had to be configured, which required plug-ins, access data, and the right authorizations for the tools.

Source code management, continuous integration, and package and container registry tools are ideally just three tools. However, there are often many more tools that make the toolchain even more complex: a separate tool for continuous delivery, various security tools for different programming languages, helpers for requirements and test management, and conventional ticket systems such as Jira. All of these tools also had to be configured and integrated with each other.

The larger and more complex the environment was, the more challenging it was to maintain the entire system. For example, Jenkins is not architecturally designed to scale horizontally. This applies to scaling of both the number of projects and the number of pipelines due to ever larger teams. In addition, the numerous different tools had to be updated regularly, and with each update, it was necessary to check whether the connection to the various tools was still working properly. Problems were common, so updating a tool required careful planning and extensive testing. Updates were often suspended or postponed, and new functions could not be used as a result. And you don't need security updates anyway, do you?

13.1.2 Step 1: Multiple Toolchains to Increase Maintainability

You can probably see what I'm getting at: The more tools there are in a toolchain, the more complex it becomes and the more expensive the setup is. In many companies, such a toolchain has grown historically and has therefore become unmaintainable, making it almost impossible to implement innovations.

In such companies, this situation was usually not pleasant for the employees, and it often led to dissatisfaction in their day-to-day work. Time and again, something that was necessary for the daily workflow did not work. Time and again, something was forgotten because the required data simply wasn't visible.

The more automation solutions were introduced, the greater the load on the systems became. At some point, many companies commonly decided to break up the toolchain: Each team or department was given its own infrastructure for the toolchain. This meant that there would no longer be many large monoliths, but many small ones.

This brought some advantages:

- One major advantage was that each large block could be significantly reduced in size. Therefore, potential deadlocks for updates could be reduced, and a failure would not paralyze the entire company, only individual teams and departments. Instead of a central Jenkins server, which was often the first to become unmaintainable, many smaller servers were set up here that had to be managed separately. During maintenance work, the systems could be updated one by one. This reduced the restrictions caused by downtime during upgrades to a much smaller group.
- Furthermore, such a procedure allowed errors to be found at an early stage that would also occur in the systems of other departments and could therefore be prevented.
- In addition, each team could then choose its own tool. This has been seen in many companies: Depending on requirements, the personal taste of the team leader, or simply coincidence, new tools were introduced and isolated solutions were created that did not meet the company's standards. In addition to Jenkins, there were also Teamcity, CircleCI, and other solutions that were sometimes more and sometimes less maintained and used.

However, the disadvantages quickly became clear:

- Because each team used its own tool stack, it was difficult to switch to other teams, as access had to be configured not only between the tools but also for the individual people. It was not possible to communicate directly or even help out quickly, as team members had to wait a long time for access rights and then get to know the entire toolstack.
- In addition, there was often a complete lack of overview within the organization: What worked? What was not working? What did productivity look like? Which libraries were used by which projects?
- The larger the company was—and the more carelessly the toolchain was approached—the more source code management systems proliferated through the infrastructure. It was often impossible to get an overview of the codebase because nobody could keep track of what was stored in all these systems. Once someone was faced with the task of untangling such a jumble, they first needed access rights to all

systems, often without knowing the right contact person, and then had to see which projects are still being actively developed and operated. And what happens when different codebases for the same project are found?

I heard a story from a large company whose network was searched for GitLab instances. A three-digit number of instances were found, mostly with outdated versions and a few users each. Just by centralizing these instances, the work became much more effective.

The worst thing about this approach was that it often did not solve the actual problems: The configuration was often still fragmented and could not be automated, and there was no exchange of information or overview. Anyone switching to another project would have to deal with a whole new zoo of individual tools at the start; everything was stuck together with digital gaffer tape. Nobody looked at the big picture.

13.1.3 Step 2: Standardized Tools, but Still with a Lot of Duct Tape

I have presented steps 0 and 1 as consecutive steps, but this is a generalization. Not every company started directly with step 0, and not every company went from step 0 to 1—many jumped directly to step 2.

In step 2, companies tackled the proliferation of different solutions. The general approach shifted so that there were no longer several solutions for source code management in parallel, but only a single solution. The same naturally also applied to the continuous integration server, even if there were still several instances of Jenkins, for example.

The difference between steps 1 and 2 primarily concern two things: From then on, companies decided to focus on explicitly selecting the best tool for each DevOps stage and on setting up a central team to manage these tools and their integration.

Such teams often had two tasks: First and foremost, they were to manage both the tools and their infrastructure. Unsurprisingly, the same applies here: The more tools there were, the more complex the setup was. Instead of the uncontrolled growth seen in steps 0 and 1, a certain amount of engineering effort was required to manage the infrastructure. For example, although a Jenkins server may have still been operated for each team, the work was much more centralized. Instead of giving the teams complete power over such systems, the focus was on taking this work away from the teams.

The selection of tools therefore started to become standardized within many companies so that everyone could work more efficiently. After all, the software development teams were supposed to develop software and not manage their DevOps toolchain.

When selecting the tools, the question arose as to whether to focus on best of breed, best in suite, or something else?

Best of Breed

In the best-of-breed approach, a separate tool is used for each problem: Jira for project management, Bitbucket for source code management, Jenkins for CI/CD, Artifactory as a package registry, Harbor as a container registry, and a number of other tools.

The basic idea is not wrong: You choose the best tool for the respective purpose. But then all the disadvantages come into play, which have already been discussed in detail in the previous sections.

Best in Suite

The best-in-suite approach is in direct competition with the best-of-breed approach. Here, the search is on for a whole suite of different tools that cover all tasks. The focus is no longer on which individual tool is best, but on the overall result.

Both variants have their advantages and disadvantages. In general, I believe that an integrated solution is the better choice. Think of modern office suites, for example, or something as simple as email. For some time now, email servers have been seamlessly linked to a variety of other tools, such as a contact book and a calendar. It is only this totality and the smooth flow of information between the individual tools that makes the difference; using the best calendar and the best contact book is not much use if everything has to be synchronized manually.

You may remember the presentation of the first iPhone by Steve Jobs. He announced at the time that there would be three separate products: a widescreen iPod with a touch display, a telephone, and an internet communication device. In the end, however, everything turned out to be one and the same device: the iPhone.

That was the right approach: It's better to offer everything from a single source and integrate functions sensibly. And yet, of course, despite the fact that smart phones come with amazing cameras, SLR cameras still have their right to exist and are used by many people; the fact that there is a suite that suits most people does not mean that special tasks are not better done with special tools.

13.1.4 Step 3: DevOps Platforms

This brings us to step 3: the DevOps platforms.

The main task of modern DevOps platforms is to master the complexity of the toolchain. Tools have to work together, and they have to be configured and maintained. DevOps platforms are a solution to many of these problems. The emphasis here is on "many"—DevOps platforms are not a solution to all problems. Although this should be the goal of the journey, a lot of development work on these platforms still needs to be done, which will take years.

What are DevOps platforms anyway? The basic idea is that the entire DevOps software development lifecycle is managed on one platform. Instead of having to jump from one tool to the next, you have one platform that contains every function you need.

Reduced Complexity

One thing should already be clear: The fewer tools are that are used, the fewer tools need to be configured and managed and the lower the complexity of the entire toolchain.

But it also makes work easier for the user—be it a developer or an admin. Developers can track their code across multiple work steps and tools without losing context. In the event of a failed deployment, an admin can link each log message to a line of code and the corresponding commit. More visibility means more insight.

Uniform Surface

A uniform surface means that there is a standardized interface so that nobody has to log in to several tools to see all relevant information.

What exactly is relevant also depends on the position. The product owner works mainly in project management, while the developer spends most of their time in source code management. Nevertheless, all information is always available and visible to everyone; when teams are working with separate tools and platforms, it is difficult to see what was planned and what had already been done. Since everything is visible to all roles in one platform within one interface, collaboration is easier.

In practice, I often see people consolidate and form into silos due to inconsistent tools with complicated interfaces and a lack of access rights. But, of course, laziness also contributes to a certain extent: Maybe development and operations are already using the same tools together, but only the security team has an insight into the security dashboard—not particularly helpful!

> **Note: Reflection**
>
> It is important that you and your company recognize the complexity of the DevOps toolchain and want to break it down. Many companies have already recognized the problem, but many have not.
>
> I've worked with German companies in the manufacturing industry and that work with large machines. In large halls with long assembly lines and complex production routes, every little optimization that can save employees time is carried out. That is important. Unfortunately, however, efforts to optimize processes are not taken in the software production line; way too often, employees run around in circles multiple times before taking the next step. And this production line is precisely where leadership from above is needed so that progress can also be implemented in the software.

13.2 DevOps Platforms at a Glance

There are essentially four different DevOps platforms on the market, which can be used to their full potential only when you purchase an enterprise license for the full version. These are GitLab—whose developers introduced, coined, and, to a certain extent, invented the term *DevOps platform*—GitHub and Azure DevOps from Microsoft, and the toolstack from Atlassian. All these platforms aim to be full DevOps platforms, but none of them really supports the whole toolchain in the way you would expect in an enterprise. It's just a huge task with many pitfalls.

I want to emphasize one thing first: You will reach your goal with any of the four platforms. All platforms are under active development and are constantly changing. (This is why I have avoided screenshots and step-by-step instructions in this book as much as possible. For such information you can refer to the documentation for the respective platform).

The brief overview that follows should not be used as a decision-making criterion; once again, I advise you to evaluation the options on the basis of your specific requirements.

13.2.1 GitLab

GitLab is primarily known for two things: source code management and continuous integration, to an extent. These are also the functions with which GitLab was first developed, although the continuous integration component was initially a separate project. In the early days of GitLab, both components were merged, which resulted in the DevOps platform it is today.

GitLab now offers solutions for all stages of the DevOps lifecycle. This starts with project management, continues with source code management and continuous integration and package management, and extends to continuous delivery, security, and monitoring solutions. Users are not forced to use all functions, as each one can easily be replaced by an alternative.

Not all solutions from the stages are perfect, but the basics are covered almost everywhere. Special strengths lie in source code management, continuous integration, continuous delivery, and security.

The nice thing about GitLab is that the company is rather independent. The product is open core, and many functions are available in the free version, which is available under an open-source license. A new release with a number of functions is published every month. The platform can be easily hosted in your own infrastructure.

Overall, GitLab works very transparently. Individual function blocks, divided into the individual DevOps stages, are transparently available on the maturity page, including the features planned for the future (*https://about.gitlab.com/direction/maturity/*).

13.2.2 GitHub

Although GitHub is a few years older than GitLab, it was only much later that it adopted the holistic platform approach. Similar to GitLab, GitHub also began with pure source code management and simple project management features. The components for continuous integration and continuous delivery (GitHub Actions) were added much later and are still relatively new.

GitHub is best known as a platform for hosting open-source projects, which are available on *GitHub.com*. Unlike GitLab, however, GitHub itself is not open source. GitHub was acquired by Microsoft at the end of 2018 and has since been operated as a more or less independent platform.

For a long time, GitHub users had to take care of their own continuous integration infrastructure yourself, which was docked to GitHub. Continuous integration servers such as Travis-CI or Jenkins were often found there. It wasn't until 2018 that GitHub Actions was introduced, which integrates CI/CD functions directly into GitHub.

GitHub relies on a large marketplace. This is both a curse and a blessing, just as it was with Jenkins. Missing functions can be added, but new external dependencies are created, and a lot of time has to be spent on configuration, as the marketplace functions do not come directly from the manufacturer. This is actually exactly the point that should be avoided by using a DevOps platform.

An enterprise subscription is also required for GitHub in order to benefit from all functions. In addition to *GitHub.com*, GitHub also offers enterprise solutions for self-hosting, which, however, have a different range of functions than the website does. The features of website usually appear much later on the GitHub Enterprise server.

The functional scope of GitLab and GitHub is very similar. It's a bit like cars: They all have four wheels and get you to your destination. The question is rather, How quickly and how comfortably is this possible, and who can you take with you on the journey?

GitHub sees itself primarily as a developer platform. There is actually no mention of DevOps in the self-description. This is also noticeable in the range of functions: Typical operational information cannot be found in GitHub.

13.2.3 Azure DevOps

In addition to GitHub, Microsoft also has Azure DevOps. You may still know it under the name Visual Studio Team Services, which it was known as until 2018.

Azure DevOps is an offering from and for Microsoft's cloud offering. Accordingly, the toolchain can be easily integrated into the Microsoft Azure world.

In contrast to GitHub or GitLab, Azure DevOps is relevant only in the corporate environment. It basically offers a similar scope to GitHub and GitLab but with a stronger focus on Microsoft technologies.

It will be exciting to see what the future of Azure DevOps will look like, as two products from Microsoft are now essentially targeting the same service. There is some overlap between the two platforms; for example, GitHub Actions evolved from Azure Pipelines, and both have the same technological basis. Accordingly, there are always rumors that Azure DevOps will be discontinued in favor of GitHub. But as long as nothing has been communicated directly by Microsoft, this is just speculation for now.

13.2.4 Atlassian

Atlassian is listed as a leader in Gartner's Magic Quadrant, but in my opinion, you have to be a little cautious with this assessment. The classification is justified, because Atlassian's tools are very important and are market leaders in their fields. However, unlike GitLab, GitHub, and Azure DevOps, the toolstack offered by Atlassian is a collection of tools; you have to install and manage all the tools separately. However, the amount of configuration you need to do to allow the tools to interact is negligible.

Particularly important tools in the Atlassian toolstack are the project management tool Jira and the wiki tool Confluence, which are widely used and are constantly being developed further. The source code management tool Bitbucket is relatively mature, and no major changes are expected. Most companies are currently migrating away from Bitbucket.

Other tools, such as Opsgenie for incident management in the event of failures or Bamboo and Bitbucket Pipelines for CI/CD pipelines, also tend to play a subordinate role in practice. They all have to be installed and managed separately, so you can't really speak of a comprehensive suite.

However, Atlassian also sees the advantages of this approach and has recently launched its own platform, Open DevOps (*https://atlassian.com/solutions/devops/*). This package is mainly controlled by Jira but offers integrations for other tools: GitHub, GitLab, or Bitbucket for source code management, Jenkins, JFrog, Bamboo, and others for CI/CD, as well as Opsgenie and Confluence.

13.2.5 Other Platforms

In addition to GitLab, GitHub, Azure DevOps, and the Atlassian toolstack, there are also some other similar solutions to consider. These include the major cloud providers AWS and GCP, which each have their own tools to offer. However, great caution is required here, as these are little-used isolated solutions that are strongly geared toward their respective clouds. They are very, very closely tied to their respective providers. AWS, for example, had a service called CodeCommit for hosting Git repositories, which AWS announced would be deprecated in August 2024.

JFrog, the company behind Artifactory, also has its own platform (*https://jfrog.com/platform/*). It is a combination of several tools from JFrog. This platform is not widely used.

> **Note: Reflection**
>
> Nowadays, there is hardly any way around GitHub or GitLab. Both are good platforms for developing and rolling out products and projects. However, I am deliberately not saying "products for operating applications" here, because there is still significant potential for improvement.
>
> In my daily conversations with people in IT, the direction is clear: Many want to migrate away from Bitbucket and Jenkins and toward GitLab or GitHub in order to modernize and simplify their development infrastructure. Although GitHub and GitLab are not yet fully usable everywhere for the entire DevOps lifecycle, the path is pointing in exactly this direction.

13.3 Summary

Realistically, everyone is heading in the direction of DevOps platforms, even if some have not yet realized it. Although GitLab and GitHub have been on the market for several years, it is clear that we are still at the beginning of a long journey. Many features that go beyond source code management, CI/CD, and security are missing; only the basic features have been implemented so far. However, everyone is focused on significantly improving collaboration within teams using DevOps principles.

It is also important to remember that a DevOps transformation does not suddenly become a success through the use of one of these DevOps platforms. The following still applies: First adapt the people to the changed circumstances, then adapt the processes, and only then choose the right tools so that the DevOps idea can unfold its full power.

Chapter 14
Beyond Culture and Tools

The DevOps concept is also having an impact on other areas, which is giving rise to new buzzwords. For example, many new DevOps-based terms and new technical functions that have an impact on DevOps have emerged.

First, we discuss the role of artificial intelligence (AI) in DevOps: What is important here? What is changing? And is it all just hype? In addition to AI in DevOps, there are numerous other topics, such as DataOps, MLOps, and AIOps. This raises further, fairly similar questions: What are they, and how do they relate to DevOps?

We also discuss how DevOps should be viewed in job descriptions and in the hiring process. There have always been discrepancies here that need to be clarified.

14.1 The Role of AI in DevOps

In IT, there are often topics that have a lot of hype around them but ultimately do not establish in practice as initially expected. For example, there has been a lot of focus on cryptocurrencies and blockchain in recent years, but not many groundbreaking changes have occurred.

The hype surrounding AI began at the end of 2022.

Let's look at the relevant terms without the hype. AI stands for *artificial intelligence*. Machine learning (ML) is part of AI (see Figure 14.1).

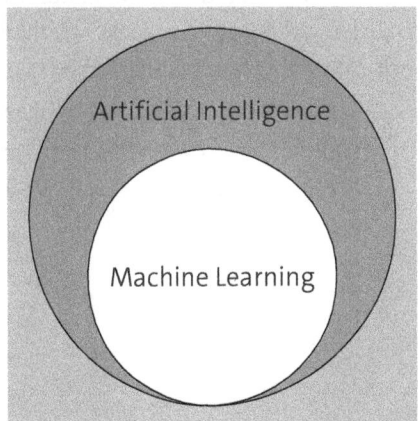

Figure 14.1 ML as a Part of AI

In ML—as the name suggests—the machine is trained to recognize patterns and predict outputs when it is given input parameters. This requires training data; neural networks work in the background, and a lot of mathematics and statistics are needed to train the ML algorithm.

An ML algorithm solves specific problems and does not attempt to be an all-encompassing AI. The search for a *general AI* is another topic, but the terms are often thrown around in the media. I will often simply use the term "AI" in this section.

There is no question that AI can do an incredible amount. Unfortunately, the hype surrounding it is also (too) big. Many people who consider using AI are not looking at their problems and challenges objectively and soberly; instead, they are instead clamoring for a solution without having examined the specific problem in detail.

You will certainly have tried ChatGPT before—as I have.

ChatGPT is an AI chatbot from the company OpenAI that has been trained with the help of a lot of data to be intelligent to a certain extent. It knows many programming languages and has an answer to almost every question.

Generative AIs such as ChatGPT make use of *large language models* (LLMs). Put simply, LLMs are trained with a huge number of parameters and datasets. This allows the models to create sentences that match the input with a very high probability. These models can be used not only for human languages but also for writing source code.

14.1.1 Making Work Easier with AI-Supported Code Generation?

Like most people, I was amazed by the capabilities of ChatGPT. "This thing can actually be helpful in the development of software!" I heard these statements from many managers in particular. And what's more, if you can develop software faster, then you can create added value for users much faster and thus make more money—perhaps you'll even need fewer software developers!

After a little experimentation with tools like ChatGPT, some strengths and weaknesses quickly became apparent. A simple example is asking ChatGPT to create a script that retrieves data from an API and outputs it in a specific data format. If you ask the right questions and enter the right prompts, and if the API is public, the result often looks good at first glance. Take the ChatGPT-created script shown in Listing 14.1, for example:

```
import requests
api_key = 'Your-API-Key'
url = 'https://api.weather.com/v3/weather/forecast'
params = {
  'api_key': api_key,
  'location': 'Berlin,DE',
  'format': 'json'
}
```

```
response = requests.get(url, params=params)
if response.status_code == 200:
  weather_data = response.json()
# Here you can process the data
else:
  print('Error retrieving data: ', response.status_code)
```

Listing 14.1 Example Script Produced by ChatGPT

However, if you then try to execute the script, you'll realize that the example looks good but, in practice, does not work at all. This is often the case with using ChatGPT to write simple scripts.

Several times I have received scripts from ChatGPT that had many API fields that should have existed but did not exist at all. ChatGPT hallucinated functions and made suggestions that did not meet the requirements. After a few iterations, I was finally able to get an executable script. However, I could have written the script myself in the meantime.

> **Note: Prompt Engineering**
>
> To work properly with AI tools, it is essential to know how to set the right work requirements so that the right code is generated. There is already a term for this: *prompt engineering*.
>
> Communicating correctly with AI is just as much an art as requirements engineering is. Even with real people, it is difficult enough to find out what is actually wanted and needed. Anyone involved in software development can tell you a thing or two about how customer requirements are misunderstood or not properly documented. Requirements are often implemented incorrectly by the development team because information is missing or incorrect. When an additional level of communication is added by incorporating AI, the potential for misunderstandings increases.

However, the fact that the AI does not generate perfect code straight away is the lesser problem. For software development, it is more important that the AI initially *only* helps with writing the code. As you have hopefully learned from this book, software development as a whole is much more than just producing lines of code. Much more time is needed to read and understand the code, to communicate within the team, to test the code, and to manage the whole toolchain around the process.

> **Note: Automatically Create Tests**
>
> Nobody can currently predict what impact tools like ChatGPT will have on software development. I think the effects are likely to initially be felt less in the writing and creation of code or even in software architecture and more in the many diligence tasks involved: in creating documentation, commenting on code, and especially in testing.

14 Beyond Culture and Tools

> Writing unit tests can easily be outsourced to AI, and computers are very good at performing repetitive tasks accurately over and over again. (You can find a good overview of the use of AI for test automation at *https://lambdatest.com/blog/using-chatgpt-for-test-automation/*.)
>
> However, this does not change the fact that it is still necessary for humans to check whether the resulting product actually meets the requirements. Integration and functional tests should still be the responsibility of real intelligence. Any thought process that is required as part of writing code must be carried out, regardless of whether the code was written by an AI or a human. Accordingly, reviews are even more important when working conscientiously on a project.

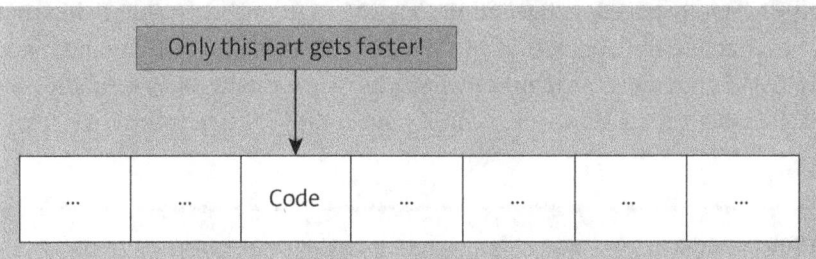

Figure 14.2 AI's Limited Role in Software Development

So let's assume at this point that writing program code should only take half as long. I am deliberately referring here to writing program code and not to software development in general, as the latter involves much more than just writing code.

The question that arises is, "Even if you assume that you write code 50% faster, does that really give you that much more time to write even more code?"

What is often missing—and what I have repeatedly stated here in this book—is a holistic view of the entire software development process. The fact that AI is used to write or generate code faster has little impact on the other aspects of software development. And it is precisely this view that is lacking in those who simply chase after the hype instead of pausing for a moment to look at the whole process.

14.1.2 More Code Leads to Higher Review Requirements!

Hand on heart: Would you use code in production that was created by ChatGPT? If yes, then you may need to explain why you trust AI more than human developers.

Let's look at testing and reviewing as the first and probably most important point: Code that has been generated by AI in particular tends to require even more time to be invested in extensive testing and review. And this means not only that you should tend to distrust generated code, but also that more code has been generated. More time is therefore inevitably needed for testing and reviewing. Because, as I said, ChatGPT and similar tools invent things that seem coherent at first glance but don't do what they are supposed to do. ChatGPT is not a programming professional or even a fact machine; it is only a statistical model that is very good at creating things that look plausible with a high probability. You have to check whether the content of the result is correct and whether the generated code is performant at all. Does it lead to performance problems with a certain data size? And what about security vulnerabilities? Are there subtle logic errors that are not found by the tests?

And even if the model's algorithms work correctly, it can still generate vulnerable and faulty code, because the models were trained with code written by humans that may not have been developed according to best practices or with security in mind. "Garbage in, garbage out" will be a major problem for all ML models for the foreseeable future.

When you find problems and security vulnerabilities in the code you have written, the work really just begins. While with a normal program you usually have history to consult to understand where the errors come from, with AI-generated code you are initially faced with a puzzle. A lot of time has to be invested in understanding and correcting problems. This requires at least as much in-depth knowledge of the programming language used, the libraries, and the source code of the project as if you were writing the code yourself from scratch. This knowledge is still indispensable. The whole process is made even more difficult by having to constantly change content for the reviews. This also costs a lot of time.

14.1.3 AI-Supporting Features

For me, it is obvious that profound knowledge of software development and the use of programming languages will continue to be needed. This knowledge and experience will not be completely replaced by AI tools in the foreseeable future. But that doesn't mean that these AI tools can't make work much easier and significantly change processes and workflows in the coming years.

The last few years were full of news and announcements about AI. New projects and start-ups that want to use AI to revolutionize software development sprouted up everywhere. It's still too early to make an assessment, but you can already recognize a few trends and get an idea of where the journey is heading. I think that there will be one or two AI-supported solutions for every stage of the DevOps lifecycle. For example, in the foreseeable future, there might be features that simplify the writing of pipelines by reading the source code of the project and automatically generating a suitable pipeline.

AI can also help by automatically analyzing failures in pipelines and creating root cause analyses. The AI could then even provide advice for addressing failures or make suggestions on how pipelines could be optimized.

In project management, AI could be used to create summaries of projects to make work easier or to discover contradictions in project descriptions. Small indications that something is not right could already be very helpful.

Team onboarding and the transition to new technologies can be mastered more easily with the help of AI; for example, new team members could use AI to experiment with code more quickly. The best way to learn a new programming language and its practical applications is to engage in pair programming sessions. With the support of AI, you can also work more efficiently in these sessions.

AI can also help to better understand security vulnerabilities and other security-related issues to make the *Sec* in DevSecOps even stronger. Developers and DevOps teams are often faced with the problem of a reported security issue that they don't know how to handle. What is the opened CVE actually about? The affected code is in fact outdated and there are only a few tests. Will the bug fix work for all scenarios? Or will the production environment blow up in our faces? In this case, an automated AI solution that contextualizes the vulnerability and explains how it is exploited can help. Perhaps in a few years it will be possible to not only find security problems based on statistical models but also to assess and minimize the risks. Such a stochastic approach could help to better assess and minimize risks, especially in the case of large, complex codebases where literally no human can keep track of everything.

> **Note: AI Tools in Practice**
>
> While I was working on this book, most AI tools were at a very early stage. It is not worth describing the specific range of functions at this point; a lot will happen here in the next few years, and I do not dare predict which of these projects will be successful in the medium and long term.
>
> In addition to ChatGPT, Copilot and Copilot X from GitHub and GitLab Duo are probably the best known. The viability of these offerings will depend on how well they integrate into the platforms and extend across the entire lifecycle.
>
> If you look beyond GitHub and GitLab, you will also find smaller companies. Sourcegraph, for example, offers Cody, an AI coding assistant (*https://about.sourcegraph.com/cody/*). Another exciting company is the Israeli start-up Tabnine whose software can also be hosted on your own infrastructure (*https://tabnine.com/*). The major cloud providers also offer their own tools, such as CodeWhisperer from Amazon and Vertex AI from Google.
>
> However, all tools have one thing in common: They incur additional costs, so their use must be well planned and evaluated.

14.1.4 Data Protection and Privacy

Of all questions relating to what is technically feasible, you must not neglect data protection and confidentiality. Will your input be used directly to further optimize and train the model? Does the code you provide as context for prompts flow directly into the training material? (This refers mainly to the lines above and below, if not the whole project, so that relevant code can be generated and suggested for auto-completion.)

Can another user then retrieve this context window by providing appropriate prompts? Where does the information on which the output of the tools is based come from?

Those responsible at GitHub, for example, were heavily criticized for ignoring the licenses of the projects for the training of Copilot. The open-source licenses, which determine the reusability of the project and, thus, the code, exist for this very reason. The future will show what legal consequences this may have. For example, follow the legal dispute against GitHub brought by American lawyer Matthew Butterick (*https://githubcopilotlitigation.com/*). There will also be similar proceedings in the EU.

> **Caution: Security Risks Due to AI Tools**
>
> I would like to use two examples to demonstrate what can go wrong:
>
> - The fact that ChatGPT can be used not only to create new code but also to optimize old code has not escaped the attention of Samsung developers. They uploaded numerous lines of code for ChatGPT to revise and improve. Since ChatGPT uses user input to further develop the system, at least in the free version, it was possible to access the code that the Samsung engineers had dumped into ChatGPT by asking clever questions (*https://techradar.com/news/samsung-workers-leaked-company-secrets-by-using-chatgpt/*).
> - Most major companies (such as Apple, Amazon, and Samsung) have therefore decided that ChatGPT may no longer be used for corporate tasks, as there are fears that confidential data will be sent via the ChatGPT company OpenAI to its largest investor, Microsoft. With other AI tools, data protection is (at least allegedly) more restrictive, and ChatGPT also promises that input will not be used if the paid version is used.

A fundamental problem here is that most AI services are available only as SaaS. If you want to use them, you inevitably have to send the data to them. There are hardly any alternatives for self-hosting, as high-performance systems with a lot of GPU power are required for both the training and use of AI systems.

Companies in the EU pay particular attention to where the data of a cloud service is stored. The keywords here are the CLOUD Act and data residency in particular. However, this does not solve the problem that private, internal company data has to be passed on and processed.

Here, too, things are likely to change in the future. Until then, however, you should keep such problems in mind.

14.1.5 It's the Overall Concept That Counts!

The great strength of DevOps is that it attempts to consider the entire value stream of software development. This is also a great advantage when dealing with the new (and perhaps revolutionary) AI helpers: Yes, the art of writing code will change, and in the future many developers will have simple functions generated for them instead of writing them themselves. But this is only a small part of the value stream. Always keep an eye on the big picture and ask yourself whether automatically generated code really solves your biggest problems. Doesn't the shoe pinch much more in other areas? Don't other processes and workflows need to be accelerated? How can AI help in those areas instead?

In the future, we will have many AI helpers, both large and small, throughout the development process (see also Section 14.2.3 on AIOps, which appears later in this chapter), whose interaction will be the key to success. GitLab Duo and Copilot X from GitHub are certainly moving in this direction: comprehensive offerings that take the entire lifecycle into account.

Unfortunately, the reality is that many companies are still struggling with DevOps principles. By this I don't just mean the team structure and the culture of collaboration, but also how things are done at a technical level.

Managers often do not have much confidence in the regular rollout of changes to production systems. In addition, the people who write the software are often not trusted. And in these cases, it doesn't help if code written by an AI is also to be rolled out.

Before you start and consider AI-supported programming as a solution to many problems, you must first correct all other points in the value stream. If every rollout of a change is difficult, with errors and downtime because nothing is automated, then there is no point in generating code at lightning speed with AI. Code changes will ultimately only pile up further and cause more problems and frustration. If you already don't trust the code that humans have written, then it won't get any better with AI—on the contrary, it will get worse!

Don't get me wrong—the advantages of AI-supported programming is undeniable. However, the advantages in the overall context of software development are much smaller than you might expect. And this is especially true if the rest of the value stream is still very old-fashioned. The bottom line is that AI can help with software development, but it takes much more than simply giving developers access to ChatGPT and letting them create code automatically. The focus must be on the big picture in order to identify and exploit potential improvements. If DevOps is properly implemented in the organization, the foundation is laid for effective integration of AI capabilities.

> **Note: Reflection**
>
> Depending on when exactly you are reading this, a lot has probably already changed in the AI space. AI in DevOps is becoming increasingly important, but you should not immediately chase the hype. Intellectual property and data protection issues need to be considered. There is still a lot going on here! It is important that you approach the topic soberly and always keep the big picture in mind.

14.2 DataOps, MLOps, and AIOps

The success of the DevOps concept has sparked the emergence of many other similar methodologies outside of software development. These terms also tend to end in "Ops," which can be very confusing when you're first looking into DevOps, it's not clear whether you want to replace DevOps in these ways, add to it, or completely change it.

So let's take a brief look at DataOps, MLOps, and AIOps, all of which have a different focus. None of them replace DevOps; instead, they extend DevOps principles to tasks beyond pure software development.

14.2.1 DataOps

DataOps applies DevOps principles, both the cultural and technical aspects, to data-driven solutions. It involves the work of mathematicians and statisticians who analyze and process the data that is collected in a company.

The lessons and experiences of software development can be applied to speed up data analysis and automate processes such as data extraction, preparation, actual calculation, and illustration in dashboards and reports. Figure 14.3 shows the individual steps of DataOps in the form of a pipeline. Automation and a focus on the reproducibility of tasks should play just as big a role in DataOps as in software development.

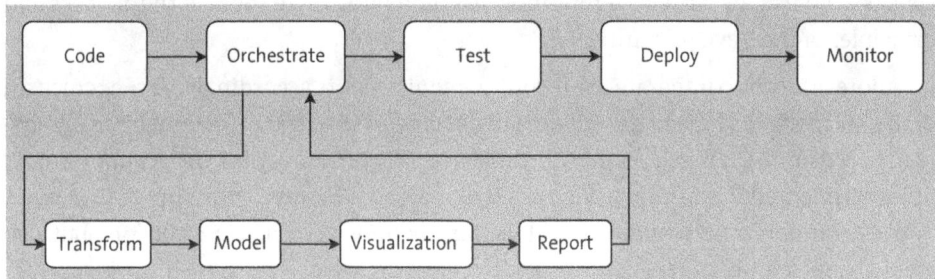

Figure 14.3 DataOps Pipeline

If DataOps is not used, many of these actions are performed manually, such as moving data back and forth in order to run the analysis in the next step. This is not only laborious but also error prone and expensive, as such approaches involve huge datasets.

And lack of communication can also be just as troublesome as it is in development projects. Validation, modeling, and integration of data into the workflow are generally done by different teams. If they simply throw their results over the wall to the next team, your processes are just as bad as they are in the worst development projects. The solution is a proper DataOps pipeline that is used in the same way as a CI/CD pipeline: The work with the data is formulated in code that is versioned in repositories. All changes have to go through a code review, and constant testing ensures QA—just like in a good software project.

14.2.2 MLOps

The ML in MLOps stands for machine learning, as explained earlier in the chapter. ML itself is the technical term that is less fraught with marketing hype than AI, as there is usually a specific problem that the machine is trained to solve using ML algorithms.

ML is essentially made up of three components: the data, the model, and the source code. A ML algorithm is defined in the code. The data is then used to train the model, which is ultimately used to make predictions or generate new material via statistical evaluations.

MLOps is a technical addition to the DevOps principle, even if the prefix "Dev" is not part of the name. This term is not about how AI can be used to make DevOps work easier, but it's about how the process of ML itself is organized.

This concerns, for example, the training of the models and the management of the training data, as a common problem with ML is that the models are trained only once and then somehow packaged and rolled out in the software. Instead, the model should be constantly improved with new data. It is also necessary to constantly evaluate how well or poorly the experiments are performing. This is obviously not possible without clever automation. Once again, it is therefore a matter of applying DevOps principles to other areas.

If you want to use ML techniques in your projects, the first step is to pick up the AI specialists, who usually use a different toolstack, and familiarize them with the tasks and principles of the DevOps culture.

It is more difficult to integrate the technical requirements accordingly, as the toolchain of the ML model and the huge amounts of data required to train it cannot be easily covered in classic CI/CD pipelines. As ML models consist of binary files, they cannot simply be experimented with in review environments, as is the case with pure source code. You also cannot retrain models at will, as this not only would take far too long but also would be incredibly expensive.

This is why new platforms and tools are currently being developed that are specifically geared toward the work of ML projects. Take a look at MLFlow (*https://mlflow.org/*), for example, which covers the entire lifecycle.

14.2.3 AIOps

While working on this book, I came across the term *AIOps* for the first time and immediately asked myself whether it was something like MLOps.

On closer inspection, however, I quickly found that AIOps concerns the use of AI for pure IT operations (i.e., using intelligent tools to make everyday tasks related to the operation of the IT infrastructure easier). The idea behind AIOps is that it can help identify problems at an early stage with the help of many different data sources. This data includes events, logs, and metrics that are used to train the system. For example, if an

ML model has learned what the normal traffic of a network looks like, it can sound the alarm for anomalies faster than even the most attentive admin could. In this way, security problems can be detected if there are accesses in the network that do not correspond to previous patterns.

In the future, it will hardly be possible to manage complex infrastructures in particular without such tools—in much the same way that hardly anything in industry works without predictive maintenance. However, it is difficult to say which commercial products (especially from the observability sector) will be successful in the long term. Check what you really need and what actually makes your operations easier and safer.

> **Note: Reflection**
> DataOps and MLOps are topics that will continue to develop in the future. The extent to which they will be useful for you will, of course, depend heavily on your role. It is highly likely that further functions will be incorporated into existing platforms such as GitHub and GitLab so that even more use cases can be covered, even if independent companies and tools such as Databricks (*https://databricks.com/*) and Hugging Face (*https://huggingface.com/*) exist.

14.3 DevOps as a Job

In Chapter 1, I briefly touched on the fact that the term *DevOps engineer* is actually wrong. Although "DevOps" in a job title can make sense, it is often used incorrectly.

If you have read this book carefully, you will have understood that DevOps is essentially a culture that can work only if the people involved accept it. Only once the people accept the culture can the processes supporting the culture be introduced, and only then can the tools supporting the processes be integrated.

Therefore, it should be clear by now that DevOps is not a role. Nevertheless, you (unfortunately) often find job offerings that are looking for DevOps people. When asked about their jobs, people often say that they "do DevOps." I always ask critical questions here, as everyone understands the term differently.

14.3.1 The Question of DevOps Engineers

Calling yourself a DevOps engineer can sometimes be correct and sometimes not, depending on the role description.

The person who is responsible for the CI/CD pipeline and the entire DevOps toolstack is often referred to as the DevOps engineer. I think this is a mistake in many cases. In principle, every team member should be able to adapt the CI/CD pipeline and also make the necessary changes to the rest of the toolchain. What should not happen is that only one person builds everything and nobody else understands the setup. Unfortunately, this scenario is common, but it is completely contrary to the actual DevOps concept, which is about collaboration between the different roles.

In larger companies, however, it can make sense to have DevOps engineers. They serve as the point of contact across teams for all topics relating to DevOps. Of course, they offer support not just on cultural issues but, above all, on the technical toolstack.

Such employees should be *enablers* for the other teams. Their tasks can consist of assisting the teams when specialist knowledge is required or when specific, very complex problems need to be solved. In principle, the DevOps engineer should ensure that the teams can work independently and are not dependent on them. A good DevOps engineer makes himself redundant.

This does not apply only to the individual DevOps engineer but also to the entire team pattern. A standalone DevOps team, as shown in Figure 14.4, makes no sense between separate development and operations teams—unless the task of this team is to abolish itself.

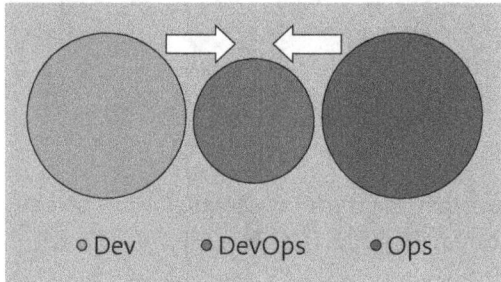

Figure 14.4 DevOps Team Topology

However, it is normal and understandable for a team to have some people who are more and some people who are less familiar with the pipelines and the setup. But the same also applies to development work and operational tasks: Knowing everything is almost impossible. However, an attempt should always be made to have a good knowledge base throughout the team.

In addition to the right DevOps teams and the right DevOps engineers, the term *DevOps* can also be appropriate for consultants or trainers who advise and train teams on DevOps topics.

In Section 14.3.3, we cover the technical DevOps learning path. This learning path is usually one of the core tasks of a DevOps engineer. However, it always depends on the exact context. I often see ads looking for senior DevOps engineers and find both types of descriptions: the role as I have described it, as well as the fake DevOps engineer role that is part of an almost pure development team.

Job descriptions like the latter are good indicators that DevOps principles are not followed on the team. In most cases, these roles are also expected to take on operations tasks on the development team. With this in mind, keep a critical eye on the use of the term in job descriptions and, during job interviews, ask exactly and, if necessary, several times what the position will entail.

However, one thing is important to emphasize once again: Specialization in certain subject areas is completely normal and often unavoidable. Not everyone is a born developer, and not everyone is interested in all the details of the business. It's the healthy mix that makes the difference, and the company needs to see this and manage it accordingly.

14.3.2 Soft Skills

Before I cover the technical aspects of the DevOps learning path, it is extremely important to talk about soft skills. Contrary to their name, soft skills are not soft at all, but are quite hard to learn and apply.

Understanding Problems

One of the most important skills is to correctly understand problems. This means that you need to recognize connections and the big picture so that problems can be tackled at the root. Don't build temporary solutions, don't get bogged down in details, and don't over-engineer.

Many people work on solutions without having fully understood the problems. Especially if you have to keep an eye on different teams and requirements as a DevOps engineer, it is very important to analyze problems properly; otherwise, you will only be correcting and adapting a lot of little things instead of making real progress.

Communication

Understanding problems is closely linked to communication. Efficient and effective communication is essential for success, especially when interacting with various other employees.

As a DevOps engineer, you are the bridge between different teams and roles. This means bringing together different people with different goals. Good communication is also important when explaining possible solutions to problems. Always try to understand the perspective of the person you are talking to and, ideally, listen more than you talk.

Collaboration

Good collaboration means mutual respect, patience, and trust. You also have to work with your teammates productively in the event of conflicts and differing goals, and you have to look for solutions together.

The three soft skills are important alone but also together: Bringing the different roles together is important in order to understand problems, which only works with close and careful communication.

14.3.3 The Technical DevOps Learning Path

Although DevOps culture is more important than the tools, the proper use of the DevOps toolchain cannot be ignored. My goal in this section is to give you insight into the tools and techniques you need to know to succeed as a DevOps engineer. Figure 14.5 provides an overview of the learning path.

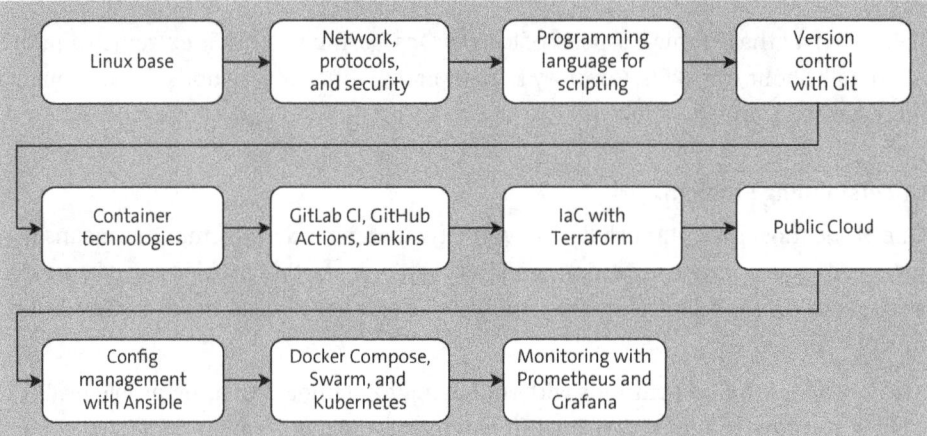

Figure 14.5 Technical DevOps Learning Path

In principle, it would of course be great if you could build up expert knowledge with practical experience in all tools and techniques. However, this is not only unrealistic but also virtually impossible, as there are simply too many tools and they are constantly changing. What was best practice yesterday may be a bad hack tomorrow. Today's market leaders may find themselves in a dead end tomorrow.

Your focus should be more on what sound basics are useful for your situation and what you should start with in order to be able to put as much as possible into practice. This is less about covering every single niche and more about meeting the broad demand for knowledge in the job market so that you can demonstrate solid knowledge.

Linux Is the Basis

In order to follow a comprehensive learning path, a solid knowledge of Linux is of fundamental importance. Linux forms the basis for all subsequent steps and is almost indispensable, as almost everything that follows in the DevOps learning path builds on knowledge of Linux. In the context of DevOps and especially in cloud native, Linux knowledge is essential in many respects. In contrast, Windows is more of a niche operating system in this area.

The first step is to understand the basic operating principle of the Linux system, as this knowledge will later be essential for creating container images and for operating containers. Comprehensive knowledge of how to use the shell is particularly important. This applies to both interactive use and mastery of scripting languages. In fact, it does not matter which specific shell is used, as the basic functions are very similar in most cases. As a rule, the Bash shell is the most common and recommended choice.

The specific Linux distribution you familiarize yourself with is less important. It is much more important to familiarize yourself with package managers and to understand how they interact in an overall system. You should also understand the difference between Linux distributions for host systems and for creating container images.

Host systems are the systems that host containers. Debian-based systems such as Ubuntu or distributions that are compatible with Red Hat Enterprise Linux (RHEL) are important here. However, DevOps is not limited to containers. If you have a solid knowledge of Linux, it is not too difficult to familiarize yourself with the basics of Alpine Linux or other minimal distributions.

Networks, Protocols, and Security

In the world of DevOps, general network and infrastructure topics are closely linked to operating systems. In addition, you should have a comprehensive understanding of basic concepts in the areas of virtualization solutions, file systems, and CPU and memory management. The latter is particularly relevant when it comes to running applications, especially when you are faced with CPU and memory limitations.

A solid understanding of processes, sockets, and networks is also essential. You might think that these topics have become irrelevant in the era of cloud and container infrastructures, but they still play an important role in this environment. Examples include security groups and network rules, which are crucial for security and communication in the cloud.

This area also includes a sound knowledge of protocols. SSH is essential and is one of the basic skills in the world of Linux. You should also understand the basics of HTTP and SSL/TLS. Dealing with REST interfaces in general is particularly relevant, as it is often necessary to retrieve data from services via REST interfaces. This knowledge is also very important for scripting and automation.

Programming Languages for Scripting

It is crucial to be proficient in a programming language, especially if you are in a role that is primarily concerned with supporting the overall CI/CD infrastructure for different teams. A scripting language can help you tremendously in this regard. The first step is often learning shell scripting, as I mentioned earlier. However, you quickly reach your limits with shell scripting when you face more complex tasks.

Shell scripting is undoubtedly important, but you should never try to solve everything with a shell script just because you can. Good knowledge of shell scripting is demonstrated by knowing when it is better not to use it.

The next step is to learn Python. Bash scripting is mainly needed for pipelining, debugging systems, and creating container images. As soon as your tasks become more complex, Python will be a great help. Typically, Python is used more for infrastructure-related scripting and less for supporting the applications themselves.

Python is particularly useful when you need to access APIs to collect data for reporting. New challenges surrounding APIs pop up regularly, and basic knowledge of a programming language like Python can help you overcome them. Python is well suited to facing such challenges because the barrier to entry is relatively low and maintenance is comparatively simple.

Regarding the continuous delivery part of DevOps, a sound knowledge of Go (also known as Golang) can be helpful. Go is a widely used language in the world of cloud native, and many applications are written in Go. This is another step beyond Python, as it allows you to dive deeper into the system.

Version Control with Git

When it comes to version control, you only really need Git. Other tools are almost irrelevant. You may still find Subversion in one place or another, but these are usually older systems that need to be replaced anyway. Other modern alternatives to Git (such as Mercurial) are rarely found and are generally negligible.

Your knowledge of Git should include not only the basic commands but also the ability to troubleshoot problems when they occur. A good indicator that you are using Git effectively is that you are able to diagnose and resolve issues without the need to re-clone the repository. Instead, you should be able to specifically correct the faulty state.

Knowledge of different development workflows is also important. It is not enough to know the Git commands; you should also know when certain workflows are best suited and when certain workflows are not suited. This is particularly relevant for anyone who works with code—be it the source code of the application or the configuration files in IaC setups.

When it comes to hosting Git repositories, there are now relatively few differences between the pure Git repository hosting platforms such as GitHub and GitLab, at least in terms of managing the source code. The structure and organization of these platforms may differ, but the basic functions are similar.

In summary, the following knowledge of Git should be known:

- Creation and pushing of commits
- Creation and merging of branches
- Elimination of merge conflicts

- Working with one and several remote repositories
- Contributing to external repositories
- Rebasing of changes to feature branches
- Reliable understanding and application of various workflows

Container

It is extremely exciting and important to familiarize yourself with the topic of containers. An initial thought might be, "Well, it's all about Docker, isn't it?"

Containers and Docker are often equated and confused with each other. In fact, there are some differences and peculiarities that even trip me up time and again.

It is crucial that you understand the following concepts and know the differences between them:

- OCI
- Container Runtime Interface
- Container engines

It is astonishing how limited DevOps professionals' understanding of these concepts often is, even in environments that use containers productively. Knowledge of creating container images and the associated security aspects goes hand in hand with these concepts: You should understand what layers are in container images, how to keep them lean, and how to avoid security vulnerabilities.

The general concepts that can also be found in Kubernetes are already visible at the container level. This includes the handling of stateless containers, the management of storage, and related aspects.

There are essentially two use cases to consider: containers for CI/CD pipelines and containers for productive deployments. Although the basic principles are similar, they have different requirements, especially in terms of lifespan.

It is therefore crucial to learn the basics of container infrastructures. This includes not only Docker but also other tools such as Podman, in order to fully understand and apply the concepts. Without this knowledge, you shouldn't even look at Kubernetes.

CI/CD with GitHub, GitLab, and Jenkins

In terms of the CI/CD landscape, there are three primary platforms that you should keep an eye on: GitHub Actions, GitLab CI, and Jenkins. Of course, there are many others, such as TeamCity, Bamboo, and Travis CI. You should be familiar with at least one of these tools. Jenkins is listed here more for historical reasons. It is on a downward trend, and many organizations are migrating away from it, as I explained in Chapter 6, Section 6.6.1.

If you are just starting out in the world of DevOps, you should therefore focus more on GitHub Actions and GitLab CI. Both platforms regularly publish improvements that can help speed up and optimize processes.

Since both GitHub and GitLab are DevOps platforms, you should not limit yourself to understanding how to write efficient pipelines; instead, take a holistic approach that shows how everything is connected. This includes topics such as the reuse of pipelines, the implementation of compliance requirements, and the integration of pipelines in areas such as deployment and monitoring.

You should also gain a deeper understanding of security. This is relevant on two levels. First, you should understand how the given platform can ensure the security of the pipeline operation and infrastructure, which is essentially guaranteed by the platform. From an application development perspective, the platform's security features for checking and analyzing applications are of practical importance, as discussed in Chapter 11, Section 11.4.

If you look at job postings, you will notice that these are exactly the skills that are often in demand. This is not surprising, as they are an integral part of the technical implementation of DevOps.

Infrastructure as Code with Terraform

I see IaC as an optional topic. The previous points I have mentioned focus on the software development lifecycle, which is not deeply related to the foundation on which the applications run. However, as new infrastructures are constantly being created in the cloud-native world, you should also have a basic knowledge of IaC.

Although there are various IaC tools, as highlighted in Chapter 9, Section 9.5.1, Terraform is the main player. In order to be able to use Terraform and the cloud sensibly, basic Linux knowledge, network knowledge, and extensive experience in dealing with CI/CD tools are both helpful and necessary.

Cloud

There are many different cloud providers, and they all have their pros and cons. The best known and most common public clouds are clearly AWS, Microsoft Azure, and GCP. I'm not a big fan of the claim that you need to know all the features of one or even all of the cloud providers.

The functions are constantly changing: New services are constantly being added and others are being discontinued. Although many companies want true multicloud setups, they are never actually implemented properly because such setups are complex and expensive. In the end, more virtual machines are often set up automatically in the cloud anyway, and they are populated with software if Kubernetes is not used. There are some differences between the cloud providers, but in the end you can achieve your goal with any of them.

It's more exciting to look at which managed services you can use—and which you should stay away from. Setting up a real high-availability database cluster is quite complex and requires in-depth database knowledge. It may be a good idea to use a cloud service, but the disadvantages of this decision should also be considered. In some cases, there are no migration paths for version upgrades for the database services in question.

The following always applies to the cloud: You should know what you should and should *not* use. Sometimes vendor lock-in is okay, and sometimes it is not.

Configuration Management with Ansible (or Puppet)

Once you know how to deploy infrastructure on a cloud system using IaC tools such as Terraform, it is then important to have the skills to equip this infrastructure with software. Kubernetes is not used everywhere, and often classic virtual machines or even physical hardware are used. In such cases, configuration management tools such as Puppet or Ansible are used in the toolchain.

If you have no experience with configuration management tools, I recommend learning Ansible. The learning curve is rather flat, and getting started is quite easy. Linux knowledge is also required here to keep the entry hurdle low.

Of course, the more complex the problems that need to be solved are, the more complex working with Ansible becomes. Although Puppet is used in many companies, I do recommend that beginners learn ansible instead, as it is both more important and simpler.

Kubernetes

Container technologies were mentioned in the previous sections, but Kubernetes was also briefly touched on. Kubernetes itself is very comprehensive; there are also many more technologies from the Kubernetes toolstack. There are two perspectives to consider: the operation of Kubernetes clusters and the use of Kubernetes clusters for more than just production environments.

The path to Kubernetes should not start directly with Kubernetes. This can quickly become frustrating! Once you have understood the concepts of containers and the construction of images and are able to apply them, it may be advisable to first make a stopover at Docker Compose and gain an understanding of orchestration of containers. Another intermediate step can be familiarizing yourself with Docker Swarm, which goes one step further than Docker Compose.

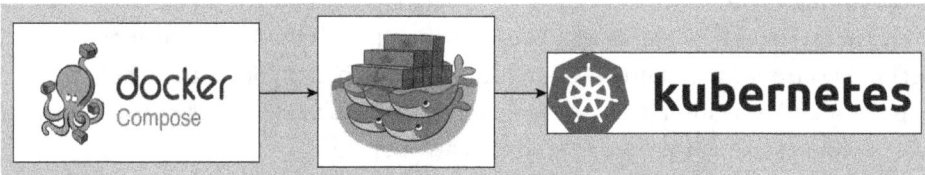

Figure 14.6 Docker Compose, Docker Swarm, and Kubernetes

Docker Compose can be used to start several containers that can talk to each other. The infrastructure requirements and the knowledge needed for this are the preliminary stage of a real cluster setup.

By using Docker Compose, you can quickly learn how container orchestration tools work. This includes points that are relevant to the operation of containers, such as health checks, the mounting of volumes, and the general short lifespan of containers that are repeatedly deployed.

If you take a closer look at Kubernetes, you will almost inevitably have to specialize: Are you more interested in the operations or development side of Kubernetes (i.e., using container orchestration in operations or building containers in development)?

The operations and especially administration sides of Kubernetes are usually in the hands of the platform engineering specialists. This primarily applies to larger companies and larger clusters, as this is a full-time job.

However, there is a big difference between using a managed Kubernetes instance from the public cloud and managing a Kubernetes cluster in your own infrastructure yourself, such as via `kubeadm`. The latter is significantly more complex to maintain.

> **Note: K3s as a Playground**
>
> If you want to learn how to use Kubernetes on the administration side, I recommend K3s (*https://k3s.io/*). This is a lightweight Kubernetes distribution that can also be installed and managed on "thin-chested" devices. It's a good way to practice deploying applications. This is how I have built up my Kubernetes knowledge over the past few years.

The container orchestration platform OpenShift from Red Hat is also widely used.

If you have a development background and are not interested in the administration and scaling of Kubernetes, then the role of Kubernetes in the CI/CD pipeline is much more exciting.

Ideally, you should know how to deploy applications to a Kubernetes cluster and how to configure and use ConfigMaps, services, ingress, secrets, and so on. Start with simple use cases to deploy applications and familiarize yourself with all the functions. You will inevitably have to do some debugging, which will teach you a lot.

The learning curve for Kubernetes is very, very steep. Good knowledge of Linux and container systems is essential here. You quickly end up on the second layer of tools and techniques from the Kubernetes environment, which complicate the whole thing—especially at the beginning. These are mainly tools such as Helm and Kustomize. After that, GitOps tools such as Argo CD and Flux also pop up.

Monitoring with Prometheus and Grafana

The topic of monitoring is basically its own big topic. If your background is on the development side, basic knowledge of monitoring should be enough. In many cases, you will find the use of classic monitoring tools such as Icinga and Check MK in the companies advertising DevOps job.

In modern infrastructures, it is helpful to know and be able to use the basic concepts of Prometheus. This also includes knowing how to use the standard metrics and how to provide your own metrics if required.

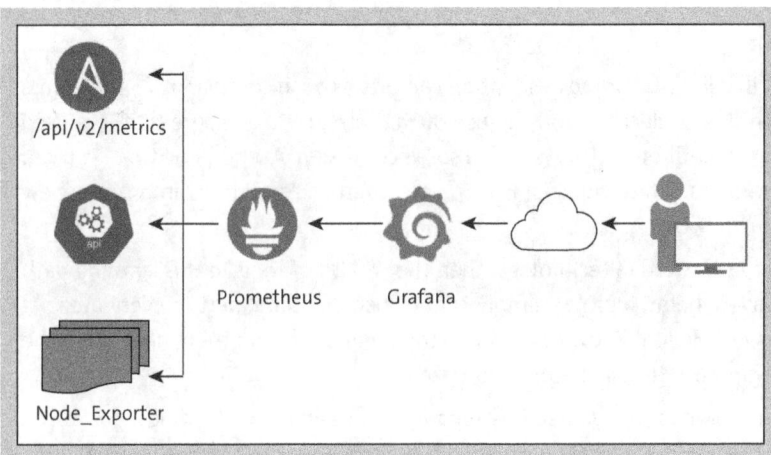

Figure 14.7 Prometheus and Grafana

Grafana helps with visualization, as it allows you to quickly create simple dashboards with a flat learning curve. The monitoring of complex container infrastructures (i.e., the connection to Kubernetes), is particularly important. You should therefore be familiar with how to define and retrieve the metrics of Kubernetes deployments via labels, for example.

Note: Observability for Kubernetes

If you would like to learn more about this, I can recommend the "Kubernetes Observability Workshop" by my colleague Michael Friedrich. In this 3.5-hour workshop, he addresses precisely these aspects: *https://o11y.love/learning-resources/*.

Influx can also be found as an alternative to Prometheus, also in combination with Grafana. In some cases, Influx and Prometheus can also be found running in parallel with different use cases.

What Else Is Missing?

Now we come to the end of our outline of a possible learning path for prospective DevOps engineers. There is much more you could dive into. The most important thing is to ensure that your knowledge of the basics is solid. Excellent Linux knowledge, both of the system itself and of the command line, the basics of network technology, and knowledge of Git will help you effectively use all tools used in DevOps environments.

And don't forget that every tool is fleeting to a certain extent. In a few years, a preferred toolchain may look completely different again. That's why a solid technical foundation is very important, because it means that other issues such as log management or secret management can be solved with simple steps. This is at least true if you don't have to learn the basics over and over again.

> **Note: Reflection**
>
> Using DevOps in a job title or job description can often cause confusion. If you are in a role that can influence decisions on the use of the term, please ensure that it is used carefully for the benefit of all. This will save some confusion. At the same time, you can also use the term as a filter mechanism to see whether candidates in an interview know what you are working on.
>
> If you are more interested in technology, then the wealth of tools in this learning path will probably overwhelm you. You simply cannot become an expert in every area. As soon as you have excellent knowledge of one tool or another and then move on to the next tool, the various skills will become outdated again.
>
> Make sure that you position yourself as broadly as possible. The technical basics are very important—they don't change. Equally, basic concepts are more important than the actual implementation.
>
> If you see DevOps in a job title you are applying for, ask carefully what the company means by DevOps. Strange explanations could quickly bring up red flags that will tell you that you should look for a different job.

14.4 Summary

The DevOps world continues to move forward more than 10 years after the birth of the term "DevOps." This is good and necessary, because it is always necessary to iterate, adapt, and observe what is needed and what is perhaps not.

The Author

Sujeevan Vijayakumaran is a senior solutions engineer at Grafana Labs. Previous, he worked at GitLab, where he helped large corporations from Germany, Austria, and Switzerland transition to a DevOps culture. He cohosts the German technology podcast, TILpod, and enjoys giving talks at open-source conferences—not only on technical topics, but also on good teamwork, efficient communication, and everything else that is part of the DevOps culture.

Index

A

A/B tests	279–280
Acceptance tests	164
Agile Manifesto	62
Agile software development	24, 61
Agile waterfall	61
AIOps	299, 398
Alert	270
Alerting	284
Ansible	253, 255, 407
Apache JMeter	168
Application performance monitoring	276
Application security	197, 307, 331
Approval rules	98
Architecture	50
Argo CD	218
Artificial intelligence	389
Atlassian	386
Atlassian stack	150
Audit trail	344
Automated tests	33
Availability	300
AWS Elastic Beanstalk	237
Azure DevOps	150, 385

B

Best in suite approach	382
Best of breed approach	382
Big bang integration	77, 116
Bitbucket	83, 386
Blameless post-mortems	245
Block storage	237
Blue-green deployments	200
Bottom-up approach	348
Branch Source plug-in	139
Bug reports	277
Buildah	196
Build integrity	330
Build management	
basics	119
problems	113
scaling	128
Build server	118
Business continuity	359
Business monitoring	281

C

CALMS model	24
automation	26
culture	26
lean	27
measurement	27, 69
metrics	267
sharing	28
Canary deployments	201
Cattle not pets	226
Change failure rate	365
Chaos engineering	258
Chaos Monkey	259
Chaos Toolkit	259
ChatGPT	390
ChatOps	247
Check_MK	284
CI/CD components	145
Cilium	297
Circle CI	151
CI server	126, 134
Cloud computing	233
commuinity cloud	234
hybrid cloud	234
multicloud	234
private cloud	234
public cloud	234
Cloud-native	240
Code	
management	81
owners	98
problems	73
Code Climate	167
Code of conduct	103
Code review	78
accelerate	95
approval rules	98
basics	91
between dev and ops	93
between dev and QA	94
in the development team	92
simplify	97
CodeWhisperer	394
Cody	394
Colocation	229
Communication solutions	247
Compliance	307, 335

Index

 approvals 344
 audit trail 344
 define guidelines 336
 fully automated 339
 manual 338
 pipeline 340, 342
 process 338
Compliance policies 320
Concurrent Version System (CVS) 378
Configuration drift 250
Container 194
Container image scanning 325
Container runtime security 325
Container scanning 325
Continuous delivery 173, 179
Continuous deployment 179
Continuous integration 113, 122
Continuous integration/continuous development (CI/CD) 27
Continuous integration and continuous delivery (CI/CD)
 landscape 405
Continuous learning 34
Conway's law 67, 75
Copilot 394
Cortex 290
Counter 287
Crossplane 252
cubeadm 208
Cubelet 211
Culture 16
Cycle time 62

D

Database as a service 238
Database reliability engineers 264
Databricks 399
DataOps 397
Declarative pipelines 131
DefectDojo 323
Dependabot 324
Dependencies 312
Dependency proxy 197
Dependency scanning 323
Dependency tree 312
Deployment frequency 363
DevOps engineer 399
DevSecOps 314, 394
Distributed tracing 295
Docker 194
Docker Compose 408

Dockerfile 194
Documentation 80, 115
DORA metrics 362, 368
Drift detection 326
Dynamic application security testing (DAST) ... 327

E

Edge cases 33
Employee satisfaction 107
Epics 120
Error budgets 303
Error culture 352
Error tracking 294
Example company 47

F

False positives 320
Fault injection 263
Feature flags 70, 203
Feature toggles 203
Feedback 32
Feedback channel 351
Fingerpointing 34
Firewalls 333
Fluentd 297
Flux 218
Follow-up actions 246
Foreman 229
Forrester 378
Fuzz testing 329

G

Gartner's Magic Quadrant 377
Gauge 288
General AI 390
General Data Protection Regulation (GDPR) 308
Git 83, 404
 passwords 328
 platforms 83
 workflows 84
Gitea 83
Git Flow 85, 88
Gitflow 89
GitHub 83, 385, 395, 406
GitHub Actions
 installation 146
 permissions management 147
 pipelines 147

Index

GitHub Flow 85–86
GitLab 83, 381, 384, 406
   runner 142
   server 142
GitLab CI/CD
   architecture 142
   integrations 143
   permissions management 144
   pipelines 144
GitLab Duo 394
GitLab Flow 87
GitOps 216
Golang 404
Golden image 189
Grafana 292, 409
Graylog 297
Grype 326

H

Harbor 197
HashiCorp Configuration Language (HCL) . 250
HashiCorp Nomad 215
Health checks 232
Helm 215
Histogram 288
Honeycomb 299
Hypervisor 229

I

Icinga 283
Idempotence 250
Identity management 333
immutable (operating system) 231
Incident management 284
InfluxDB 291
Infrastructure
   abstract 228
   cattle not pets 226
   failures 224
   hardware 222
   problems 222
   provisioning 228
   server utilization 223
Infrastructure as a service 235
Infrastructure as code (IaC) 182, 404
Infrastructure team 56
Inner sourcing 102, 104, 317
Integration tests 163
Interactive application security testing (IAST) ..
   327

Issue board 30
Iteration 361

J

Jaeger 296
Jenkins 136, 379, 405
   agents 136
   compliance 342
   controller 136
   permissions management 137
   pipelines 140
   plug-in hell 138
Jenkins Configuration as Code plug-in 137
Jenkinsfile 139
JFrog Artifactory 379
Jira 65, 71
Jobs 127, 399

K

k3s 209
Kibana 297
KICS 326
Kubernetes 206, 407
   annotations 214
   ConfigMaps 212
   control plane 210
   deployment 212
   HorizontalPodAutoscaler 213
   labels 214
   pods 212
   ReplicaSet 212
   service 212
   sets 212
   worker node 210

L

Large language models (LLM) 390
Latency 273
LaunchDarkly 205
Lead time 27, 62, 364
Learning path 402
Lifecycle 42
Lift and shift 226, 238
Linkerd 297
Linter 167
Linux 402
Litmus Chaos 259
log4j 331
log4shell 312

Index

Logging 274, 296
Logstash 297
Loki 297

M

Machine learning 389
Maintainability 379
Manifests 256
Marketplace 139
Merge hell 121
Merge train 99
Metrics monitoring 272
Microservices 240
MinIO 237
Misunderstandings 36
MLFlow 398
MLOps 398
Mob programming 101
Monitoring
   availability 300
   business monitoring 281
Monorepositories 108
Multistage builds 195

N

Nagios 283
Need-to-know principle 76
Nexus 196
Nightly build 123

O

Objectives and key results (OKRs) 356
Object storage 237
Observability 268
   engineering 276
   platforms 298
Obstacles to integration 51
Onboarding 80, 114
Open Container Initiative 196
OpenGitOps 217
OpenShift 209
Open source 103
OpenTelemetry 279, 296, 299
OpenTofu 251
Operating level agreements 301
Operating services 221
Operating system 230
Operational performance 362
   reliability 366

Operations team 53
Opsgeny 284
Orchestration 205, 232

P

Packaging 193
PagerDuty 284
Pair programming 86, 99–100
Performance tests 167
Perses 294
Persistent volumes 212
Pipeline 123, 126
   best practices 132
   building blocks 133
   compliance 340
   declarative 130
   efficient authoring 132
   GitHub Actions 147
   GitLab CI/CD 144
   Jenkins 140
   scripted 130
   tests 160
Platform 377
Platform as a service 236
Playbooks 253
Podman 196
Predictive maintenance 399
Processing time 28
Profiler 279
Project management 64
Project planning 61
Prometheus 285, 409
   labels 289
   scalability 290
PromLens 290
Prompt engineering 391
PromQL 289
Pulumi 251
Puppet 255, 407
Python 404

Q

QA team 52
Quality assurance 153
Quality gate 204, 319

R

Rancher 209
Red Hat Enterprise Linux (RHEL) 403

Index

Registry 196
Release management 173
Reliability 263, 366
Reliability engineering 262
Renovate Bot 324
Reproducible builds 130
Requirements analysis 49
Requirements engineering 49, 164
REST interfaces 403
Retrospectives 354
Roles 399
Rollbacks 199
Root cause 246
Rubber duck debugging 100

S

Sapling 108
Scaling 213
Scrum 64
Secret detection 328
Security 307
   dashboards 321
   dealing with errors 319
   development process 318
   inner sourcing 317
   pull and merge requests 322
   quality gates 319
   reporting 311, 321
   scanner 322
   shift left 316
   team structure 315
   user accounts 333
Security team 57
Semantic versioning 192
Semgrep 326
Sentry 294
Service accounts 334
Service discovery 286
Service level agreement (SLA) 301
Service level indicators (SLI) 303
Service level objectives (SLO) 302
Service meshes 297
Shell scripting 404
Shift left 316
Single source of truth 82
Site reliability engineering 263
Soft skills 401
Software as a service 239
Software bill of materials (SBOM) 332
Software development and delivery experience 351

Software development lifecycle 24, 42
Solarwinds 330
SonarLint 167
SonarQube 167, 327
Sonatype Nexus 379
Source code management 90
Splunk 297
Squash 99
SSH 283
Stages 127
Staging environment 175
Standardized tools 381
Static application security testing (SAST) ... 326
Static code analysis 166
Subversion 74, 81, 378
Supply chain security 307, 329
Systems thinking 29
System tests 163

T

Tabnine 394
Tanzu 209
Teamcity 151
Team exchange 355
Team for the transformation 357
Team structuring 38
Team topology 357
Technical debt 33, 48, 79
Technology 17
Telegraph 291
Telemetry 279
Terraform 250, 406
Test-driven development 169
Testing 153
Tests
   acceptance tests 164
   automate 164
   coverage 165
   duration 166
   integration tests 163
   pipeline 160
   unit tests 162
   with AI 391
Thanos 290
The Three Ways 28, 267
The Twelve Factor App 241
Threat model 337
Tickets 65, 102
Time series 272
Time to restore service 366
Toolchain 378

Top-down approach 348
Tracing 278
   OpenTelemetry 279
Traefik 211
Transformation 350
Transitive dependency resolution 324
Trivy 326
Trunk-based development 85
Tuning 320

U

Unit tests 162
Unleash 205

V

Value stream 359
Value stream mapping 368
   charter 372
   lead time 373
   percentage complete and accurate ... 373
   process time 373
   workshop 373
Version management system 83
Version number 192–193
Vertex AI 394
Virtual machines 229
Visibility 134, 268
Vision and strategy 350

W

Waterfall model 48, 61

Z

Zed Attack Proxy 327

- Get hands-on practical experience with Git

- Understand branches, commands, commits, workflows, and more

- Learn to use GitHub, GitLab, and alternative Git platforms

Bernd Öggl, Michael Kofler

Git

Project Management for Developers and DevOps Teams

Get started with Git—today! Walk through installation and explore the variety of development environments available. Understand the concepts that underpin Git's workflows, from branching to commits, and see how to use major platforms, like GitHub. Learn the ins and outs of working with Git for day-to-day development. Get your versioning under control!

407 pages, pub. 10/2022
E-Book: $44.99 | **Print:** $49.95 | **Bundle:** $59.99

www.sap-press.com/5555

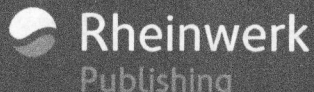

- Get hands-on practice with Docker, from setup to orchestration
- Work with Dockerfiles, Docker Compose, GitLab, and Docker Hub
- Learn about project migration, security, Kubernetes, and more
- Consult and download practical code examples

Bernd Öggl, Michael Kofler

Docker

Practical Guide for Developers and DevOps Teams

Learn the ins and outs of containerization in Docker with this practical guide! Begin by installing and setting up the platform. Then master the basics: get to know important terminology, understand how to run containers, and set up port redirecting and communication. You'll learn to create custom images, work with commands, and use key containerization tools. Gain essential skills by following exercises that cover common tasks from packaging new applications and modernizing existing applications to handling security and operations.

491 pages, pub. 01/2023
E-Book: $44.99 | **Print:** $49.95 | **Bundle:** $59.99

www.sap-press.com/5650

- Your practical guide to using Kubernetes, from managing containers to deploying Kubernetes applications with Helm

- Master the basics, including pod and container management, YAML, ReplicaSets, ConfigMaps, and storage

- Configure service and Ingress networks, load balancing and auto-scaling, application self-healing, and more

Kevin Welter

Kubernetes

Practical Guide for Developers and DevOps Teams

Unravel the complexities of Kubernetes with this hands-on guide! Start with an introduction to Kubernetes architecture and components such as nodes, Minikube, and kubectl commands. Follow tutorials to set up your first clusters and pods, and then dive into more advanced concepts like DaemonSets, batch jobs, and custom resource definitions. Perform resource management, set up autoscaling, deploy applications with Helm, and more!

401 pages, pub. 09/2024
E-Book: $54.99 | **Print:** $59.95 | **Bundle:** $69.99

www.sap-press.com/5964

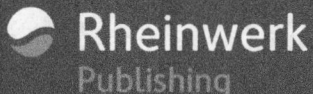

- Learn to program your own AI applications—even if you've never coded before!
- Get started without code using the KNIME platform and then expand your skills by programming with basic Python
- Work with neural networks, transfer learning, anomaly detection, reinforcement learning, and more

Metin Karatas

Developing AI Applications

An Introduction

It's time to get practical about AI. Move past playing around with chatbots and plugging your data into others' applications—learn how to create your own! Walk through key AI methods like decision trees, convolutional layers, cluster analysis, and more. Get your hands dirty with simple no-code exercises and then apply that knowledge to more complex (but still beginner-friendly!) examples. With information on installing KNIME and using tools like Auto-Keras, ChatGPT, and DALL-E, this guide will let you do more with AI!

402 pages, pub. 06/2024
E-Book: $39.99 | **Print:** $44.95 | **Bundle:** $49.99

www.sap-press.com/5899

- Learn to work with scripting languages such as Bash, PowerShell, and Python
- Get to know your scripting toolbox: cmdlets, regular expressions, filters, pipes, and REST APIs
- Automate key tasks, including backups, database updates, image processing, and web scraping

Michael Kofler

Scripting

Automation with Bash, PowerShell, and Python

Developers and admins, it's time to simplify your workday. With this practical guide, use scripting to solve tedious IT problems with less effort and fewer lines of code! Learn about popular scripting languages: Bash, PowerShell, and Python. Master important techniques such as working with Linux, cmdlets, regular expressions, JSON, SSH, Git, and more. Use scripts to automate different scenarios, from backups and image processing to virtual machine management. Discover what's possible with only 10 lines of code!

470 pages, pub. 02/2024
E-Book: $44.99 | **Print:** $49.95 | **Bundle:** $59.99

www.sap-press.com/5851

Torsten T. Will

C++

The Comprehensive Guide

If you need to know C++, look no further! This comprehensive guide has everything you need to master the modern C++23 language, from syntax fundamentals to advanced development concepts. Follow practical code examples as you learn object-oriented programming, work with standard library containers, program concurrent applications, and more. Don't just learn how to code—learn how to code better with expert tips and guidance on the rules of compact, secure, and efficient code.

1089 pages, pub. 10/2024
E-Book: $64.99 | **Print:** $69.95 | **Bundle:** $79.99

www.sap-press.com/5927

- Your all in one guide to modern C++

- Work with the C++ language, from basic syntax and functions to more advanced features such as pointers and macros

- Learn to use the standard library and containers